The flying circus of physics
WITH ANSWERS

Jearl Walker

DEPT. OF PHYSICS
CLEVELAND STATE UNIVERSITY

The flying circus of physics WITH ANSWERS

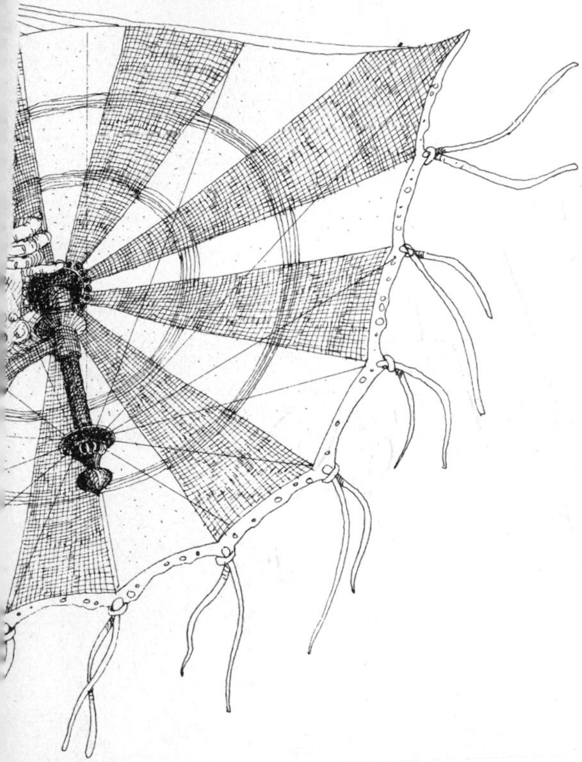

John Wiley & Sons
NEW YORK SANTA BARBARA LONDON SYDNEY TORONTO

Library of Congress Cataloging in Publication Data:

Walker, Jearl, 1945–
The flying circus of Physics.
Bioliography
1. Physics–Problems, exercises, etc.
I. Title.
QC32.W2 530 75–5670
ISBN 0-471-02984-x

Printed in the United States of America

10 9 8 7 6 5 4 3 2

for elizabeth

Preface

These problems are for fun. I never meant them to be taken too seriously. Some you will find easy enough to answer. Others are enormously difficult, and grown men and women make their livings trying to answer them. But even these tough ones are for fun. I am not so interested in how many you can answer as I am in getting you to worry over them.

What I mainly want to show here is that physics is not something that has to be done in a physics building. Physics and physics problems are in the real, everyday world that we live, work, love, and die in. And I hope that this book will capture you enough that you begin to find your own flying circus of physics in your own world. If you start thinking about physics when you are cooking, flying, or just lazing next to a stream, then I will feel the book was worthwhile. Please let me know what physics you do find, along with any corrections or comments on the book.* However, please take all this as being just for fun.

Jearl Walker

My grandmother's house
Aledo, Texas, 1977

*Physics Department, Cleveland State University, Cleveland, Ohio 44115.

Acknowledgments

I should in no way give the impression that this book was written by me alone. Lots of people contributed, helped, argued, criticized, encouraged, and understood. Since I was a graduate student at the University of Maryland when I wrote the book, I must thank Howard Laster and Harry Kriemelmeyer for their willingness to support a graduate student with such an offbeat idea. Dick Berg, also at Maryland, contributed many ideas and hours of discussion. Sherman Poultney not only gave me several good problems but also was understanding when my dissertation occasionally (frequently) fell victim to my book. My wife, Elizabeth, typed and edited the manuscript. Art West, who was also a graduate student at the time, gave very valuable and detailed suggestions on the semifinal version. However, it was Joanne Murray who toiled through my morass of the English language and read and edited many versions of the manuscript. I am especially indebted to her. I also thank Don Deneck, Edwin Taylor, George Arfken, Ralph Llewellyn, and A. A. Strassenburg who thoughtfully read the manuscript and offered many very valuable suggestions.

Jearl Walker

Contents

The flying circus of physics
WITH ANSWERS

1

Hiding under the covers, listening for the monsters

vibration

friction

resonance

1.1
Squealing chalk

Why does a piece of chalk produce a hideous squeal if you hold it incorrectly? Why does the orientation of the chalk matter, and what determines the pitch you hear?

Why do squeaky doors squeak? Why do tires squeal on a car that is drag racing from a dead stop?

1 through 3.*

resonance

vibration

friction

1.2
A finger on the wine glass

Why does a wine glass sing when you draw a wet finger around its edge? What exactly excites the glass, and why should the finger be wet and greaseless? What determines the pitch? Is the vibration of the rim longitudinal or transverse? Finally, why does the wine show an antinode in its vibrational pattern 45° behind your finger?**

124, p. 154.

Exceptionally good references: *Weather* (a journal), Jones (82), Bragg (159).

*The numbers following the problems refer to the bibliography at the end of the book.

**An antinode is where the vibrational motion is maximum.

coupled oscillations

1.3
Two-headed drum vibrations

If a two-headed drum, such as the Indian tom-tom, is struck on one head, both heads will oscillate although they may not both be oscillating at any given instant. Apparently the oscillation is fed from one to the other, and each periodically almost ceases to move. Why does this happen? Wouldn't you have guessed that the membranes would oscillate in sympathy? What determines the frequency with which the energy is fed back and forth?

124, p. 149; 126, p. 474.

harmonic motion

1.4
Bass pressed into records

If I turn down the volume on my record player and just listen to the sound coming directly from the stylus, I can hear high frequencies whenever they occur in the music, but there is almost no bass. Amplifiers take this weaker bass into account and amplify the low frequencies much more than the high. Is there any practical reason for reducing the strength of the bass pressed into records?

143.

oscillations

shearing

1.5
Whistling sand

In various parts of the world, such as on some English beaches, there are sands that whistle when they are walked on. A scraping sound seems plausible, but I can't imagine what would cause a whistle. Do the sand grains have some unique shape so that the sand resonates?

81, p. 145; 144, Chapter 17; 145, p. 140; 146 through 150; 1483.

oscillations

shearing

1.6
Booming sand dunes

Even more curious is the "booming" occasionally heard from sand dunes. Suddenly, in the quiet of the desert, a dune begins to boom so furiously that one might have to shout to be heard by his companions. The clue to this may lie in the accompanying avalanche on the leeward (downwind) side of the dune. Then again, there is nothing unusual about such avalanches for that is precisely how the dune itself flows across the desert floor. Under some conditions could one of these avalanches cause a large vibration of the sand and thus produce the booming?

144, Chapter 17; 146; 150.

vibration

standing waves

1.7

Chladni figures

Chladni figures are made with a metal disc supported at its center and sprinkled with sand. As a bowstring is drawn across an edge, the sand jumps into some geometric design on the plate (Figure 1.7). Why? Nothing to it, you say? It is just simple standing waves set up on the plate by the bowing? Well, then tell me why, using the same bowing motion, you get one design with sand and another design with a finer dust? You can even mix them up beforehand; they'll separate into their own designs as you bow the plate.

81, pp. 129–131; 82, pp. 172–180; 124, pp. 61–62; 127, pp. 172–176; 128, pp. 130–131; 130 through 138; 139, p. 207; 141, pp. 178–190; 142, pp. 88–91; 1529; 1551.

Figure 1.7
Bowing a plate to get Chladni figures. (Some of these may require the support of the plate in places other than the center.)

string vibration

1.8

Pickin' the banjo and fingering the harp

Why does the banjo produce a twangy sound and the harp a soft mellow sound? One difference between the two instruments is that the banjo is plucked with a pick but the harp is plucked with a finger. How does this make a difference?

82, pp. 283 ff; 128, pp. 92–93; 145, p. 89.

string vibration

resonance

1.9

String telephone

How does the string telephone that you played with as a child work? How does the pitch heard in the receiving can depend on the tightness and density of the string and the size of the can? Approximately how much more energy is transmitted with the string telephone than without it?

82, pp. 103–104.

string vibration

friction

1.10

Bowing a violin

Plucking a string, as a guitar player does, seems a straightforward way to excite vibrations in it.

But how does the apparently smooth motion of bowing excite the vibrations of a violin string? Does the sound's pitch depend on the pressure or speed of the bowing?

82, pp. 219–221, 291–300; 124, pp. 98 ff; 126, pp. 453–456; 127, pp. 101–103; 128, pp. 93–94; 145, pp. 89–99; 151, pp. 90–93; 152, pp. 167–170; 153; 1552.

Figure 1.10

string vibration

1.11

Plucking a rubber band

If you tighten a guitar string, you raise its pitch. What happens if you do the same with a rubber band stretched between thumb and forefinger? Does its pitch change when it is stretched farther? No, the pitch remains fairly unchanged; or, if it does change, it becomes lower rather than higher. Why is there a difference between rubber bands and guitar strings?

154; 155, pp. 186–187.

vibration

phase change

1.12

The sounds of boiling water

When I heat water for coffee, the sound of the water tells me when it has begun to boil. First there is a hissing that grows and then dies out as a harsher sound takes over. Just as the water begins really to boil, the sound becomes softer. Can you explain these sounds, especially the softening as the water begins to boil?

157; 158, p. 295; 159, pp. 88–89; 160, p. 168.

vibration

1.13

Murmuring brook

At some time in your life you've probably spent a sunny afternoon lying in the grass, listening to the murmur of a brook. Why do brooks murmur? Why do waterfalls and cataracts roar?

What is responsible for the spritely sound of a just-opened soft drink? Look into a clear soft drink and try correlating the noise with the creation, movement, or bursting of the bubbles.

145, p. 140; 159, pp. 129–130; 161 through 163.

stress

phase change

1.14

Walking in the snow

Sometimes snow crackles when you walk in it, but only when the temperature is far enough below freezing. What causes the noise, and why does its production depend on the temperature? At approximately what temperature will the snow begin to crackle?

164, p. 440; 165, p. 144; 166.

absorption

1.15

Silence after a snowfall

Why is it so quiet just after a snowfall? There aren't as many people and cars outside as usual, but that alone doesn't explain such quietness. Where does the energy of the outdoor noise go? Why does the snow have to be fresh?

A similar sound reduction occurs in freshly dug snow tunnels in Antarctic expeditions: the speakers must shout to be heard if they are more than 15 feet apart. Again, what happens to the sound energy?

165, p. 134; 167.

1.16

Ripping cloth

Why is it that when you tear a piece of cloth faster, the pitch of the ripping is higher?

Figure 1.18
"Listen. There it is again. 'Snap, crackle, pop.'"

1.17

Knuckle cracking

What makes the cracking sound when you crack your knuckles? Why must you wait a while before you can get that cracking again?

168.

1.18

Snap, crackle, and pop

Why exactly do Rice Krispies* go "snap, crackle, and pop" when you pour in the milk?

1.19

Noise of melting ice

Plop an ice cube or two into your favorite drink, and you'll hear first a cracking and then a "frying" sound. What causes these noises? Actually, not all ice will produce the "frying" sound. Why is that?

Icebergs melting in their southward drift also make frying noises that can be heard by submarine and ship crews. The sound is called "bergy seltzer."

169.

*A breakfast cereal from the Kellogg Company.

acoustic conduction

1.20

An ear to the ground

Why did Indian scouts in the old Westerns fall to their knees and press their ears against the ground to detect distant, and unseen, riders? If they could hear the distant pounding of hooves through the ground, why couldn't they hear it through the air?

124, p. 21.

propagation

1.21

Voice pitch and helium

When people inhale helium gas, why does the pitch of their voices increase?

One should be very, very cautious in inhaling helium. One can suffocate with the helium while feeling no discomfort because there is no carbon dioxide accumulation in the lungs. Never, never inhale hydrogen or pure oxygen. Hydrogen is explosive and oxygen supports burning. Even a spark from your clothes can lead to death.

170, p. 219; 171, pp. 16–17.

speed of sound

1.22

Tapping coffee cup

As you stir instant cream or instant coffee into a cup of water, tap the side with your spoon. The pitch of the tapping changes radically as the powder is added and then during the stirring. Why?

Tap the side of a glass of beer as the head goes down. Again, the pitch changes. Why?

You may have a tendency to answer that the foam or the powder damp the oscillations caused by the tapping, but even if that is true, would that change the pitch or only the amplitude?

159, p. 158; 173.

speed of sound
and temperature

1.23

Orchestra warmup and pitch changes

Why does the pitch of the wind instruments increase as an orchestra warms up? Why does the pitch of the string instruments decrease?

124, pp. 49–50; 126, p. 498; 172.

interference

1.24

Bending to the ground to hear an airplane

I have read that if I put my head close to the ground while listening to an airplane fly by, the pitch of the airplane's noise may seem to increase. Similarly, if I stand by a wall near a waterfall, I may hear, in addition to the normal sound of the waterfall, a softer background sound. The closer I stand to the wall, the higher the pitch of the extra sound. In either case, why would I hear a sound whose pitch depends on the nearness of my ear to the solid structure?

82, pp. 98–100; 145, p. 59; 174 through 180.

interference

waveguides

1.25

Culvert whistlers

Stand in front of a long concrete culvert and clap you hands sharply. You will hear not only the echo of your clap, but also a "zroom," which starts at a high pitch and drops to a low pitch within a fraction of a second.* What's responsible for the "zroom"?

181; 182.

1.26

Music hall acoustics

Why are concert halls generally narrow with high ceilings? If echoes are undesirable shouldn't the walls and ceiling be close to the listener? That way the listener will not be able to distinguish the direct sound from the reflected sound. What is the minimum time difference between two sounds that the listener can, in fact, distinguish? Why does a hall full of people sound much better than an empty hall?

If echoes are to be eliminated, why aren't the walls and ceilings covered with material that will absorb the sound? Granted that the hall's beauty might be destroyed, it still appears that halls are designed so as not to eliminate all sound reflections. In fact, the walls and ceilings may be covered with nooks and crannies that reflect the sound in every which way. On the other hand, a hall with no reflections is said to be acoustically dead.

124, Chapter 13; 127, pp. 531–540; 128, Chapter 10; 142, Chapter 14; 145, pp. 279–293; 152, Chapter 9; 158, pp. 609–616; 170, pp. 265–266; 171, pp. 44–50; 183, pp. 123–180; 184, Chapter 14; 185, Chapter 11; 186, Chapter 8; 187, pp. 291–300; 188 through 195; 1528.

*Crawford (181) has described these as being analogous to atmospheric whistlers (see Prob. 6.31).

reflection

focusing

1.27

Acoustics of a confessional

Some rooms are especially noted for their strange acoustics; some even provide a focusing of the sound. Such focusing was apparently used in the "Ear of Dionysius" in the dungeons of Syracuse where the acoustics somehow fed words and even whispers of the prisoners into a concealed tube to be heard by the tyrant.

For an example in recent times, the dome covering the old Hall of Representatives in the Capitol building (Washington, D.C.) would reflect even a whisper from one side of the chamber in such a way that it would be audible on the opposite side. More than once this was rumored to have embarrassed representatives whispering party secrets to their colleagues.

The Cathedral of Girgenti in

Sicily provided even more severe embarassment. Its shape is that of an ellipsoid of revolution, so that sound produced at one focus of the ellipsoid is nearly as loud at the other focus. Soon after it was built one focus was unknowingly chosen for the position of the confessional.

The focus was discovered by accident, and for some time the person who discovered it took pleasure in hearing, and in bringing his friends to hear, utterances intended for the priest alone. One day, it is said, his own wife occupied the penitential stool, and both he and his friends were thus made acquainted with secrets which were the reverse of amusing to one of the party (141).

139, p. 194; 141, p. 48; 197, Chapter 11.

propagation

refraction

1.28

Sound travel on a cool day

Why does sound travel farther on a cool day than on a warm day? This is especially noticeable over calm water or a frozen lake. The range of sounds in the desert, on the other hand, may be noticeably limited.

81, pp. 34-35; 82, p. 107; 124, p. 17; 127, pp. 322-325; 142, pp. 117-118; 185, pp. 309-311; 186, pp. 66-67; 187, p. 137; 207, pp. 50-52; 209, pp. 24-25; 210, p. 600; 211, pp. 474-475; 212; 213, pp. 49-52.

1.29

Silent zones of an explosion

During World War II it was often noticed that as one would drive toward a distant artillery piece, the roar of its fire would disappear at certain distances (Figure 1.29). Why were there such silent zones?

Sound travel over large distances is also curious. For example, during World War I people on the English shore could hear gunfire from installations in France. What conditions permit such an enormous sound range?

150; 165, pp. 135 ff; 187, p. 137; 214, p. 2; 215, pp. 23-25; 216; 217, pp. 9-14; 218; 219, pp. 291-293; 313, pp. 71 ff.

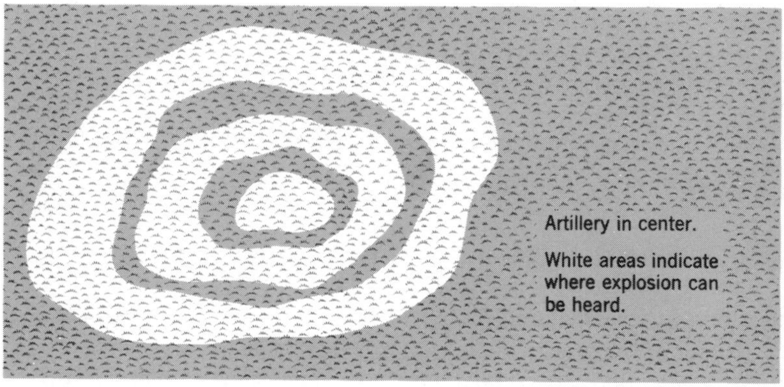

Figure 1.29

reflection

Rayleigh scattering

1.30

Echoes

I am sure you can explain echoes—they are reflections of the sound waves by some distant object, right? Then explain why some echoes return to the speaker with a higher pitch than that of the initial sound. Also, why does a high-pitched sound usually produce a louder, more distinct echo than a low-pitched sound? How close to the reflecting wall can you stand and still hear an echo?

81, p. 31; 82, pp. 86-87; 127, pp. 311-313; 142, p. 132; 164, p. 426; 182; 198, pp. 147-154; 206.

Rayleigh waves

intensity and distance

reflection

1.31

The mysterious whispering gallery

It was Rayleigh who first explained the mysterious whispering gallery in the dome of London's St. Paul's Cathedral. In this large gallery there is a peculiar audibility for whispers. For instance, if a friend were to whisper to the wall somewhere around the gallery, you would be able to hear his whisper no matter where you might stand along the gallery (Figure 1.31*a*). Strangely enough, you will hear him better the more he faces the wall and the closer he is to it.

Is this just a straightforward reflection and focusing problem? Rayleigh made a large model of the gallery to find out. He placed a birdcall at one point along the model gallery and a flame at another point. When sound waves from the birdcall impinged on the flame, the flame would flare, and so the flame was his sound detector. You are probably tempted to draw the sound rays shown in Figure 1.31*b*. But before you put too much faith in them, suppose a narrow screen were to be placed at some intermediate point along the inside perimeter of the metal sheet (as shown in Figure 1.31*c*, but exactly where along the perimeter doesn't matter). If your idea about the rays is correct, the flame should still flare because the screen is out of the way, right? Well, as a matter of fact, when Rayleigh inserted a screen, the flame did not flare. The screen must somehow have blocked the sound waves. But how? After all, it was only a narrow screen placed seemingly well out of the way of the sound rays. This result gave Rayleigh a clue to the nature of the whispering gallery.

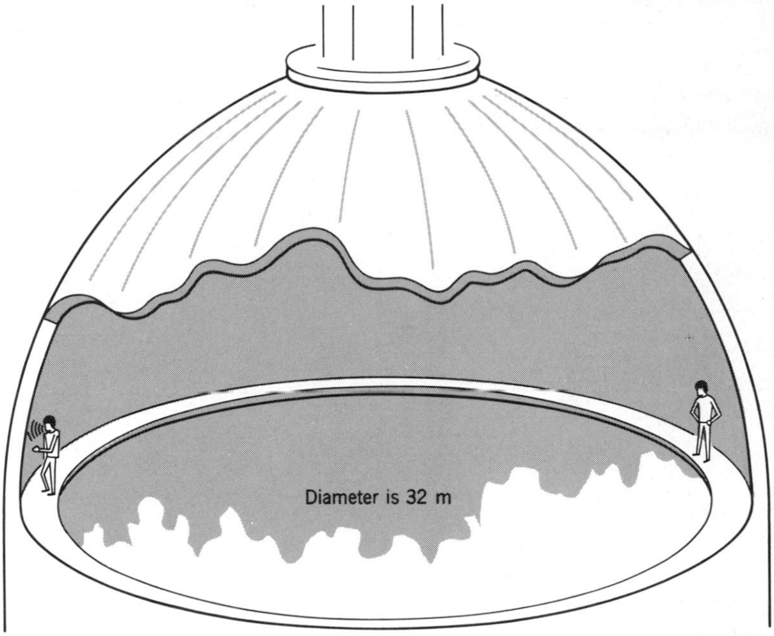

Figure 1.31a
Cutaway view of whispering gallery.

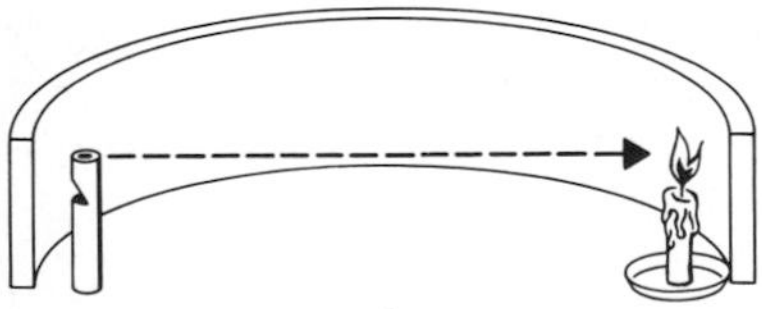

Figure 1.31b
Rayleigh's model of whispering gallery. Birdcall causes flame to flare.

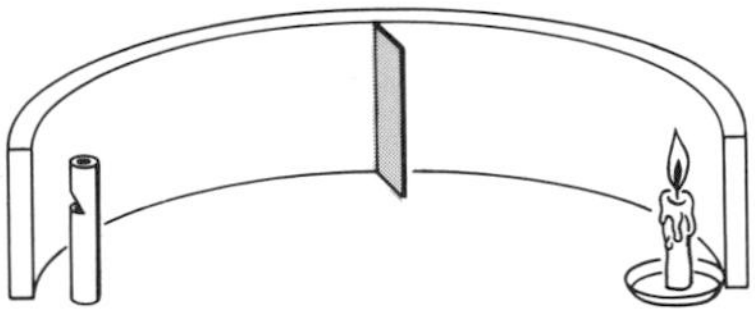

Figure 1.31c
With thin screen placed near the wall, birdcall cannot make flame flare.

81, pp. 32–33; 82, pp. 87–92; 127, pp. 315–316; 198, pp. 126–129; 199 through 205.

interference

1.32
Musical echoes

What causes the musical echo you can sometimes hear when you make a noise near a fence or a flight of stairs? Can you calculate the pitch of the echo?

81, p. 32; 127, pp. 313–314; 145, p. 13; 164, pp. 426–427; 182; 206; 207, pp. 47–48; 208.

turbulence

refraction

1.33
Tornado sounds

My grandmother could always forecast a tornado by the deathly silence that would suddenly fall before the appearance of the tornado. Why the silence? When the twister hit, there would be a deafening roar much like a jet plane's. Why the roar? Finally, there are reports that in the tornado's center it is, again, deathly quiet. Can this be true? Wouldn't you hear at least the furious destruction taking place outside the center?

165, pp. 144–145; 223, pp. 67, 83; 224 through 226.

reflection

Rayleigh waves

1.34
Echo Bridge

The whispering gallery effect may be responsible for some of the sounds you can hear beneath a bridge arch. If you stand near the wall of such an arch (Figure 1.34) and whisper faintly, you will hear two echoes; a loud handclap will yield many echoes. Can you account for these echoes? They can result either from normal reflections off the water or from the whispering gallery effect, or from both.

82, p. 87; 202; 203.

Figure 1.34
Echo bridge.

refraction

1.35

Sound travel in wind

Why is it easier to hear a distant friend yell if you are downwind rather than upwind? Is it because, as is commonly thought, there is a greater attenuation in the upwind direction?

81, pp. 33–34; 82, pp. 107-108; 124, pp. 17–18; 127, pp. 322-325; 142, pp. 119-121; 185, pp. 11–13, 311; 186, pp. 66-67; 187, p. 137; 207, pp. 52-53; 210, pp. 599-600; 212; 213, pp. 52–55; 222.

propagation

1.36

Brontides

Throughout history there have been tales of mysterious sounds from the sky, rumblings, and short cracklings when the sky is perfectly clear and there are no obvious noise sources. These noises—called brontides, mistpoeffers, or Barisal guns—are heard virtually everywhere: over flatland, over water, and in the mountains. In a study of 200 Dutch mistpoeffers it was found that the sound came most often in the morning and afternoon, less often at noon, and hardly ever during the night. In some places they are far from rare. For example, near the Bay of Bengal they are heard so frequently that the people ascribe the sounds to the gods. In other places, however, they are now probably dismissed as sonic booms.

One is tempted to identify these mysterious sounds as distant thunder, but thunder is normally not heard at distances greater than 15 miles*. Besides, these sounds are heard on clear days. Can you think of other possible explanations?

164, p. 442; 227; 1611, Section GS.

*See Prob. 1.38

diffraction

1.37

Shadowing a seagull's cry

For an example of quiet "shadows" behind objects, let me offer the following story (see Figure 1.37).

> In the spring the seagulls resort in large numbers to the moss to lay their eggs and when the young birds are able to fly, the air is filled with their shrill screams. There is a road at a little distance from the nests and by the side of the road there is sometimes a row of stacks of peat. The length of one of these stacks is many times as great as the wave-length of the scream of the birds and consequently a good sound shadow is formed. Opposite the gap between two stacks the sound is unpleasantly loud; opposite the stack itself there is almost complete silence, and the change from sound to silence is quite sudden (234).

Would there be a quiet region if seagulls cried in a deep voice rather than their shrill one?

128, p. 18; 234, p. 103.

Figure 1.37

refraction

1.38

Lightning without thunder

Often a lightning stroke appears unaccompanied by thunder. In fact, thunder is rarely heard beyond about 15 miles from the lightning flash. Why? Is 15 miles really such a great distance for sound to travel? No, artillery fire and explosions can certainly be heard beyond 15 miles. Why not thunder as well?

82, pp. 114–116; 142, p. 118; 164, pp. 441–442; 219, pp. 304-305; 220, p. 196; 221.

diffraction

1.40

Cracking a door against the noise

If I close my door, which leads to a very noisy hall, my room is kept quiet. If I open the door wide, though, it is hard to think with all the noise. How about cracking the door just a little? That certainly should be almost the same as closing it all the way. Yet, I try it and discover the noise to be almost as bad as with the door wide open. Why does even a small crack make such a disproportionate difference in the noise level of my room?

128, p. 19; 155, p. 177.

1.41

Feedback ringing

There was an era in rock and roll when feedback was used extensively to give a psychedelic quality to the music. A guitar player would play facing into his own speaker, and the speaker output would be picked up and reamplified by his electric guitar. That same type of ringing can be heard if a radio announcer holds a radio tuned to his own station near his microphone. In either case what causes the ringing?

refraction

1.39

Submarine lurking in the shadows

Though sonar systems are powerful enough to detect submarines at very large distances, they are usually limited to only several thousand meters (in the tropics to even less than that). Consider, for example, a sonar unit and a sub at about the same depth (Figure 1.39). For some reason other than just absorption, sound radiated toward the sub never reaches it; the sub is said to be in a shadow area and won't be detected. What causes those shadows?

171, pp. 86-89; 185, p. 235; 217, pp. 16-19; 228, pp. 376-379; 229 through 232.

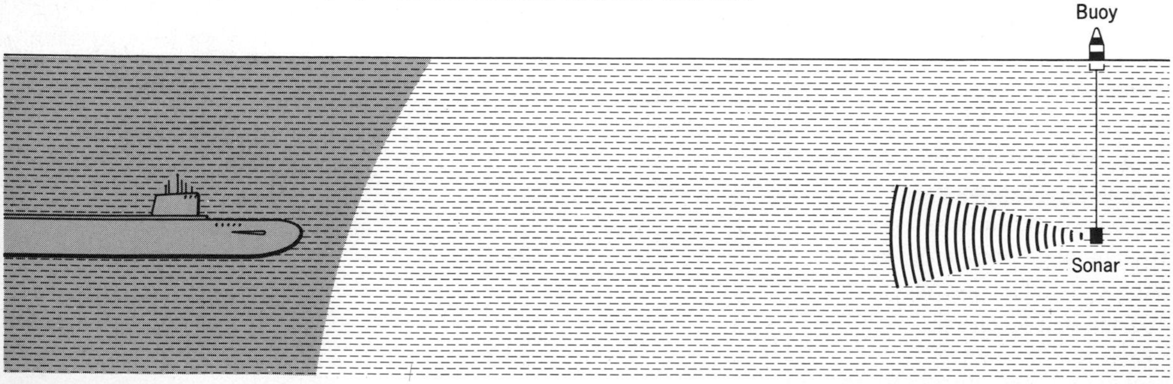

Figure 1.39

diffraction

1.42

Foghorns

Foghorns should be designed to spread their sound over a wide horizontal field, wasting as little as possible upward. Doesn't it seem strange, then, that rectangular foghorns are oriented with the *long* sides of their openings *vertical* (Figure 1.42). Isn't that orientation precisely the wrong one?

142, pp. 124-125; 145, p. 167; 159, pp. 159-160; 235, pp. 78-79; 236.

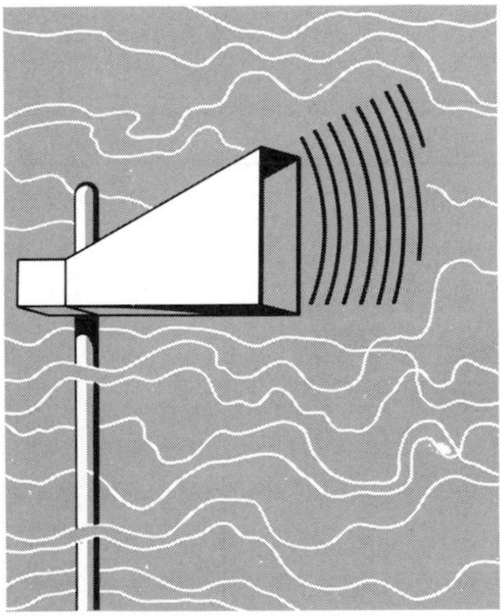

Figure 1.42

diffraction

1.43

Whispering around a head

You can hear a friend's normal voice reasonably well whether he is facing you or turned away. Why is it that you can hear his whisper only if he is facing you, even if the whisper is as loud as his normal voice?

159, pp. 85-86; 198, p. 127; 237, p. 188; 238, pp. 47-48; 239, p. 220.

resonance

1.44

End effects on open-ended pipes

Why is there an antinode of air movement (and a node of pressure) at each end of an open pipe when standing sound waves are set up inside? Since there is a node at a closed end, there should be an antinode at an open end, right? Can you actually show why there is an antinode there? As a matter of fact, the antinode is not precisely at the open end, and where it really is depends on several parameters of the pipe (width, for example). Will this departure from simple theory effect the practical use of pipes in such things as organs?

82, pp. 136-139; 126, pp. 493-496; 127, pp. 181-182; 145, pp. 163-165; 240; 241.

resonant oscillation

1.45

Getting sick from infrasound

Infrasound (sound of a subaudible frequency) can make you nauseous and dizzy. . . it can even kill you. Now that its danger is being recognized, infrasound is being discovered in many common settings: near aircraft, in cars at high speeds, near ocean surfs, in thunderstorms, and near tornados, for example. It may even warn animals and some especially sensitive people of an impending earthquake. Why does infrasound affect people and animals this way? In particular, how can it cause such things as internal bleeding?

171, pp. 139-147; 1489 through 1491; 1534 through 1536.

vibration

cavitation

resonance

1.46

Noisy water pipes

Why do the pipes sometimes groan and grumble when I turn on and off my water faucet? Why doesn't it happen all the time? Where exactly does the noise originate: in the faucet, the pipe immediately behind it, or a turn in the pipe somewhere down the line? Why is there rumbling only with certain flow rates? Finally, why can the problem be alleviated by adding a vertical pipe of trapped air to the water pipe?

183, p. 46; 251; 252.

resonance

vortex motion

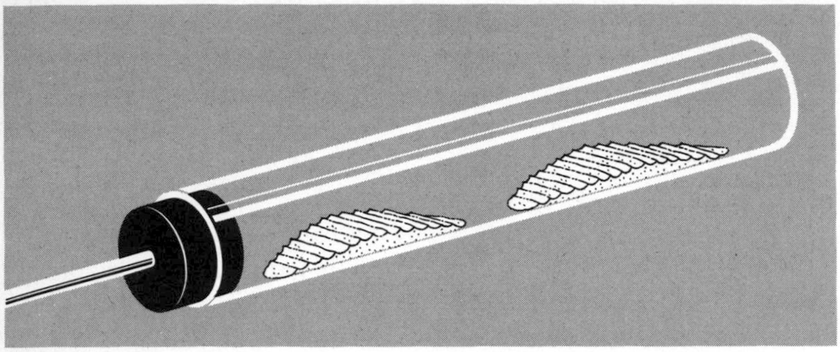

Figure 1.47a
The dust is left in piles and ripples when the rod is stroked.

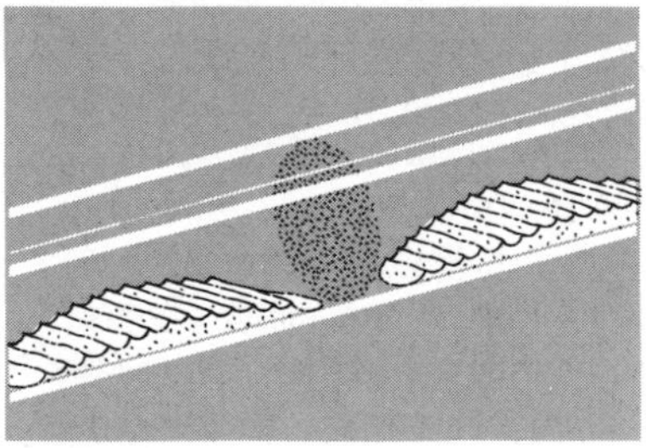

Figure 1.47b
With a loudspeaker as an exciter, thin discs of dust form across the tube's cross section.

1.47

Piles and ripples of a Kundt tube

The Kundt tube has long been a simple demonstration of acoustic standing waves, but can you really explain how it works? It consists of a long glass tube containing some light powder (cork dust or lycopodium powder, for example). The tube is corked at one end and sealed at the other with a brass rod (Figure 1.47a). When the rod is stroked with a rosin-coated chamois, not only does the rod squeal, but also the dust in the tube collects in periodic piles along the tube. Standing sound waves must do this to the dust, but how? Moreover, if one of the piles of powder is examined closely, it is found to contain a series of ripples. If standing waves make the piles, what makes the ripples?

If the rod is replaced by a pure tone loudspeaker, discs form in between the piles (Figure 1.47b). Each disc resembles a very thin barrier extending across the tube. What generates them?

82, pp. 208–214; 124, pp. 113–114; 127, pp. 188–191, 255–258, 472; 128, pp. 22–23; 130; 141, pp. 244–253; 145, pp. 220–222; 207, pp. 151–156; 243 through 250; 1517.

resonance

cavitation

1.48

Pouring water from a bottle

As water is poured from a bottle, the pitch of the pouring noise decreases. As water is poured back in, the opposite change in pitch occurs. Why?

resonance

1.49

Seashell roar

What causes the ocean roar that you hear in a seashell?

82, pp. 196–197; 141, pp. 253–254; 150; 238, pp. 57–58, 65.

resonance

vibration

1.50

Talking and whispering

What determines the pitch of your voice? Why are women's voices higher than men's? Many young men go through a stage in which their voices change. What causes that? How do you switch from a normal voice to a whisper?

81, pp. 113–114; 124, pp. 75–77, 132–136; 127, pp. 207–211; 141, pp. 238–244; 142, pp. 179–181; 145, pp. 254–255; 151, pp. 175–177; 238, Chapter 7; 239; 253, p. 387; 254.

resonance

1.51
Shower singing

Why does your singing sound so much richer and fuller in the shower (Figure 1.51)?

1.52
A shattering singer

Champagne glasses can be shattered by opera singers who sing at some high pitch with great power. Why does the glass shatter, and why must a particular pitch be sung? Why does it take several seconds before the glass shatters?

1.53
Howling wind

Monster movies always have a howling wind as a background sound to the sinister deeds of the monster. How does wind howl?

150; 164, pp. 442-443.

resonance

Bernoulli effect

1.54
Twirl-a-tune

A musical toy called "Twirl-a-tune"* is a surprisingly simple toy: it's nothing but a flexible, corrugated plastic tube made much like a vacuum cleaner hose and open at both ends. When held by one end and whirled about (Figure 1.54), it produces a musical tone. At higher speeds, you get higher pitched tones; the transition from pitch to pitch is not smooth but takes place in jumps. A gathering of many twirlers can produce quite a sound, and the fairies in a particular English presentation of *A Midsummer Night's Dream* even gave a chorus of such twirling tubes to enhance their magic (1588). How are the tones made, and why are the pitch changes discrete?

The tendency will be to dismiss the questions by pointing to the standard textbook example of sound resonance in open-ended pipes. But here you will first have to understand why there is any sound at all and why the sound's frequency range depends on the whirling speed. Also, you should figure out which way the air moves through the tube. Only then can you use the textbook explanation of why only certain frequencies will be stored and built up inside the tube.

Will the centrifugal force on the tube affect the frequency of the sound?

1588.

*Avalon Industries, Inc., 95 Lorimer St., Brooklyn, New York 11206; see Ref. 1588 for other trade names.

Figure 1.54
Twirl-a-tune.

Figure 1.51

resonance

vortex formation

1.55

Whistling wires

Why do telephone wires whistle in the wind? Why did the aeolian harps of ancient Greece sing when left in the wind? In particular, do the wires or strings themselves have to move in order to produce the sound? If they move, do they move in the plane of the wind or perpendicular to it? What determines the pitch you hear?

Suppose you were to simulate the wind whistling through telephone wires by waving a fork with long, thin prongs. Which way would you wave it, in the plane of the prongs or perpendicular to that plane? Try it both ways.

What causes the sighing of trees in winter and the murmur of an entire forest? Do all trees sigh at the same pitch?

82, pp. 304-313; 124, pp. 114-116; 126, pp. 480-482; 127, pp. 218-220; 142, p. 215; 145, pp. 149-152; 150; 155, pp. 188-189; 164, pp. 443-448; 165, p. 144; 207, pp. 156-157; 256, pp. 126-128; 257, pp. 123-130; 258 through 261.

sound from vortices

feedback

1.56

The whistling teapot

Other types of whistles use an obstruction in the way of the air stream. For example, an edge tone can be produced by directing an air stream onto a wedge (Figure 1.56*a*). Similarly, a ring tone is made by placing a ring in the stream path (Figure 1.56*b*). The most familiar of all is the common teapot whistle that has a hole in the stream's way and that produces what is called a hole tone (Figure 1.56*c*). In each example the whistling sound depends on the obstructing object, but how? What really produces the whistling you hear when your teapot boils?

124, pp. 116 ff; 126, pp. 482-485; 127, pp. 220-223; 142, p. 216; 145, pp. 169-174; 151, pp. 95-97; 257, pp. 130-138; 258; 263 through 269.

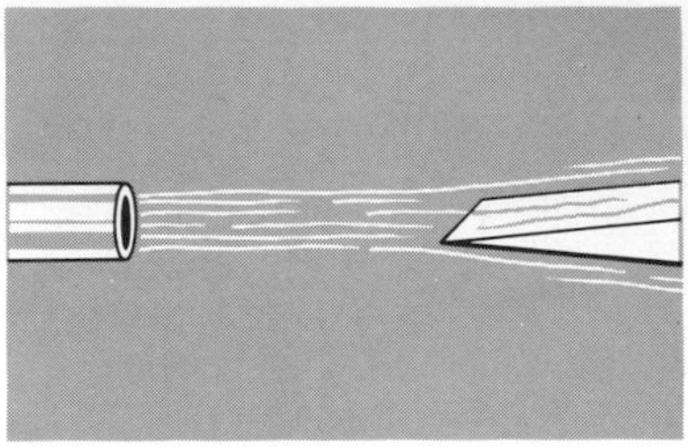

Figure 1.56a
Edge-tone setup.

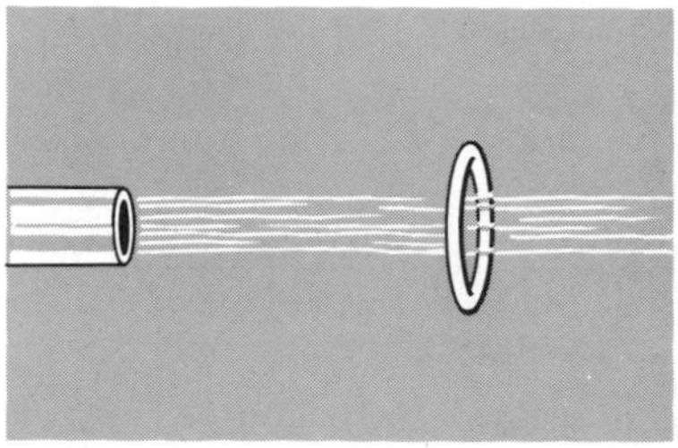

Figure 1.56b
Ring-tone setup.

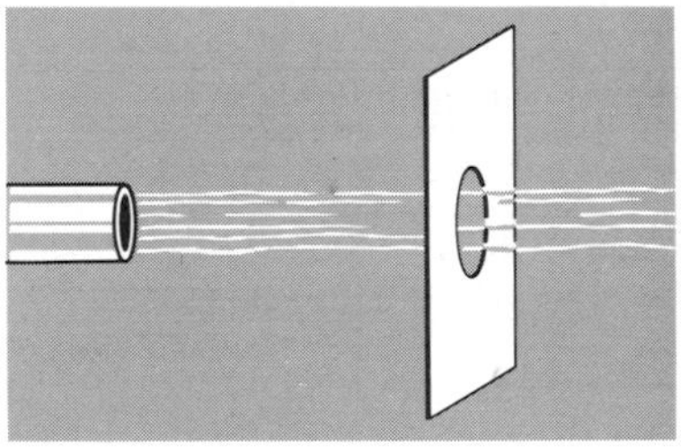

Figure 1.56c
Hole-tone setup.

resonance

sound from vortices

1.57

Blowing on a Coke bottle

Making a Coke bottle hum by blowing across its opening is an example of still another type of whistle. Not only is there an obstruction, the edge of the bottle, but there is also a cavity adjacent to the obstruction. Flutes, recorders, and organ pipes are other examples of the same kind of whistle.

Why do such devices produce particular frequencies? In other words, how do the fingering of holes on the cavity (as in the case of the flute) and the change of air pressure across the obstruction determine the different frequencies that can be made? In the case of the Coke bottle, does the bottle's mouth size affect the frequency? How about shape? Suppose I partially fill a bottle with water, excite it with tuning forks to find its resonant frequency, and then tilt the bottle. The internal shape changes, of course, but does the resonant frequency?

142, p. 163; 151, pp. 95–97; 159, pp. 74–75; 170, pp. 218-219; 258; 310; 1553.

resonance

1.58
Police whistle

How does an American police whistle work? As above, there is an edge across which air is blown and there is an adjacent cavity. There is also a small ball in that cavity. What does the ball do for the whistling? Why won't the whistle work underwater?

258.

1.59
Whistling through your lips

Finally we come to the most common whistle of them all, although perhaps the most difficult to explain: whistling through your lips. What's responsible for this sound? Can you whistle underwater?

82; 258.

1.60
Gramophone horns

Remember the old gramophones with their cranks and big horns? Why did they have horns? Did the horns concentrate sound in a certain direction? Why did they use an expanding horn and not just a straight tube? The point was that if the sound box's diaphragm coupled directly to the room's air without the intervening horn, there was poor sound emission. What can an expanding horn do in coupling the sound box with the air?

124, pp. 212–214; 186, pp. 208-209.

vibration

acoustic impedance

power

1.62
Sizes of woofers and tweeters

Why is the woofer (low frequency speaker) so much larger than the tweeter (high frequency speaker) in most hi-fi systems?

128, p. 148; 187, pp. 272–273, 280; 228, pp. 174–175.

sound from vortices

1.61
Vortex whistle

The vortex whistle (Figure 1.61) produces sound when you blow through a tube that juts out from a round cavity. Apparently vortices are set up in the cavity and, when they emerge from a central hole, a whistling sound is made. Unlike the common "police whistle," the vortex whistle produces a frequency that depends on the pressure with which the whistle is blown. Hence, by varying the pressure, you can play tunes on it. What produces the whistling sound, and why does the frequency depend on the pressure?

258; 262.

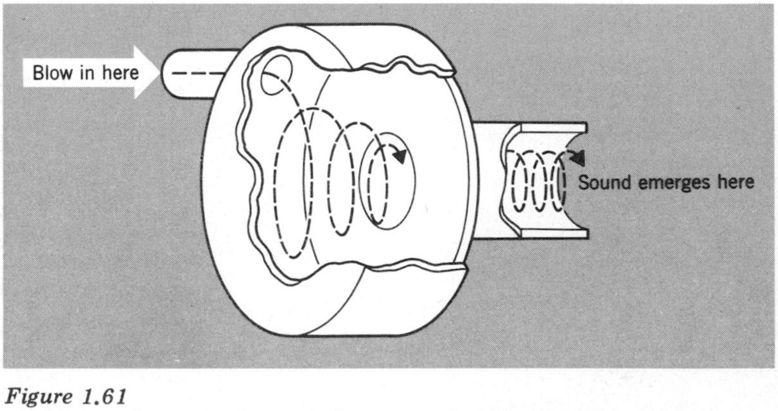

Figure 1.61
Vortex whistle (after Bernard Vonnegut, J. Acoustical Soc. Amer., 26, 18 (1954).

spherical and plane waves

intensity versus range

impedance

1.63
The cheerleading horn

How does a cheerleader's horn make the yell louder in one direction? Do multiple reflections inside the horn limit the direction of spreading? This may seem reasonable, but considering the size of the horn compared to the wavelengths of the sound, how can there possibly be such a concentrating effect due to internal reflections? So, again, why is the yell louder in the direction of the horn?

127, pp. 205-207; 142, p. 111; 145, pp. 239-240; 159, pp. 12-13; 213, p. 47; 235, p. 78; 242.

combination tones

nonlinear response

1.64
Bass from small speakers

Isn't it surprising that telephones, high fidelity earphones, and small transistor radios can reproduce bass (see Prob. 162)? The speakers in them are so small, yet one does hear bass from them. The horns on early gramophones shouldn't have been able to handle low frequencies either. In both cases, why is bass heard?

82, pp. 256-261; 124, pp. 29-32, 84-87, 165, 214; 127, pp. 400-402; 128, pp. 31-32, 56-59; 151, pp. 117-120; 152, pp. 105-108; 170, pp. 40-42; 184, pp. 403-406; 209, pp. 179-187; 228, pp. 253-256; 237, pp. 66-68; 256, pp. 231-245; 270, pp. 129-133; 271, pp. 50-52; 272, Chapter 7, pp. 411-413; 273 through 279.

Doppler shift

1.65
Screams of race cars, artillery shells

Why does the pitch of a race car's scream change as the car speeds past you? Surely the noise thrown forward is no different from that thrown backward.

On the battlefield men can predict the danger of an incoming shell by the scream it makes. Not only do they listen for the change in loudness but also the pitch and its change. What does the pitch tell them?

128, p. 19.

Doppler shift

ranging

1.66
Bat sonar

To find their way and to locate insects, most bats emit a high frequency sound and then detect the echo. What does a bat actually do with the echo? Does it emit a sound pulse and time its return, thus finding the distance to a reflecting object? Does it detect the Doppler shift (frequency shift) if either it or the object is moving? Or does it locate the object by triangulation of the return sound, much as we perceive depth with binocular vision? Maybe it is even more complicated because some bats chirp, that is, each sound pulse sweeps from about 20 kHz down to 15 kHz. How can such a chirp be used to extract more information about the object?

What is the smallest insect that a bat can detect using a constant frequency pulse of 20 kHz?

142, pp. 353-354; 280 through 284; 1493 through 1497.

Brownian motion

hearing

1.67
Hearing Brownian motion

Hearing involves detection of air pressure variations, right? Well, the air next to the eardrum is continually fluctuating in pressure. How large are those fluctuations on the ear drum, and are they large enough to be heard? If they are, then why don't you hear them? Shouldn't there be a continuous roar in your ears?

311.

acoustic power

signal-to-noise ratio

1.68

When the cops stop the party

Some cocktail parties are quiet; others are loud. Can you roughly calculate the critical number of guests beyond which the party becomes loud? You might take the transition point as when the background noise on your listener becomes as great as your volume on him.

Suppose the guests are called to attention by the hostess, and then, afterwards, allowed to resume their conversations. About how much time will pass before the party becomes loud again?

285.

shock wave

1.69

V-2 rocket sounds

If you were being fired on with artillery shells, you would first hear the shell's scream, then its explosion, and finally the roar of the gun. But in the V-2 rocket attacks on London during World War II, the first two sounds came in reverse sequence: first the explosion, and slightly later the rocket's whine. Why the difference?

142, p. 153.

hearing

1.70

Cocktail party effect

How can you distinguish the words of one person when there is a lot of background noise? If you tape a friend talking to you at a loud party, it's likely that on tape you won't be able to hear him at all, much less understand the words. Why the difference?

171, p. 62; 238, pp. 15-16; 286.

acoustic conduction

hearing

1.71

Taping your voice

If you've ever taped your own voice, you were probably surprised by how thin it sounded when you played it back. Other people recorded their voice and their playbacks sounded fine to you. But yours. . .it just wasn't right. What was wrong?

312.

acoustic conduction

hearing

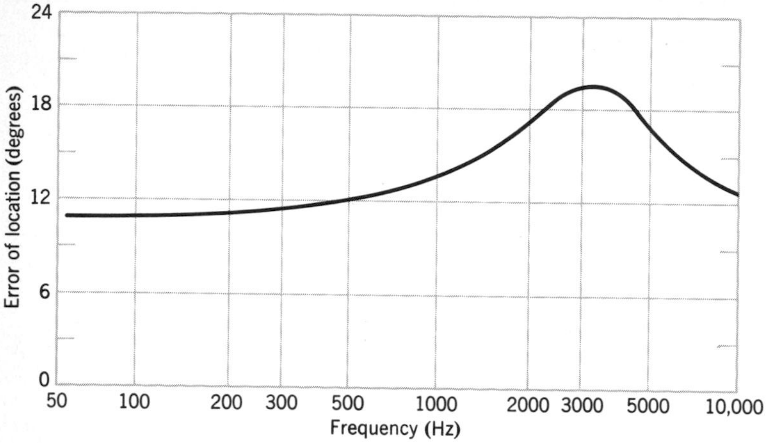

Figure 1.72
[From Konishi (Ref. 1554) after Steven and Newman (Ref. 1555).]

1.72
Fixing the direction of a sound

Since you have two ears instead of one, you can locate the direction of a sound as well as just hear the sound. If you were to plot your ability to fix the direction of a pure tone versus the frequency of that tone, you would find that your ability is reasonably constant with frequency except in the region of 2 to 4 kHz (Figure 1.72). Why does your ability get worse in that particular region, whereas it is better for both higher and lower frequency tones?

1554; 1555.

shock waves

refraction

1.73
Sonic booms

What causes the sonic booms produced by supersonic aircraft? Is the boom produced only when the plane first breaks the sound barrier? Does it depend on the engine noise? Sometimes you hear not just one boom but two right in succession. Why two? Why not always two? Does the boom depend on the aircraft's altitude? Does it matter if the plane is climbing, diving, or turning? Under some circumstances the aircraft may generate a "superboom"—an especially intense shock wave. Under other conditions a boom will be made by the plane but will never reach the ground. What probably happens to it?

288 through 298.

shock wave

reflection

1.74
Sounds of thunder

When I was little my mother told me that thunder had something to do with lightning. How is thunder produced, and why does it last for such a relatively long time? Must it always boom? I've read that if you stand within 100 yards of the lightning flash you first hear a click, then a crack (as in a whip crack), and finally the rumbling. What causes the click and the crack? If you're a little further away you'll hear a swish instead of the sharp click? Why a swish?

82, pp. 114–116; 220, Chapter 6, pp. 195–199; 299, pp. 124–127; 300, pp. 162–163; 301, pp. 66-67; 302 through 305; 1617.

sound propagation

attenuation

1.75
Hearing aurora and frozen words

Is it possible to hear aurora*? There have been reports of cracklings or swishings (sounding like "burning grass and spray") coordinated with changes in the light intensity of aurora. While it is hard to imagine how sound made at such a high altitude (above 70 km) could reach an observer on the ground with any appreciable power simply because of the attenuation over such a large distance, recently an explanation along this line was proposed: electrons from the aurora would excite what are called plasma acoustic waves that would create normal acoustic waves. Regardless of the actual mechanism, however, could you hear a sound made so high? What exactly happens to the acoustic power when the sound travels downward through the atmosphere?

Another interesting explanation has been that "What one hears is one's own breath that freezes in the cold air" (Figure 1.75). When the air is calm and very cold, can you actually hear the collision of ice crystals formed from your breath? If this is ever possible, how cold must it be?

1506 through 1511; 1532.

*See Prob. 6.30.

Figure 1.75

shock waves

1.76
Dark shadows on clouds

During the fighting near the Siegfried Line in World War II, U. S. troops spotted dark shadows crossing over white cirrus clouds. These shadows were arcs whose centers lay on the German side, supposedly being caused by the heavy artillery. Why were these shadows produced? Would you expect the shadows to come singly or always in pairs? Finally, was the cloud background necessary?

142, p. 154; 306 through 308.

1.77
Whip crack

What makes the sound when a whip is cracked? Try to support any guess with rough numbers.

82, p. 30; 159, p. 184; 288; 309.

2
The walrus speaks of classical mechanics

Linear kinematics and dynamics

2.1 to 2.22

force	equation of motion
displacement	energy
velocity	momentum
acceleration	
cross section	
flux	

2.1

Run or walk in the rain

Should you run or walk when you have to cross the street in the rain without an umbrella? Running means spending less time in the rain, but, on the other hand, since you are running *into* some rain, you might end up wetter than if you had walked. Try to do a rough calculation, taking your body as a rectangular object. Using such a model, can you tell if your answer (whether to run or walk) depends on whether the rain is falling vertically or at a slant?

2.2

Catching a fly ball

If in baseball a highly hit ball—a "high fly"—is knocked to your part of the outfield, there are two things you could do. You could dash over to the correct place and wait to catch the ball, If you do, I'll ask you how you guessed where the ball was to come down. Or, you might run over at a more or less constant speed, arriving just in time to make the catch. In that case I'll ask you how you determined the correct running speed. Experience helps, of course, but you must also have an intuitive feel for the physics involved in the ball's flight. What tips you off as to where to go and how fast to run?

121.

2.3

Running a yellow light

Every driver will occasionally have to make a quick decision whether or not to stop at a yellow light. His intuition about this has been built up by many tests and some mistakes, but a calculation might reveal some situations where intuition will not help.

For some given light duration and intersection size, what combinations of initial speed and distance require you to stop (or run a red light)? What range of speed and distance would allow you to make it through in time? Notice that for a certain range of these parameters you can choose either to stop or not. But there is also a range in which you can do neither in time, in which case you may be in a lot of trouble.

123.

2.4

Getting the bat there in time

To make a hard line drive you must get the bat into the proper position for the collision with the thrown ball. How much error can you stand, both in the vertical direction and in time, and still be able to get the hit? For example, would it be all right if your timing is off by some small amount such as 0.01 second?

4.

Exceptionally good reference: Crabtree (36).

2.5
Turn or stop

It is hard to find any physics of a more real-world nature than that which involves your own death. For example, suppose you suddenly find yourself driving toward a brick wall on the far side of a T-intersection (Figure 2.5). What should you do? Use your brakes *fully*, without skidding, while steering straight ahead? Turn at full speed? Or turn while applying your brakes as well as you can?

Consider this question in parts. First, assume you can stop the car in time by braking and steering straight. Would turning instead also save you? Right now you'll probably want to think about this with ideal conditions. Later you can throw in the possibility of a skid, differences in road handling between front and rear tires, and brake fade. What if the straight-ahead option won't stop you in time? Should you bother turning, or are you doomed to an early death?

If you were to find a large object in the road, would it be better to attempt a head-on stop or to try to steer around it? Of course, the answer will depend on the object's size.

Don't answer quickly in any of these cases, for even though you may be very experienced, your intuition may be wrong. If it *is* wrong, the question becomes far more relevant.

122.

Figure 2.5
Turn or stop for the brick wall?

2.6
The secret of the golf swing

How should you swing a golf club in order to impart maximum speed to the ball? While many golfers might prefer to keep the problem in the realm of the esoteric, we should be able to consider it using some physics. What should the initial backswing angle be? When should you relax your wrists? Should the club, arms, and ball be along a straight line when contact with the ball is made?

5; 6; 1613.

momentum transfer

center of mass motion

2.7
Jumping beans

Why do jumping beans jump? There they are, lying quietly in your hand, when suddenly, every few seconds, they jump into the air. Can you convince a friend they violate conservation of momentum?

7; 8; 9, p. 238.

2.8
Jumping

How high can you jump? Can you calculate the height? Would you be able to jump higher if your legs were longer? Is there any initial

orientation or way of swinging the arms that would increase your jump?

How far can you jump? Some athletes bicycle their legs as they fly through the air; does this really help? At what angle is it best to leave the ground? At the angle (45°) that maximizes the range of a projectile?

Why do pole vaulters and broad jumpers charge forward for the jump but high jumpers approach the jump much slower? Shouldn't all three obtain the maximum possible speed before they leave the ground?

Can you jump as high or as far on a seacoast beach as you can on a mountain? If the height above sea level makes a difference, some caution should be exercised in comparing record jumps at various altitudes.

10 through 13.

2.9

Throwing the Babe a slow one

Pitchers sometimes threw Babe Ruth slow balls because they thought it would be harder for him to hit a home run if the ball were moving slower when struck. Does this belief have any physical basis?

14, p. 274.

impulse

collisions

2.10

Karate punch

In karate classes I was taught to terminate a punch, kick, or edge-of-hand chop several centimeters inside my opponent's body. This is different from normal street fighting where there is much follow through. Which technique will produce more damage? Through a rough calculation, can you show the feasibility of a karate fighter breaking a wooden board, a brick, or a human bone with a punch?

1632.

2.11

Hammers

Should a sculptor use a heavy hammer or a light one on his chisel? Which hammer should be used to drive a nail? When would an elastic collision (that is, one with a full rebound of the hammer) be more desirable than an inelastic collision? Consider something on a grander scale, a piledriver, for example: should the piledriver be heavy or light compared to the pile? A guess is easy to make, but a calculation should back it up.

15; 16, pp. 396–399.

elasticity

2.12

Softballs and hardballs

Should you hit a hardball and a softball differently? In particular, should there be more follow through for one than for the other?

4.

2.13

Heavy bats

Why do home-run hitters prefer heavy bats? It seems it would be harder to give the heavy bat a large final speed and hence harder to hit the ball very far. Should you use the heavy bat for a bunt? Considering the range of weight normally used, does the bat's weight really make much difference?

4.

center of mass motion

2.14

Jerking chair

A body's center of mass moves only if an external force is applied, but you can get to the other side of the room in a chair without letting your feet touch the floor. If all your twistings and contortions are internal forces, what provides the external force?

power

energy

2.15

Click beetle's somersault

If you poke a click beetle lying on its back, it throws itself into the air, as high as 25 cm, with a noticeable click. That in itself is trifling, you say? But the beetle, without using his legs, hurls himself upward with an initial acceleration of 400 g and then rotates his body to land on his legs. 400 g! Even more surprising is that the power needed for this is 100 times the direct power output of any muscle. How does the beetle produce such an enormous power output? How frequently can he perform this amazing feat, and what physically determines the frequency?

21; 1485; 1531.

weight

momentum transfer

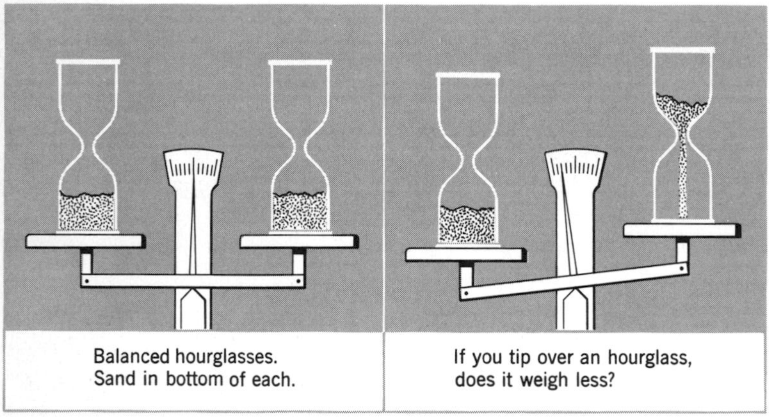

Figure 2.16
The weight of an hourglass.

2.16

The weight of time

Does the weight of an hourglass depend on whether the sand is flowing (Figure 2.16)? If some of the sand is in free fall, won't the weight of the hourglass be less?

17.

pressure

force

2.17

Pressure regulator

Have you ever used a pressure cooker? Mine has a solid cylinder on top of the lid that somehow regulates the pressure. There are three different size holes drilled into the side of the cylinder, and I pick the pressure by placing the appropriate hole over the hollow stem extending from the pan (Figure 2.17). How does it work? The pan's steam must lift the same cylinder no matter which hole I pick. Why do I get different pressures by using different holes?

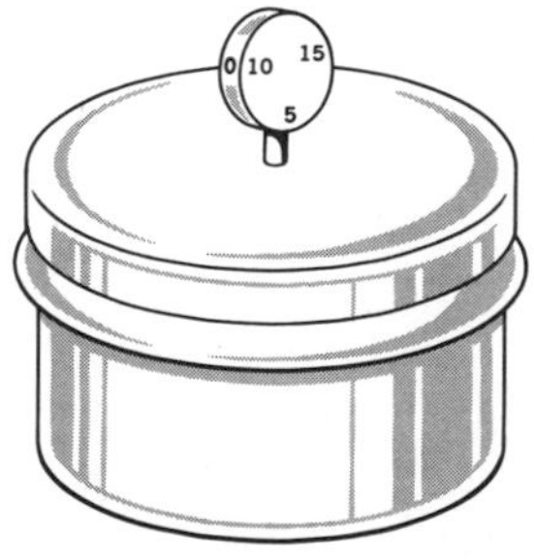

Figure 2.17
Pressure regulator.

elasticity

2.18

The superball as a deadly weapon

If a small superball,* which is a very elastic rubber ball, is dropped immediately after a large one as shown in Figure 2.18*a*, the small ball will be shot back up into the air after the two balls strike the floor. If the mass of the small ball is appropriately chosen, the other ball will completely stop at the floor and the smaller ball will rebound to about nine times its original height.

Try this as well: drop a large superball, a small superball, and a ping-pong ball as shown in Figure 2.18*b*. If the balls are appropriately chosen, the ping-pong may reach almost 50 times its original height.

18 through 20.

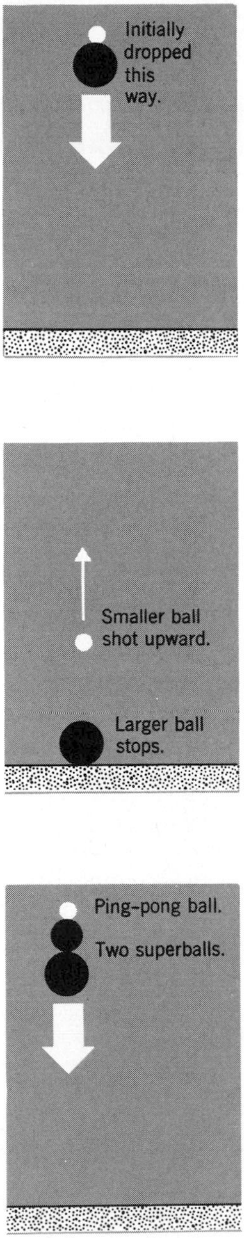

Figure 2.18
Rebounds of several superballs dropped simultaneously.

*® Wham-O Manufacturing Company, San Gabriel, California; similar balls are sold under other brand names.

Friction

2.19 through 2.22

2.19

Locking brakes

If you must stop your car in a hurry, should you slam on the brakes and lock them?

2.20

Wide slicks on cars

If you had to decide between regular-width tires with no treads and wide tires with no treads (both are called slicks), which would you choose for better braking ability?

In drag racing wide slicks are preferred for the rear tires. Why?

work

power

2.21

Friction in drag racing

In drag races there are two measurements of interest: the final speed and the total elapsed time on a quarter-mile course. To help gain traction, a sticky fluid is poured under the rear wheels before the "go" light, but apparently the track's friction really affects only the elapsed time and has little influence on the final speed. Why?

22.

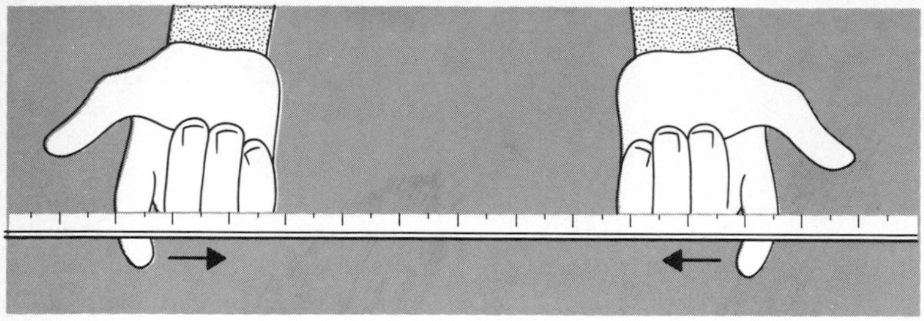

Figure 2.22
Sliding your index fingers under a yardstick

2.22
Sliding stick across fingers

Hold a yardstick horizontally on your index fingers and slide your fingers together smoothly (Figure 2.22). Does the stick slide smoothly over your fingers? No, it slides first on one finger and then on the other, and so on. Why does the sliding change back and forth?

23; 24, pp. 83–84.

angular and linear momentum conservation

action–reaction

2.25
Left on the ice

For a mean trick, your friends desert you in the middle of a large frozen pond. The ice is so very slippery that you can't walk off in a big huff, or even crawl off in a small one. How can you get off the ice?

Now let us suppose you were first placed on your back. After lying there for a while, your back is frozen to the bone and you want to turn over. How do you do it on such slippery ice?

They could have been meaner. They could have stood you up and tied you to a pole fixed in the ice (Figure 2.25). How could you turn yourself about that pole if they

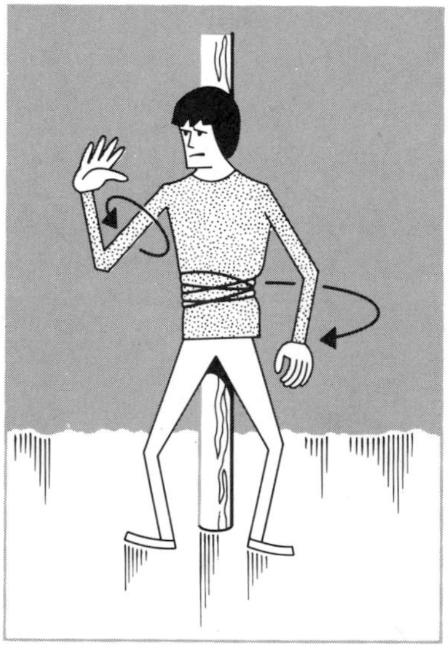

Figure 2.25

Rational kinematics and dynamics

2.23 through 2.55

angular motion — angular momentum

torque — rotational energy

moment of inertia

2.23
Accelerating and braking in a turn

Why is it unwise for you to do any significant braking when your car is in a turn? For example, suppose that while in the curve you decide you are taking it a bit too fast. What happens if you apply the brakes too hard?

Race drivers accelerate as they are leaving a curve, not while they are in it. Why?

29; 30, p. 8.

friction

torques

2.24
Starting a car

There is much debate about how to start a stick-shift car on a slippery road. Some claim you should have the car in low gear; others swear you must put it in high. Why does the gear you use matter at all? What is needed to get the car moving? Why must the initial speed be small? What advantages would any one gear have over the others? You'll have to explain how the torque applied to the wheel depends on the gear and decide when you need more or less torque.

28.

had left your hands free? The pole is too slippery to use, and the ice is too slick to turn with your feet. How then do you turn around to face the other way?

9, p. 238.

precession

center of gravity

2.26
Turning a car, bike, and train

How do you turn a bicycle? How, exactly, do you initiate the turn? On a motorcycle you turn by leaning the bike and not by turning the handle bars. Why the difference?

When a train goes around a curve it leans because the roadbed is banked to prevent the centrifugal force from derailing the train. Will the leaning also affect the turning as in the motorcycle case? Try a rough, back-of-the-envelope calculation to see. Whether or not the effect is significant or even real, the outside rail on a curve is often elevated.

Finally, do you have a similar consideration in turning high speed cars, such as the Formula 1 race cars?

16, pp. 535-536; 24, pp. 156-157; 35; 36, pp. 43-44; 37, pp. 146-147; 38, pp. 89-93; 1612.

collision

impulses

linear kinematics

2.27
Pool shots

How do you set up a "follow shot" (the cue ball follows after the ball with which it has collided) or a "draw shot" (cue ball returns after the collision). I thought that if a moving object hits a stationary object of the same mass, the first object stops.

A massé shot is one in which the cue ball describes a parabolic path (Figure 2.27*a*). (These shots are usually outlawed in most pool halls for a missed massé shot will rip up the table cover.) How must the cue ball be hit to bring this about and why, in detail, does it happen?

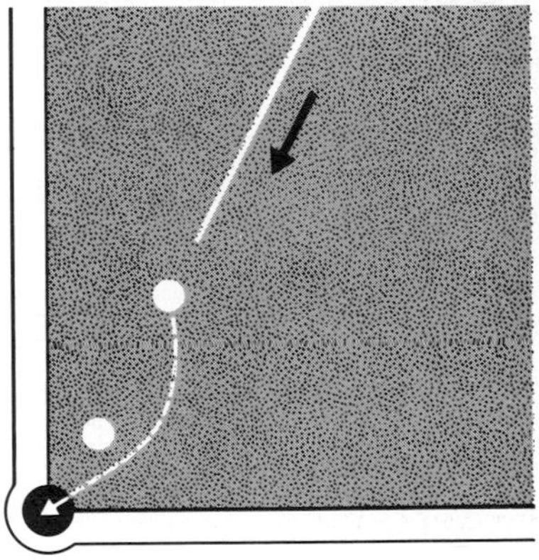

Figure 2.27a
Masse shot.

Why is the cushion higher than the center of the balls (Figure 2.27*b*)? Wouldn't you get better rebounds if the cushion were at the center's height?

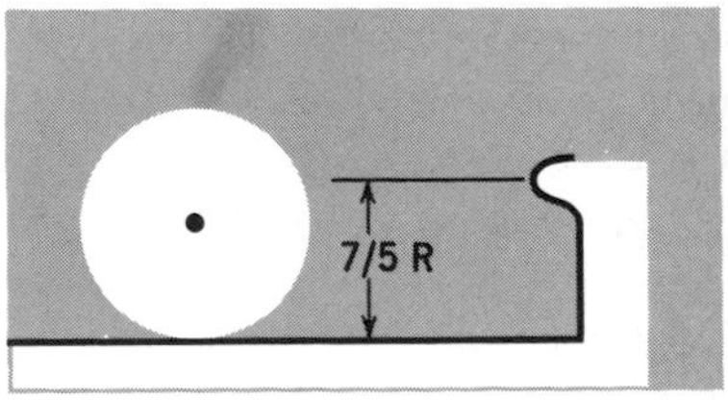

Figure 2.27b
Height of cushion.

How can English be used in a cushion shot?

14, pp. 143-146; 24, pp. 158-161, 250-251; 25; 26, pp. 183-186; 27, pp. 139-143, 268-274, 290-301.

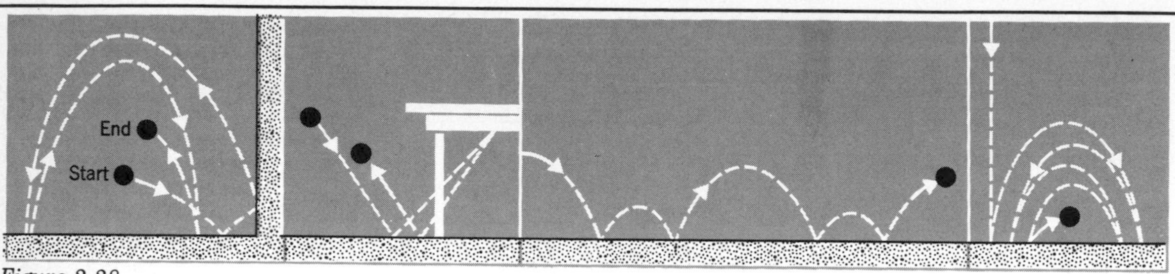

Figure 2.28
Superball tricks (after drawings copyrighted 1970 by Wham-O Manufacturing Co., used with permission).

2.28
Superball tricks

One of the most significant advances made by our technological society is the Superball. Because of its high elasticity it can perform some rather amazing tricks. Several are shown in Figure 2.28. Figure out how you set up each trick and explain how they work.*

31; 32.

*® Wham-O Manufacturing Company, San Gabriel, California; similar balls are sold under other brand names.

stability

mechanical efficiency

2.29
Bike design

Why is a modern bicycle designed the way it is? In the past there has been a great variety of designs (Figure 2.29*a*). Some, for example, had radically different wheel sizes, and some had the pedals attached to the front wheel. Is the modern bike more efficient or more stable than its predecessors?

Why does the modern bike have a curved front wheel fork? Would the bike be more or less stable with the other possible fork designs shown in Figure 2.29*b*?

35; 39; 41; 42, Vol. II, Chapter 6; 43.

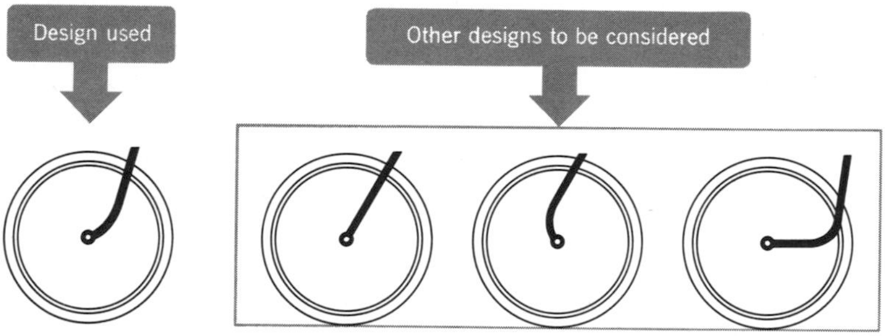

Figure 2.29b

Figure 2.29a

resonant driving force

2.30
Hula-Hoop

The Hula-Hoop* is a plastic hoop that can be kept rotating about your waist by an appropriate circular motion of your hips (Figure 2.30). The toy was first popular in the 1950s, but similar hoops rotated about the arm or leg have been used for toys and in dances for a long time. The American Indians, for example, used them in some types of hoop dances.

Think about what it takes to keep a Hula-Hoop up and going. You throw it around your waist and then trap it with your driving hula motion. Should the initial speed you give it be more than the speed at which you are going to trap it? How do you drive it around? Is the hoop's motion in phase with yours? What is the minimum speed you can use?

34.

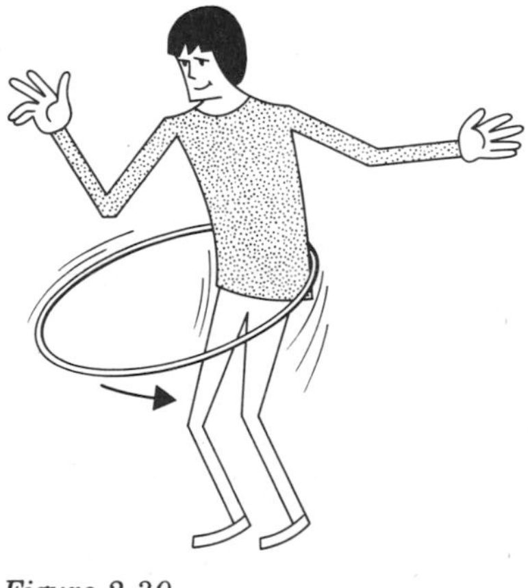

Figure 2.30
Hula-Hoop.
*® Wham-O Manufacturing Company, San Gabriel, California.

Figure 2.31
"Nobody likes to fall, Rocco—but this is ruining our image." The Saturday Evening Post.

stability

torques

2.31
Keeping a bike upright

How do you keep your balance on a bicycle? When you sense a fall, do you steer into the fall and thereby right the bike? Or does the bike itself do most of the stabilizing? It must at least contribute some stability because if it is pushed off riderless it will stay up for almost 20 seconds.

How do you balance and steer the bike when you ride without using your hands? Suppose you stand next to the bike and you lean the bike to the right. Which way does the front wheel turn and why?

24, pp. 156–157; 35; 36, pp. 43-44; 37, pp. 146–147; 38, pp. 89-93; 42, Vol. II, Chapter 6; 43; 44, pp. 122–123.

driven rotational motion

2.32
Cowboy rope tricks

How does a cowboy keep his lasso up and spinning? What minimum speed must he maintain in order to keep the lasso horizontal? Vertical?

33; 120.

moment of inertia

stability

2.33
Spinning a book

If you hold it closed with a rubber band, you can toss this book into the air spinning about any of the three axes shown in Figure 2.33. The motion about two of the axes is a simple, stable rotation. The rotation about the third axis, however, is much more complicated, no matter how carefully you throw the book. (See Figure 2.33.) Try it. What causes that uncontrollable wobbling about the third axis?

44, p. 115.

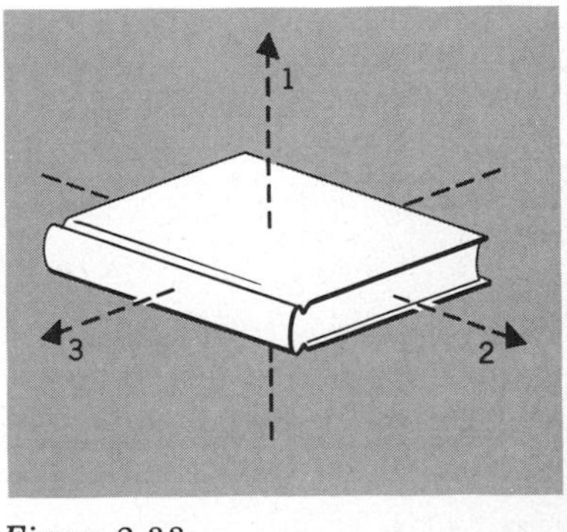

Figure 2.33

energy conservation

collision

2.34
Fiddlesticks

Fiddlesticks* is a remarkably simple yet fascinating toy. It consists of a plastic ring (of relatively large inner diameter) on a stick. Once the ring is sent spinning by a flick of your fingers, the stick is held vertically. The ring begins to drop (slower than you would expect), and as it comes down the stick, the ring spins faster and falls even slower (Figure 2.34). By inverting the stick just before the ring reaches the lower end of the stick, the process can be repeated indefinitely. Why does the spin increase as the ring falls? In fact, why doesn't the ring just fall with the full gravitational acceleration?

Now use two rings at once. Not only is this more spectacular, but a curious thing often happens. The top ring may be dropping faster than the lower ring and thus may run into the lower ring. If this occurs, the rings bounce apart, the upper ring rising (Figure 2.34). Why?

*® Funfair Products, Inc., New York, N.Y. 10016.

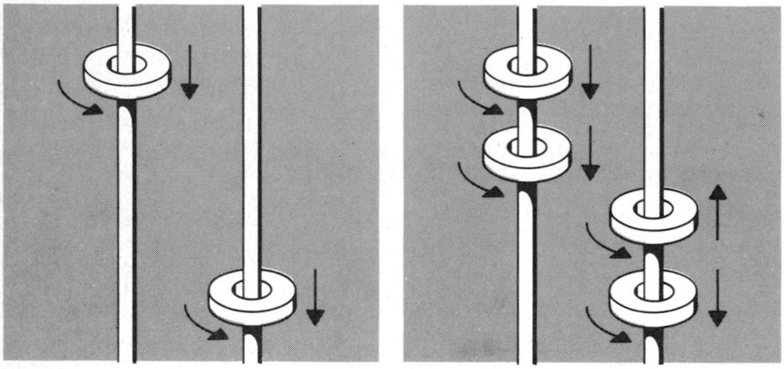

Figure 2.34

torques

center of gravity

2.35
Eskimo roll

How does a kayaker right an overturned kayak without ever leaving the cockpit?

45; 1563.

2.36
Large diameter tires

Will large diameter tires really make your car go faster?

moment of inertia

stability

2.37
Car in icy skid

If your car starts to skid on an icy road, are you supposed to straighten it out by turning the front wheels in the direction in which you want to move or in the direction of the skid? Why?

46.

2.38
Tire balancing

If your tire is balanced statically with a simple bubble leveler, will it still be balanced when it's spinning? Can you get both static and dynamic balancing with a single balancing weight added to the rim? How about two?

47.

torque

moment of inertia

2.39
Tearing toilet paper

Why, on some toilet paper dispensers, can I get a long piece of toilet paper without tearing if the roll is fat, but when the roll has been nearly used up, the paper inevitably breaks too soon, giving only short pieces? Why is just the opposite true for other dispensers?

torques

angular momentum

2.40
Skipping rocks

How does a stone skip across the water? If you skip a stone across hard-packed, wet sand, the marks in the sand provide a record of the stone's flight. You'll find the first bounce is short (several inches), the next is long (several feet), and this sequence repeats itself over and over until the stone comes to rest (Figure 2.40*a*). Why does it follow this pattern?

During World War II the skipping rock effect was used by the British in the bombing of German dams. It is very difficult to drop a bomb on a dam, especially when you are being fired upon. So, the RAF developed a bomb (cylindrical, with a length of about 5 feet and a slightly smaller diameter) which was given a backspin around its length of about 500 rpm in the plane's bomb bay before it was released over the target (Figure 2.40*b*).

When it hit the water, the bomb skimmed like a stone, bouncing in shorter and shorter jumps until it hit the dam itself. Then, instead of rebounding away, the back-spin forced it against the wall and made it crawl downwards until it exploded, on a hydrostatic fuse set for 30 feet below surface, still clinging to the dam. It was a beautifully simple idea for positioning a bomb weighing almost 10,000 lbs to within a few feet. [50]*

48; 49; 1486.

*From *The Royal Air Force in World War II*, edited by Gavin Lyall, copyright © 1968 by Gavin Lyall. Published by William Morrow and Company, Inc.

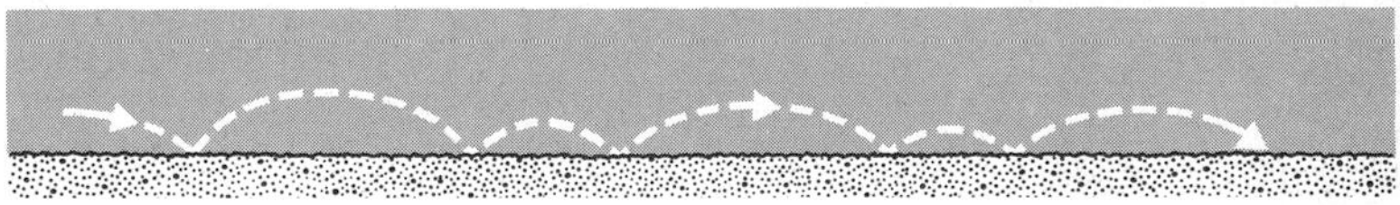

Figure 2.40a
Path of skipping rock on sand.

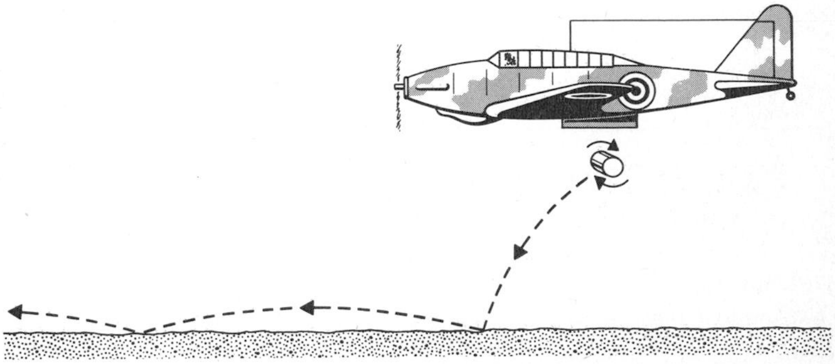

Figure 2.40b
The skipping–rock bomb.

torque

angular momentum

2.41

Car differential

When your car takes a turn, the outside wheels must move faster than the inside wheels. Since there is an inside and an outside wheel on each axle, how is this turning accomplished?

24, pp. 254–255; 52, pp. 500–501.

moment of inertia

2.42

Racing car engine mount

Some of the European racing cars have their engines mounted in the centers of the cars, rather than in the fronts or rears. The racing circuits in Europe are really just streets and therefore have lots of fast turns. Considering the torque needed to turn a car, what advantages does a center-mounted engine have over the conventionally mounted engine in this situation?

center of gravity

stability

2.43

Tightrope walk

How does a tightrope walker keep his balance? Why does a long bar help?

orbits

torques

angular momentum

2.44

Carnival bottle swing

There's an old carnival sideshow trick involving hitting a bottle with a pendulum suspended directly above it (Figure 2.44). To show your skill, you must start the pendulum so that it misses the bottle on the forward swing and then hits it on the return swing. The barker, of course, won't let you throw the pendulum over the top. Still, this trick shouldn't be too hard, should it? With a few tries you should find the arc needed to win the prize. Well, try it, and then worry about why it doesn't work and then about what would make it work.

53, p. 184.

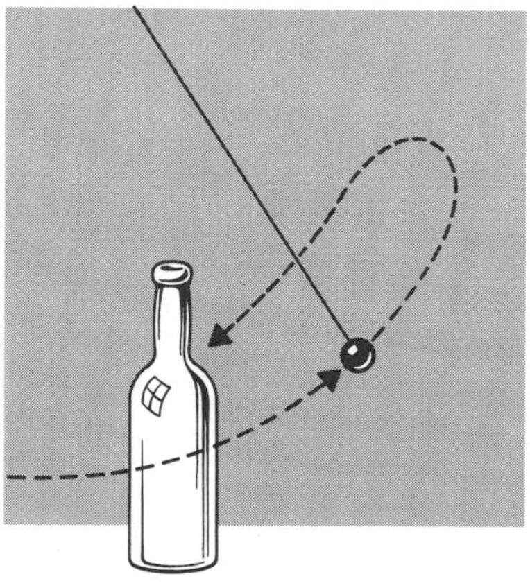

Figure 2.44
Swing the ball so as to hit the bottle on the return swing.

torques

angular momentum conservation

2.45

Falling cat

It is common knowledge that if you drop a cat upside down it will land on its feet; even tailless cats show this mysterious ability to right themselves. Now, if there is no external torque the cat's angular momentum must be constant. *Is* the angular momentum constant during the fall? If so, how does the cat turn itself through a full 180°? If the angular momentum is not conserved then somewhere, somehow, there must be a torque on the cat. But where? References 36 and 54 contain photographs of a cat turning over, and they are clear enough to provide an explanation.

9, p. 238; 36, pp. 56–57; 54; 55.

2.46

Ski turns

A ski turn can be a set of rather complicated twists and gyrations but consider the several simple elements of such a turn.

The Austrian turn requires a sinking of the whole body, followed by a powerful upward thrust and a rotation of the upper part of the body. The lower part, and hence the skis, rotate the opposite way as a result. Why? For a given upper-body rotation, how much does the lower body turn?

The normal skiing stance gives a straight skiing path, but a shift of one's body either forward or backward on the skis will force a turn. Why and which direction of shifting gives which sense of turn?

If the skis are edged (the ski's uphill edge is held into the snow so that the ski is at an angle to the snow's surface), turns are also caused by a shifting of weight, but the sense of turn is opposite to that in the normal-stance case. Why is that, and again, what forces the turn?

55; 1525.

2.47
Yo-yo

Can you tell me why a yo-yo comes back up? How about the sleeping yo-yo in which the yo-yo is thrown down and spins at the end of the string until it returns when you give the string a slight jerk. If the sleeping yo-yo touches the floor, it will walk along the floor—this is "walking the dog."

As an even better trick, put the yo-yo to sleep, take the string off your finger and hold it between your thumb and index finger. Now give that hand a slap. As soon as the yo-yo starts to climb back up, let go of the string. The yo-yo will charge up the loose string, neatly winding it up. Dazzle your friends by catching the yo-yo in your coat pocket when the string is wound up.

24, pp. 246–247; 56.

2.48
Slapping the mat in judo

When you've been thrown in judo, slapping the floor with your arm at the moment of impact will prevent injury in the fall. How does it work? The effect is probably partly psychological, but I know that for the most part it is real. When I was taking lessons, I was always hurt when I missed the slap or when my timing was off. When I slapped the mat properly, my discomfort was only mild.

angular momentum
torques
stability

2.49
Bullet spin and drift

Why are bullets given a spin as they travel down the rifle barrel? The rifle, in fact, derives its name from the rifling—the spiral grooves in the bore—that impart this spin.

If the bullet is given a counterclockwise spin as seen from the rear, it will drift to the left of the target. A clockwise spin will cause a drift to the right. Why? Can you calculate, roughly, the amount of drift for small and large guns?

16, pp. 536–537; 26, pp. 154–155; 36, pp. 53, 140–144; 37, pp. 148 ff, 274 ff; 38, pp. 117–119; 40, pp. 440–441; 64, pp. 393–394.

center of gravity
torques
stability

Figure 2.50
Leaving a leaning tower for a librarian.

2.50
The leaning tower of books

If you want to construct a stack of books leaning to the side as much as possible (Figure 2.50), what is the best way to stack them? Would you put the edge of one book over the center of the lower book?

57 through 59; 1559.

torques

angular momentum

center of gravity

stress and strain

2.51
Falling chimneys

When a tall chimney falls, it usually breaks in two at some point along its length. Why doesn't it fall in one piece? Where do you think the break will occur? Will the chimney bend towards or away from the ground after the break (Figure 2.51*a*)? You can check your answer by toppling a tall stack of children's blocks and seeing which way the stack curves as it falls.

If the chimney does not break, something even stranger may occur: the base of the chimney may hop into the air during the fall (Figure 2.51*b*). How can it do this, seemingly against gravity?

9, pp. 124–125; 60 through 63.

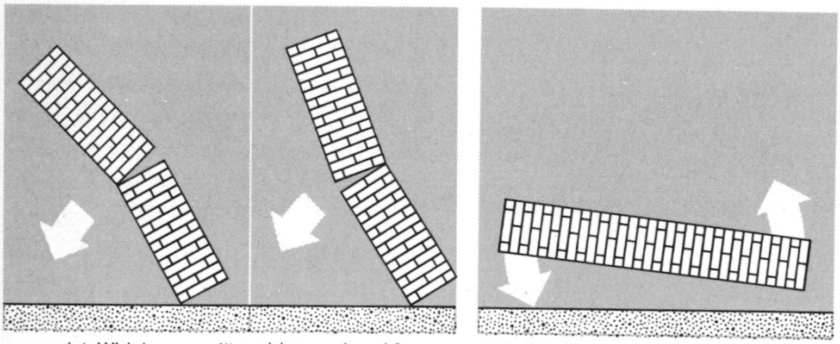

(a) Which way will a chimney break? (b) If it doesn't break, it may hop up.

Figure 2.51

forces in a rotating frame

2.52
The Falkland Islands battle and Big Bertha

During World War I, there was a famous British-German naval fight near Falkland Islands (about 50°S latitude) in which the British shots, while well aimed, were mysteriously landing about a hundred yards to the left of the German ships. The British gun sights were not faulty, for they had been set very precisely back in England. During the German shelling of Paris in the same war, a huge artillery piece called Big Bertha would pump shells into the city from 70 miles away. If normal aiming procedure has been employed, Big Bertha's shots would have missed their mark by almost a mile. What was happening to the shells?

68 through 72; 1488.

forces in a rotating frame

2.53
Beer's law of river erosion

Why does the right bank of a river in the northern hemisphere suffer more erosion, on the average, than the left bank?

24, p. 164; 72; 73.

2.54
A new twist on the twirling ice skater

The twirling ice skater has long been used as an example of the conservation of angular momentum. When she pulls her arms in, she spins faster due to the conservation of angular momentum (there are no external torques).

This is all true, of course, but I would like to explain the speeding up in terms of forces because force arguments are more accessible to the imagination than angular momentum arguments. What *is* the force that speeds up her spinning?

74.

airfoil theory

angular motion

2.55
Boomerangs

Returning boomerangs are designed to be thrown great distances and to return to the thrower. Australian natives have thrown them as far as 100 yards and to heights of 150 feet with five complete circles. The nonreturning type, which is more practical for hunting, can be thrown as far as 180 yards.

The ordinary boomerang is shaped like a bent banana. Is it essential that the boomerang have this particular shape? Can one make a returning boomerang in the shape of an X or a Y? Most boomerangs are designed to be thrown with the right hand. What is the difference between left- and right-handed boomerangs? Why does a boomerang (of any shape) return? Why does it loop around in its path (see Figure 2.55)? Finally, how does the path depend on the boomerang's orientation as it leaves the thrower's hand?

26, pp. 153-154; 37, pp. 291-296; 65 through 67; 1564.

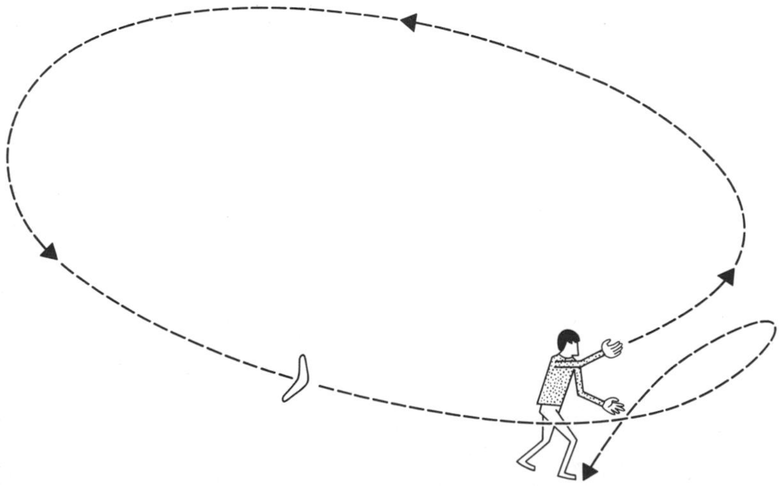

Figure 2.55
Boomerang path.

Periodic motion

(2.56 through 2.68)

angular momentum

torques

potential and kinetic energy

center of gravity

2.56
Swinging

When you swing, you must pump first to gain height and then just to keep going. How does pumping work? How do you pump if you want to start to swing from rest? Do you pump differently when you are sitting and standing? Is it possible to turn a complete circle on a well-oiled swing, or is there some limit to the height you can reach? You might want to consider a swing hung on rigid bars as well as on chain or rope. How much work do you do in pumping from rest to some maximum height?

9, p. 239; 26, pp. 245-246; 42, Vol. I, pp. 179-181; 75 through 80.

oscillations

resonance

2.57
Soldiers marching across footbridge

In 1831 cavalry troops were traversing a suspension bridge near Manchester, England, by marching in time to the bridge's swing. The bridge collapsed. Ever since then, troops have been ordered to break step when crossing such bridges. What is the common explanation for the danger, and is the danger real? Make rough calculations if you can.

81, pp. 59–60; 82, pp. 193–194; 1571.

resonant oscillations

2.59

Road corrugation

A road that is initially flat may develop a bump, and soon thereafter ripples appear down the road. In fact, the ripple itself seems to propagate slowly along the pavement. Thus many unpaved roads and even some blacktops and concrete roads look like washboards, especially after a rain leaves the depressions filled with water.

A similar pattern has been observed on trolley car and railroad tracks. A train passing over such corrugation makes so much noise that the tracks are called "roaring rails." Skiers may also find a washboard surface on their ski trails. What causes such corrugation? What determines its periodicity? Can you predict the periodicity by simulating the effect in a sandbox with a hand-held wheel?

89.

torques, angular momentum

energy change, resonant oscillations

2.58

Incense swing*

Pilgrims to Santiago de Campostella, Spain, visit the shrine of St. James to burn incense. The incense and charcoal are held in a large silver brazier hung from the ceiling. The brazier is set swinging with a small amplitude, and then it is pumped by about six men (see Figure 2.58) until it is swinging through 180°. The swinging makes the charcoal burn energetically for the pilgrims. The pumping is the interesting part: they do it by shortening the rope by about a meter each time it passes through the vertical; they release the same amount of rope when the container reaches its maximum height. How does this shortening and lengthening of the rope increase the amplitude?

*H. Pomerance, personal communication.

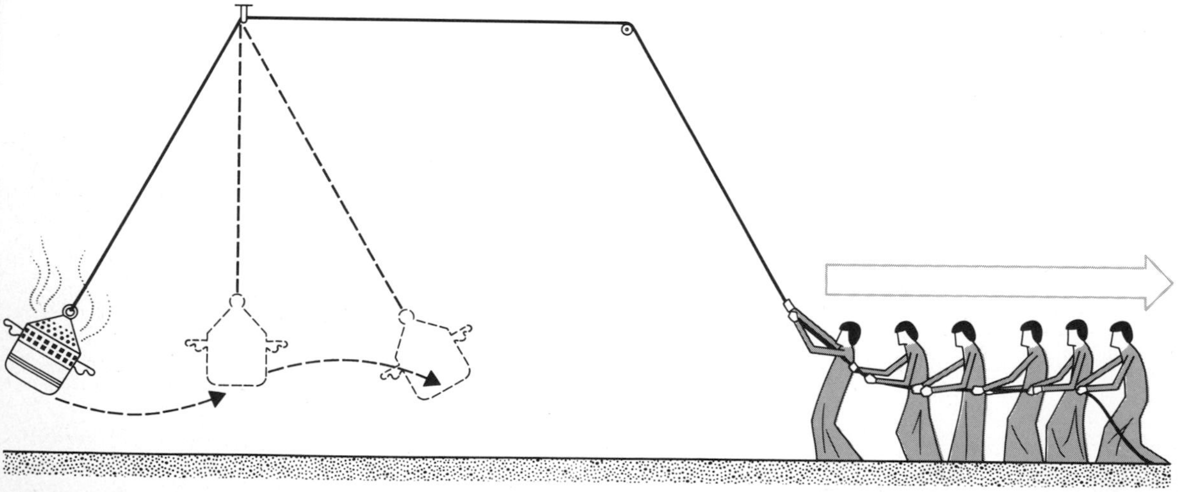

Figure 2.58

2.60
A ship's antiroll tank

A ship's rolling is normally just unsettling, but if the waves strike the ship at its resonant frequency, the rolling can be very dangerous. Consequently, some ships have carried tanks partially filled with water to diminish the danger (Figure 2.60). Such a tank had carefully chosen dimensions so that the resonant frequency of the water it held matched that of the ship. But isn't there something wrong? Since the resonant conditions *were* matched, how could the tank have managed to stop the resonant buildup of the ship's rolling?

44, p. 270; 88, pp. 202–203.

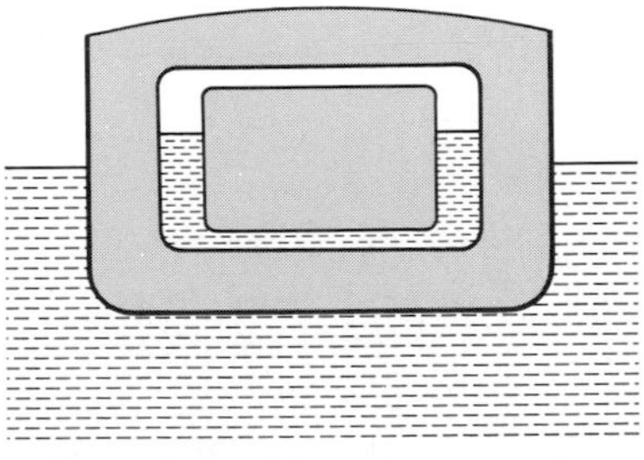

Figure 2.60
The antiroll water tank in a ship, as shown in cross section.

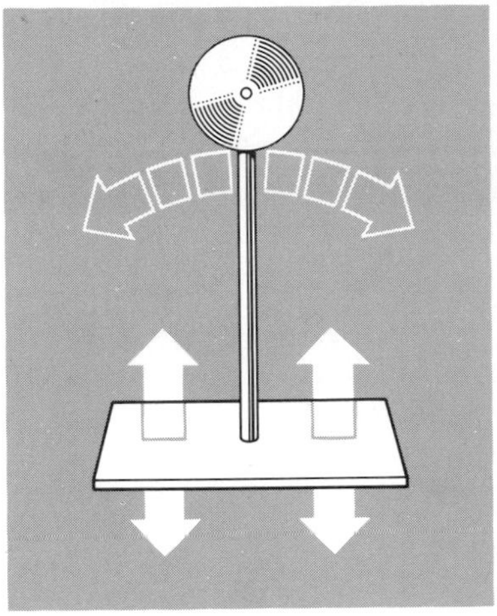

Figure 2.61
If the plate is oscillated vertically fast enough, the pendulm won't fall over.

2.61
Inverted pendulum, unicycle riders

Suppose you inverted a pendulum and tried to stand it on its end. It would be unstable and would fall over at the slightest disturbance. But if the pendulum were made to oscillate up and down fast enough (Figure 2.61), it would be stable even against small disturbances. A unicycle rider accomplishes the same thing, except that he uses a horizontal oscillation to stabilize himself. Why is there more stability in the oscillating cases? What determines the oscillation frequencies needed to gain such stability? Rather than use equations entirely, can you explain the inverted pendulum physically?

83 through 87; 795.

2.62
Spring pendulum

You are already familiar with springs and pendulums, but have you considered putting them together by suspending the pendulum bob on a spring? If you choose the spring and the bob appropriately you will have a remarkable example of sympathetic oscillation. Just as you would expect, a vertical pull sets up vertical oscillations; but soon the vertical motion dies away, and the bob begins to swing like a pendulum (Figure 2.62). After a short time it is again oscillating vertically. Somehow the energy of the system moves back and forth between the two oscillatory modes and continues to do so as long as there is energy left in the system. How must you choose the bob's mass and the spring's mass and length to obtain this oscillation exchange? Why does the exchange take place at all? With what beat frequency does the bob switch from mode to mode?

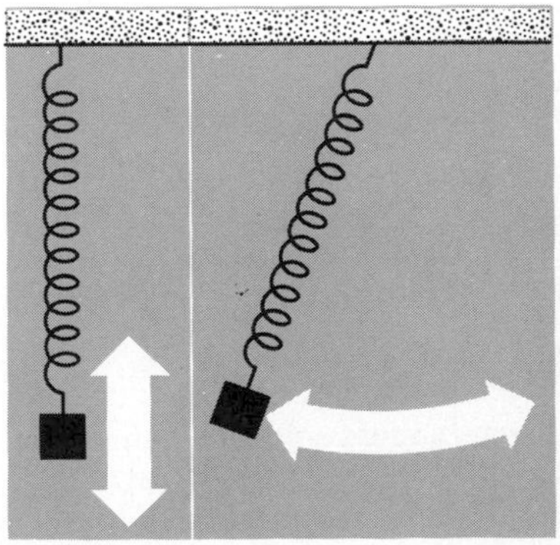

Figure 2.62

coupled pendulum motion

2.63

The bell that wouldn't ring

There's no sense in putting up a bell refusing to ring, but that's what was done at the Cathedral of Cologne. The pendulum frequencies of the bell and its clapper were so unfortunately chosen that the bell and clapper swung in phase, and of course the bell won't ring that way. Under what conditions will the pendulum motions be so matched? And when it does happen, what can you do about it, short of throwing the bell out of the belfry?

16, pp. 409–413; 37, p. 148.

Figure 2.63
"It dings when it should dong."

coupled harmonic motion

2.64

Swinging watches

Once hung on a chain, free to swing, should a pocket watch change its timekeeping rate? Many pocket watches do, even though they keep very good time if fastened down securely. If hung free on a chain by its stem (Figure 2.64), the watch will gradually begin to swing and may gain or lose up to 10 or 15 minutes a day. Why does it swing, and why does the timekeeping get messed up? Finally, why do some watches gain time while others lose time?

16, pp. 420–424; 24, pp. 114–117; 42, Vol. II, pp. 85–87, 190; 90.

Figure 2.64
Swinging pocket watch.

vibration

resonance

standing waves

2.65

Earth vibrations near waterfalls

Waterfalls pound the earth so hard that you can feel the vibration in the ground from a considerable distance. For most waterfalls one frequency of vibration is dominant, and the frequency is higher, the shorter the waterfall. In fact, the product of the frequency and the height of the waterfall is always one fourth the speed of sound in water. Why should the frequency have anything to do with the waterfall's height, and why in the world should their product be one-fourth the speed of sound?

91.

impulse

vibrational modes

2.66

Stinging hands from hitting the ball

Sometimes when you're batting a ball, your hands may get a good, healthy sting. The sting is related to what part of the bat hits the ball. Not only can such a collision cause a sting, but also it makes it much more likely the bat will break. Why are there such points on the bat, and where are they?

4.

vibration

2.67

The archer's paradox

No matter how well an arrow is aimed, when it is loosed and the feathered end is passing the bow grip, it will deviate considerably from the line to the target, perhaps as much as 7° (Figure 2.67). The archer's paradox is that a well-aimed arrow will still strike the

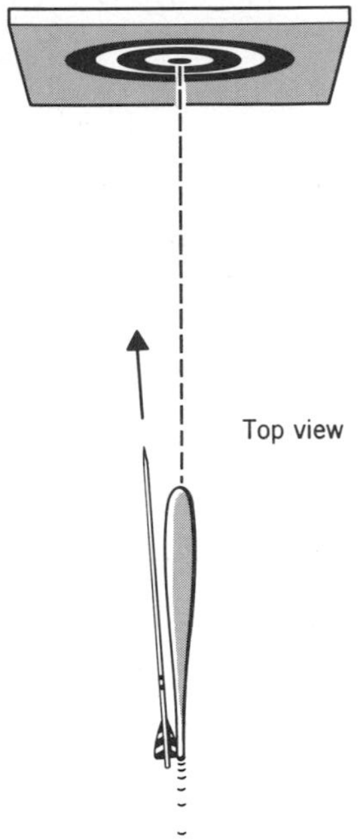

Figure 2.67
Once the arrow is loosed, it doesn't point toward the target.

target. How can this be? First of all, why is there a deviation and second, given the fact of the deviation, why does the arrow then hit the target?

High-speed photographs of the arrow show that the last time the arrow touches the bow's stock is when it is first loosed. It does not touch the stock even as the feathered end passes. If that's true, how does the arrow find its way to the target?

92.

vibrations

driven rotation phase

2.68
Magic windmill

A fascinating toy which you can easily build yourself is the rotor on a notched stick (Figure 2.68*a*). One stick has notches along its length and a small propeller on the end (on a straight pin jammed into the stick). The second stick is used to stroke the notches. Holding your forefinger on the far side of the notched stick and your thumb on the near side, run the stroking the stick back and forth notches, as shown. As you are stroking, let your forefinger press against the notched stick (Figure 2.68*a*). The propeller will turn in one direction. Now loosen your forefinger and let your thumb press against the stick, stroking the stick back and forth all the while. The propeller will turn in the opposite direction.

When you're showing this to the uninitiated, you can slyly shift from the forefinger to the thumb and make a great mystery of the change of direction of the spin. The number of lies you can feed someone about why the rotor reverses is almost unlimited—I like to attribute it to a variation in cosmic ray intensity.

The first question you should ask yourself is why the rotor turns at all. Next comes the bigger mystery of why the spin sense depends on which side of the stick you press.

If you want something flashier, put four rotors on the stick (Figure 2.68*b*). All four will turn in the same direction, so there's nothing essentially different about this. Another design, which is more difficult to explain, has two rotors mounted one behind the other (Figure 2.68*c*). Something strange does happen in this case. You can make both rotors turn to the left or both to the right or, best of all, you can make one go in one direction and the other in the opposite direction.

93 through 96; 1487.

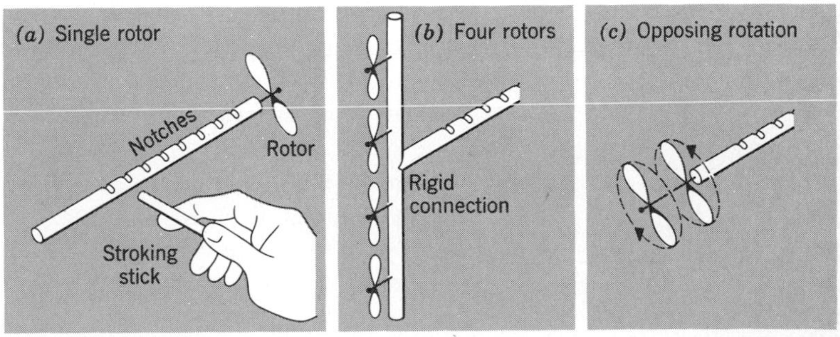

Figure 2.68
Magic windmill. (After R. W. Leonard, Am. J. Phys., 5, 175 (1937).

Gyroscopic motion

(2.69 through 2.73)

torque	precession
angular momentum	

2.69

Personalities of tops

Why does a spinning top stay up? Can you explain it using only force arguments, without invoking torque and angular momentum? The top stays up against gravity; hence, there must be a vertical force. What produces that force?

Can you also explain the personalities of individual tops? Some "sleep," that is, remain vertical; others precess (Figure 2.69) like mad. Some are always steady in their motion; others are worrisome before finally settling down to a steady motion. Some die long, lingering deaths; others depart rapidly. How do you account for these varied temperaments?

36; 37, Chapter 1.

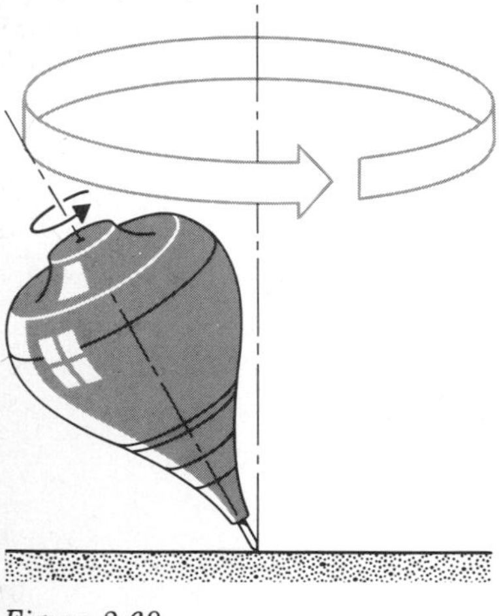

Figure 2.69
In precessing, the top's axis itself rotates about the vertical.

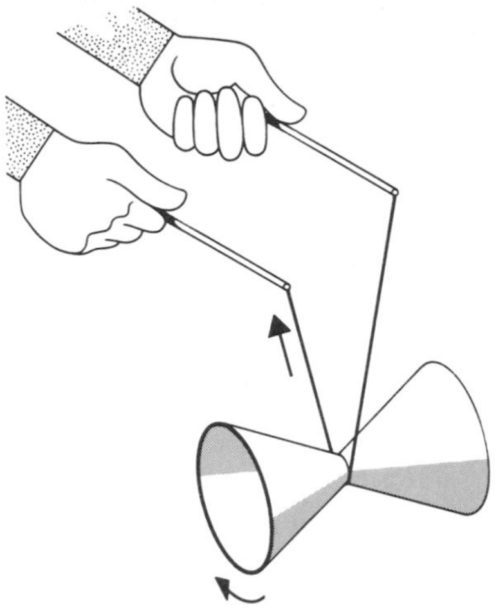

Figure 2.70

2.70

Diabolos

The diabolo, an ancient toy, is a spool made of two cones stuck together, which is spun by means of a string whose ends are tied to sticks (Figure 2.70). Spinning is initiated by first lowering the right hand, smoothly drawing it back up and thus spinning the diabolo, then quickly dropping that hand again and repeating the process until sufficient spin has been given the diabolo.

Why is the diabolo so much more stable when spinning? Even then, you may have to make corrections. For instance, suppose the near end begins to dip. What should be done with the sticks to make the spool horizontal again? Or suppose that you want the spool to turn to your left. What must you do with the sticks?

36, pp. 40-41, 120-121; 37, pp. 458-459.

2.71

Spinning eggs

In times of doubt, you can distinguish a hard-boiled egg from a raw one by spinning them. The cooked egg will stand on end and the raw one will not. Why? Another way to tell if an egg is raw or cooked is to spin it, stop it with your finger, and quickly release it (Figure 2.71). A cooked egg will sit still, but a raw one will begin to turn again. Again, why?

36, pp. 5-6, 51, 155;,37, pp. 16-17, 264-272; 108; 109, pp. 39, 57; 110, p. 123.

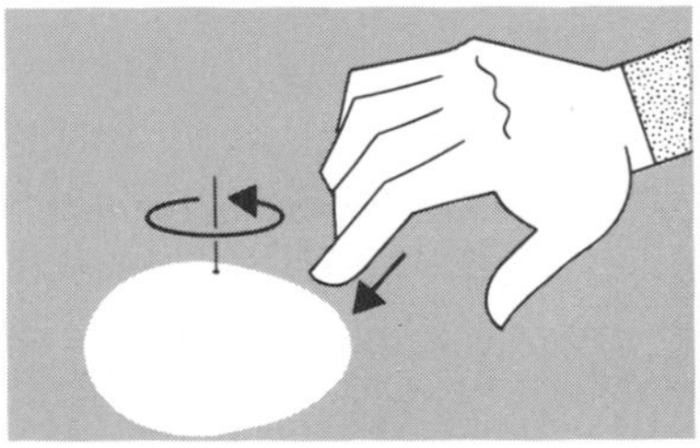

Figure 2.71
Testing for a fresh egg.

2.72

The rebellious celts

Some of the stone instruments made by primitive men in England display curious personalities when they are spun on a table. These stones, called celts, are generally ellipsoidal in shape. When you spin them about the vertical axis some behave as you would guess, but others act normally only when spun in one direction about the vertical (Figure 2.72*a*). If you spin them in the other sense, the

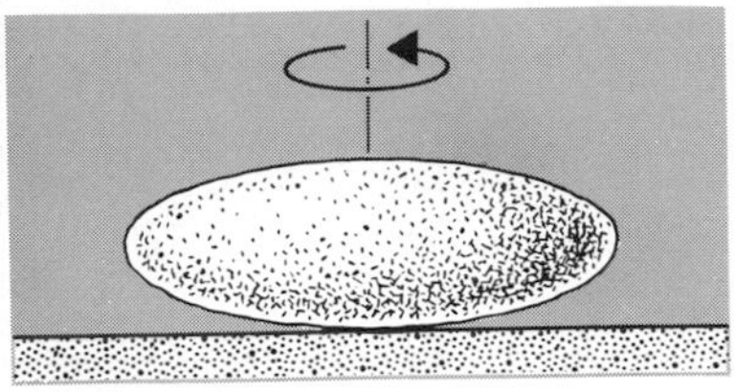

Figure 2.72a.
Spinning celt.

rebellious stones will slow to a stop, rock for a few seconds, and then spin in their preferred direction. Some stones demand one direction, others demand the opposite.

If you tap one of these stones on an end, say at point *A* in Figure 2.72*b*, it will rock for a while. But soon the rocking ceases, and the stone begins to rotate about the vertical axis.

Try to make some wooden celts displaying this rebellious nature. What causes such personalities?

26, pp. 204-205; 36, pp. 7, 54; 37, pp. 363-366.

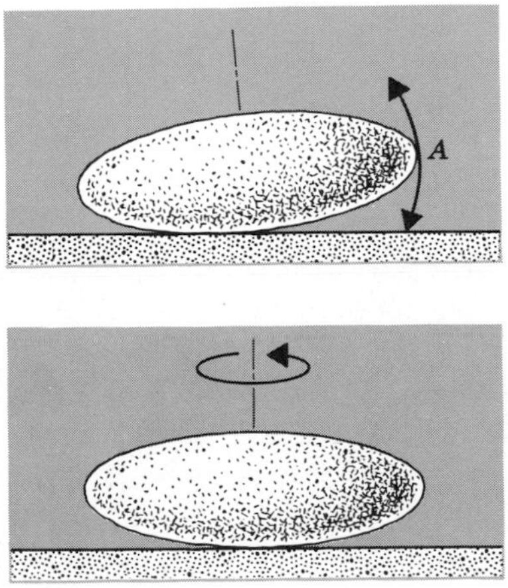

Figure 2.72b.
Celt initially set rocking at A begins to spin.

2.73

Tippy tops

There is a kind of top that really knocks me out—it is part of a sphere with a stem in place of the missing section (Figure 2.73). Given a spin on its spherical side, it will quickly turn over and spin on the stem, the heavier side thus rising against gravity. Why does it rise? What forces the top up? Isn't it completely contrary to your intuition that the spinning top is so unstable in the initial orientation and so much stabler in the final one?

The same behavior can be seen with high school and college rings having a smooth stone. Footballs and hard-boiled eggs will also raise themselves up on their points when spun in similar fashion.

36, pp. 5-6, 51, 155; 97 through 108; 109, pp. 39-57.

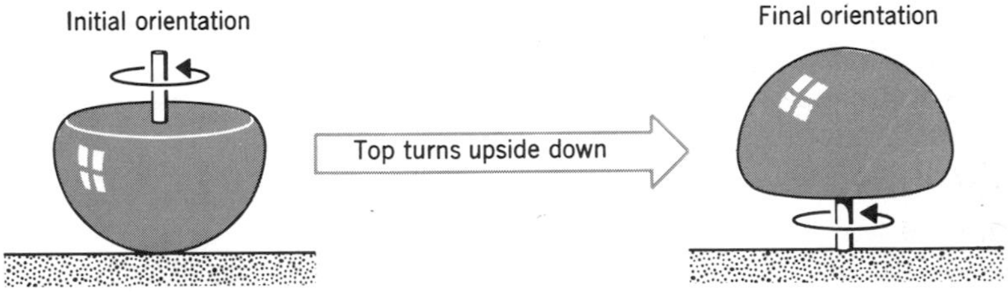

Figure 2.73

Gravitation

(2.74 through 2.79)

gravity	kinetic and potential energy
orbits	
torques	
moment of inertia	

2.74

Seeing only one side of moon

Why do we see only one side of the moon? Because the moon turns on its own axis at such a rate that as it orbits the earth it always presents the same face to us. But is this pure chance?

26, pp. 369 ff; 111.

2.75

Spy satellites over Moscow

The United States would like to see what Russia is up to, so we put up spy satellites with long-range cameras. We would really like to have a permanent satellite stay directly over Moscow 24 hours a day. Why *don't* we? Why, instead, do we put up a series of satellites whose times over Moscow overlap?

2.76

Moon trip figure 8

When the astronauts go to the moon, why is their path (earth-moon-earth) essentially a figure 8 (Figure 2.76) instead of an ellipse? In particular, does the figure-8 path require less energy?

Figure 2.76

2.77

Earth and sun pull on moon

How large do you think the sun's pull on the moon is, compared to the earth's pull? Well, after all, the sun doesn't steal our moon away, so the earth must be pulling harder, right? That's satisfying, but unfortunately it isn't true. The sun pulls more than twice as hard as the earth. So why don't we lose the moon?

117.

2.78

Making a map of India

I have read it is difficult to survey India because the plumb line one uses in surveying is pulled northward to the Himalayas and thus does not hang toward the earth's center. Is this true? How large do you think the effect is, and is it large enough to influence large-scale surveying?

13; 118; 119.

2.79

Air drag speeds up satellite

Artificial satellites don't orbit the earth forever. Eventually the earth's atmosphere, thin as it may be up there, will bring them down. But did you know the linear speed of a satellite in a nearly circular orbit will increase because of the air drag? The satellite will experience an acceleration forward along its path, and the accelerations's magnitude will be the same as if the air drag were turned around and were pushing the satellite along. How can that be?

112 through 116.

3

Heat fantasies and other cheap thrills of the night

Pressure

(3.1 through 3.9)

Boyle's law   partial pressure

atmospheric and water pressure

3.1

The well-built stewardess

LOS ANGELES (AP)—What happens to a stewardess wearing an inflatable bra when the cabin of her jet plane is depressurized?

Just what you're thinking, Herman. Inflation.

As Los Angeles Times columnist Matt Weinstock told it Friday, this set of potentially explosive circumstances occurred recently on a Los Angeles-bound flight. He gallantly withheld the identity of girl and airline.

"When she had, ahem, expanded to about size 46," Weinstock wrote, "she frantically sought a solution. Somehow she found a woman passenger who had a small hatpin and stabbed herself strategically.

"However, another passenger, a man of foreign descent, misunderstood. He thought she was trying to commit hara-kiri the hard way. He grappled with her trying to prevent her from punching the hatpin in her chest.

"Order was quickly restored, but laughter still is echoing along the the airlines."

Weinstock says it really happened.

Exceptionally good reference: Chemical Principles Exemplified," edited and written by R. C. Plumb, monthly in *J. Chem. Ed.*

Good thing they don't make these bras puncture-proof.

. . .Associated Press

Can you calculate the stewardess's pectoral measurements as a function of altitude?

3.2

Making cakes at high altitudes

Why does the recipe for a cake change when you do the baking above 3500 feet? The side of the cake mix box calls for more flour, more water, and a higher baking temperature when the mix is used at altitudes greater than 3500 feet.

316, pp. 184–186.

pressure

humidity

3.3

The Swiss cottage barometer

One of my grandmother's most fascinating possessions is her Swiss cottage barometer. She explains that when the pressure falls, a little man comes out of the cottage to warn of a coming storm. During fair weather a little woman comes out instead. How does this cottage barometer work, and does it actually measure the barometric pressure? I notice that when I place it in the bathroom it predicts bad weather far more often. Why the increase in frequency?

317, p. 201; 318, p. 209.

3.4

Wells and storms

My grandmother claims that during storms her water well is easier to pump but the water may be unfit to drink because of an increase in suspended matter. This happens, she says, whether the storm brings rain or not. Others have noticed that artesian wells generally increase in strength during storms, again regardless of the rain. Why would these wells respond to storms? Might there be an opposite effect in which a normally freely flowing well is stopped?

318, p. 143.

elasticity

surface tension

3.5

One balloon blowing up another balloon

Blow up two identical balloons, one more than the other. Take care that air doesn't leak until you've joined the two balloons by a short length of tubing as shown in Figure 3.5. What will they do when they are joined? Does the smaller balloon expand at the expense of the larger one? Intuition

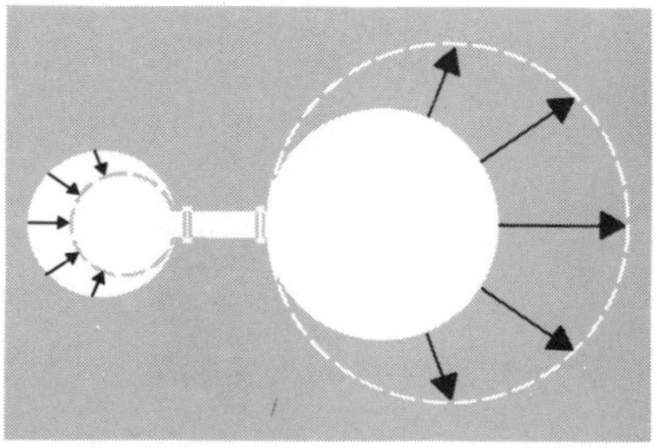

Figure 3.5

may say yes, but actually the opposite happens: the smaller balloon shrinks and the larger balloon expands. Why? The same phenomenon occurs with soap bubbles. See Boys's soap bubble book (322).

321; 322, pp. 56-57.

3.7

Emergency ascent

Suppose that while scuba diving at some great depth, say 100 feet, you had to make an emergency ascent without additional air. One lungful has to be enough for you to reach the surface, or you'll die. How would you do it? (This is not really just an academic question, for submarine crews are trained to make such emergency escapes.) Would you continuously release air as you ascend, or keep it all in? Well, although it may seem unreasonable, you had better release air or you won't make it. In fact, novice scuba divers practicing in swimming pools are occasionally killed because they neglect to exhale when practicing emergency ascents. Why?

It is said the urge to take another breath stems from the partial pressure of the CO_2 in your lungs, not the volume of the CO_2. Researchers conclude from this that the most dangerous and crucial point in your ascent will be at some intermediate point and not near the surface. Once you pass the crucial point, the urge to take another breath will relax considerably. Why is this? What is the crucial depth? How fast should you swim to the surface? Can you swim too fast? If you can, then what's a reasonable rate?

325 through 328.

Boyle's law

partial pressure

3.6

Champagne recompression

When a tunnel under London's Thames River had been completed and the two shafts had been joined, the local politicians celebrated the event at the tunnel's bottom. In the tunnel they unfortunately found the champagne flat and lifeless. When they returned to the surface, however, "the wine popped in their stomachs, distended their vests, and all but frothed from their ears [Figure 3.6]. One dignitary had to be rushed back into the depths to undergo champagne recompression " (314). What happened to the politicians?

314; 315.

Figure 3.6
The danger of subterranean champagne.

3.8

Blow-holes

You'd probably imagine that caves are full of stagnant air. Some are, but at the entrances of some, called "blow-holes" by spelunkers, a fierce wind blows constantly. Why is that? Even stranger are the breathing caves where the air blows in for a moment and then out alternately. What drives the air back and forth?

318, pp. 143-144; 319; 320.

3.9

Decompression schedule

In deep-sea-diving ascents there is always the serious threat of "bends," in which bubbles form from the nitrogen dissolved in the tissue during the dive. This can be not only painful but also paralyzing and even fatal. Consequently, the ascent is made slowly enough that the nitrogen is disposed of without bubble formation. You have seen this in movies: the diver stops at various depths in his ascent. Where do you think the longest stop is: near the surface where the diver is almost at atmospheric pressure, near the bottom, or at some intermediate point? I would have eliminated the first choice immediately, but the decompression schedule in Figure 3.9 contradicts me: the longest stops are near the surface. Why should that be? What is the deepest dive you can take without having to wait around on the way up?

323; 324.

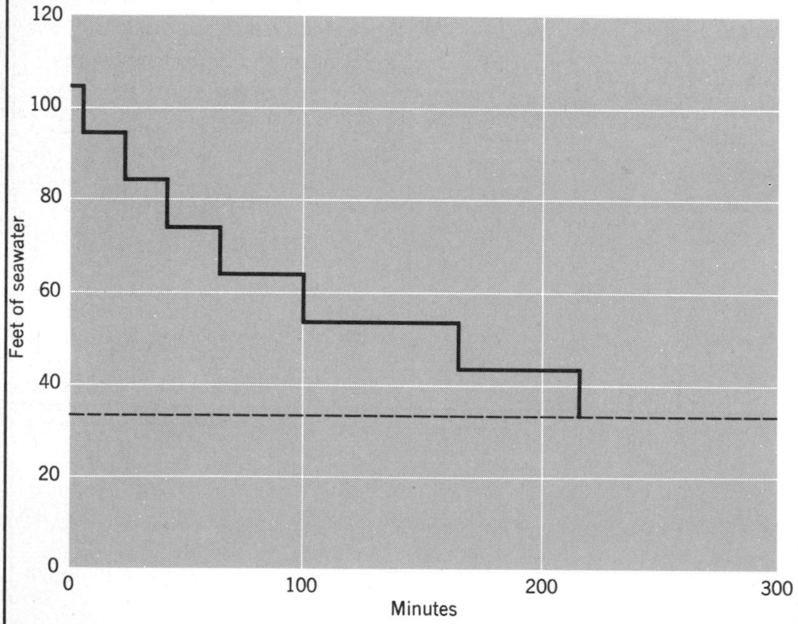

Figure 3.9
Decompression schedule as recommended by the U.S. Navy for a one-hour dive at 200 feet. Dashed line indicates the sea-level pressure. [After H. Schenck, Jr., Amer. J. Phys., 21, 277 (1953).]

Thermal expansion & contraction

3.10 through 3.15

3.10

Hot water turning itself off

When I turn on the hot water in my sink, the water's flow rate slowly decreases and the flow may even stop. The cold water won't do that, so why does the hot water behave so badly? Why does it do that only when I've first turned it on and not the second time, after I've turned it up?

thermal expansion

3.11

Bursting pipes

Why do water pipes burst in winter? If the only thing that occurs is the freezing of water next to the pipe walls, then there shouldn't be any great strain on the pipe and hence the pipe shouldn't burst. Besides, the bursting usually occurs away from the point where the water is frozen. So, again, what causes the burst? Is there any real advantage in letting outside taps drip all winter as some people do? Finally, is there any truth to the common idea that hot water pipes burst far more often than cold water pipes?

253, pp. 136–137; 338, pp. 35–36; 339.

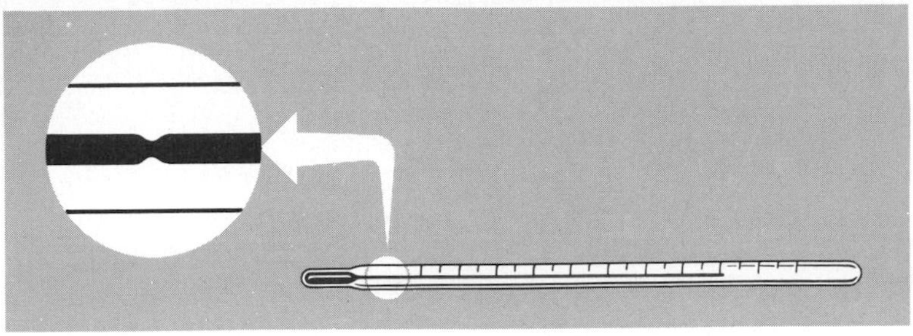

Figure 3.12

3.12

Fever thermometer

When you take your temperature, the heat of your mouth makes the mercury expand. Why doesn't the mercury level fall as soon as you remove the thermometer from your mouth? It doesn't, because a constriction in the tube prevents it from falling (Figure 3.12). But why? After all, during the expansion the mercury passed through the constriction. Why won't it do the same during the contraction?

Why does the reading drop for a moment if you stick the thermometer into hot water? (Don't overheat the thermometer so that it breaks.)

160, p. 114; 317, pp. 117–118, 129; 329, p. 50; 330, p. 41; 331, p. 6.

3.13

Heating a rubber band

Stretch an uninflated balloon and then touch it to your face. It feels warm. Now let it contract to its normal size. It feels cool. Why?

If you heat a rubber band it contracts. Why is its behavior precisely opposite that of metal? What's different about its structure? Figure 3.13 shows a rubber-band engine based on this property. The spokes of the wheel are rubber and hence will shrink when heated. The wheel turns because of the shift in the center of gravity.

155, p. 244; 332, Vol. 1, p. 44-1; 333 through 337.

Figure 3.13
A rubber-band heat engine. (Reprinted by special permission from Feynman, et al., The Feynman Lectures on Physics, Vol. 1, 1963, Addison-Wesley, Reading, Mass.)

3.14

Watch speed

Since metal expands when it's heated and a watch spring is metallic, wouldn't you think that a watch would run at different speeds in cold and in warm weather?

9, p. 82; 160, p. 125; 317, p. 129; 329, p. 43; 330, p. 90; 331, p. 23.

buoyancy

nonlinear oscillations

3.15

U-tube oscillations

If a U-tube of water is heated and cooled as shown in Figure 3.15, the water will begin to oscillate from one side to the other. (There must be open reserviors with the

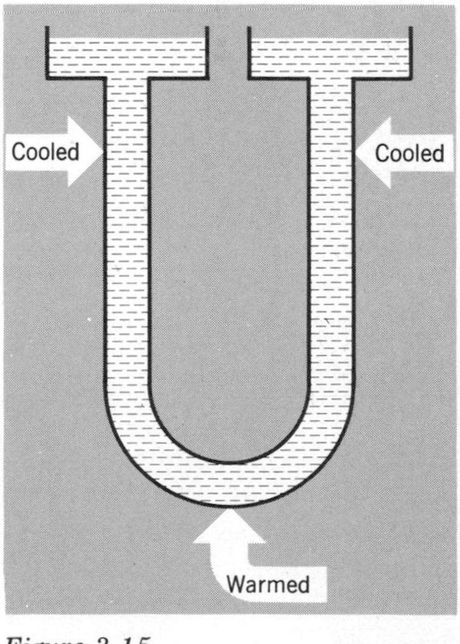

Figure 3.15
The water will oscillate from side to side if the tube is heated and cooled as shown. [After P. Welander, Tellus, 9, 419 (1957).]

air–water surface area larger than a critical size.) The change in water levels can be a millimeter or so, and the period of the oscillations can range from about 20 seconds to 4 hours depending, in part, on the cross-sectional area of the U-tube. Doesn't the symmetry of the situation make it seem curious that the water oscillates? What first starts the oscillation and what parameters determine its period?

340.

adiabatic process

3.16

Bike pump heating up

Why does the valve on a bicycle pump get hot when you're pumping up a tire? Is it because of friction from the air being forced through the valve? Well, perhaps, but if you use a gas station's compressed air supply, the valve usually doesn't get hot.

341.

condensation

3.17

West-slope hill growth

Why is it that in the United States there is often more vegetation on the westward slopes of hills and mountains than on the eastward slopes? You may even find extreme cases where the east side is barren though the west side has thick growth.

360, pp. 162–165.

adiabatic processes

3.18

The Chinook and going mad

The Chinook is a warm, dry wind that blows down from the Rockies into such places as Denver (Figure 3.18). It can be up to 50°F above the ambient temperature and may reach speeds as high as 80 mph. The mystery is how a *warm* wind could come down off a *cold* mountain. Besides, warm air should rise, shouldn't it? Legend says the warmth comes from the ghosts of Indians buried in the mountains.

Chinook-like winds are by no means confined to the Denver area. In Switzerland this wind is called the foehn; in Ceylon, the kachchan; in South Africa, the berg wind; in Southern California, the Santa Ana; and in other places, other names. They all share the properties of being dry and warm.

They also share a very controversial feature, namely, it is said that they drive men and animals mad. When these winds blow, crime rates increase, rape and murder are more frequent, there are more traffic accidents, and people just act generally more irrational. This could be an old wives tale, or there could be some truth in it. How could dry, warm winds affect a man physiologically? Is there any physical reason for the irrational behavior?

164, pp. 217-218; 343, p. 348; 344, pp. 94-98; 345 through 358.

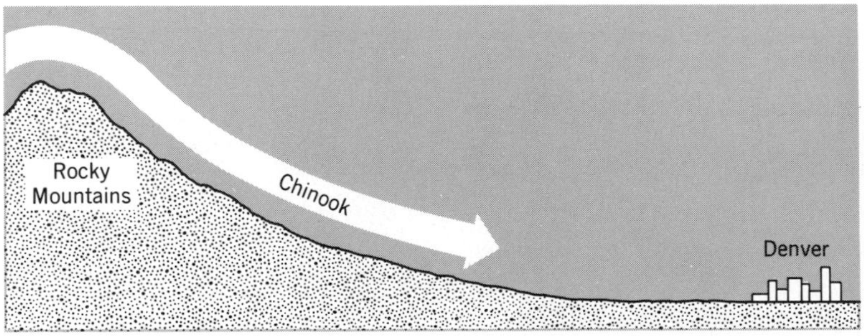

Figure 3.18
The Chinook wind blowing down off the Rockies.

adiabatic process

3.19
Coke fog

Have you ever noticed the thin fog that gathers at the mouth of a chilled champagne or soda bottle just after it's been opened (Figure 3.19)? What causes the fog?

342.

Figure 3.19
Fog at mouth of freshly opened, chilled champagne bottle.

3.20
Convertible cooling effect

On a hot day you're in luck if you've got a friend with a convertible. Driving down the road with a good breeze always does the trick against the heat. You feel cooler but a thermometer should read the same with or without a breeze, shouldn't it? Try it. With a thermometer in the back seat, measure the temperature when the car is parked and when it is moving. You'll probably find that the thermometer reads about 1/2°C lower when the car is moving. Why?

359.

adiabatic process

radiation absorption

3.21
Death Valley

Death Valley is both the lowest point on the American continent and the hottest place in the world. Temperatures there may be as high as 120°F for several days straight, and once a temperature of 134°F was recorded. Isn't there something physically wrong in its being so hot if it is so low? Since hot air rises and cold air sinks, and since the valley is surrounded by mountains with cold air on their tops, shouldn't the valley be a relatively cool place?

223, p. 200.

adiabatic process

condensation

latent heat

radiation

3.22
Mountain top coldness

Why are mountain tops cold? Isn't the solar heat per unit area on a mountain about the same as at sea level? And shouldn't cold air sink?

condensation

buoyancy

adiabatic process

3.23
Holding a cloud together

What holds a cloud together? Or, on partially cloudy days why are some parts of the sky cloudy and others not? Wouldn't you expect a more uniform distribution of the clouds over the sky?

363, pp. 44–67; 365.

3.24
Mushroom clouds

Why do ground-level nuclear and other large explosions leave mushroom clouds?

371, pp. 202–203; 372; 373.

cloud genesis

stability

buoyancy

3.25
Holes in the clouds

Mysterious circular holes have occasionally been observed in otherwise uniform cloud banks. The feeling is that these holes, which are usually quite large, are not just random arrangements of the clouds. Suggestions as to their cause have ranged from burning meteors to accidental or intentional cloud seeding. How exactly could any of these explanations account for such holes?

362, p. 91; 374 through 379.

cloud genesis

condensation

Figure 3.26a
Two types of mountain peak clouds.

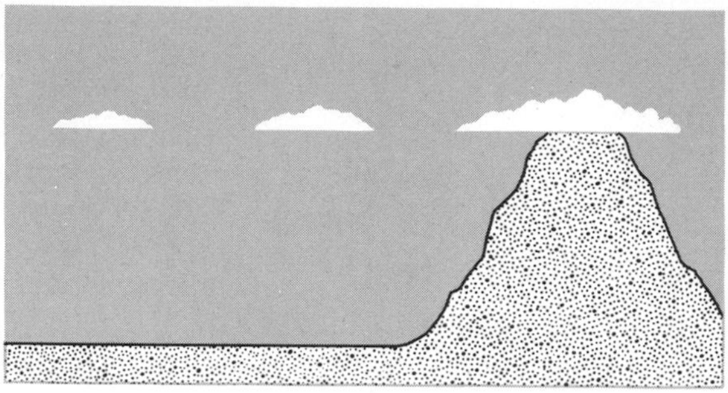

Figure 3.26b
Wavelike clouds associated with a mountain peak.

3.26
Mountain clouds

If you have ever lived near mountains you may have noticed the stationary clouds often found over mountain peaks. Two are shown in Figure 3.26*a*. What causes these formations? The wavelike series of clouds that sometimes occurs near a peak is even more intriguing (Figure 3.26*b*). What determines the spatial periodicity of these clouds?

164, pp. 301-303; 360, pp. 86-88; 361, pp. 14-21, 39; 362, pp. 64-73; 363, pp. 75-82; 364, pp. 229 ff; 365 through 370.

shock wave

condensation

3.27
Spherical cloud of A-bomb blast

In some circumstances, a nuclear blast is accompanied by a thin, spherical cloud (Figure 3.27). What causes these clouds? How fast do they expand? Will they significantly reduce the radiation produced by the explosion?

219, pp. 311-312; 371, pp. 34 ff.

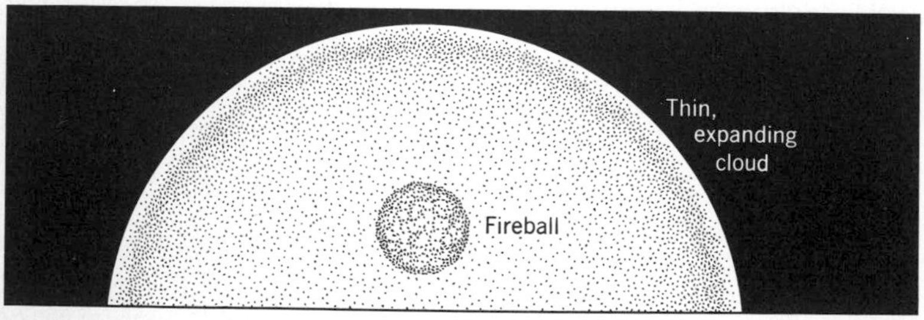

Figure 3.27

absorption

buoyancy

condensation

evaporation

3.28
Burning off clouds

When there were low-hanging clouds on an early summer morning, my grandmother would often say the sun would "burn them off" and the day would be sunny. Since they did often disappear later in the morning, I figured she was right and that the sunlight absorbed by the clouds indeed "burned them off." Was I correct?

363, p. 76; 364, pp. 273-274.

cloud genesis

stability

buoyancy

3.29
Mamma

What causes the breastlike cloud structure called mamma (Figure 3.29)? In particular, why are there sometimes bright gaps between the mammae?

362, pp. 54–56.

Figure 3.29
Mamma cloud formation.

condensation

3.30
Cause of fog

London's fogs have diminished in intensity in the last decade partially because there was a reduction in the use of open coal burning. What has open coal burning got to do with fogs? In general, what causes fogs?

388, pp. 480–510.

3.31
Breath condensation

Why does your breath condense on the window pane on a cold day? More specifically, what actually causes the water molecules to form into a drop? Why did those water drops condense in those particular places on the glass. . .what was so special about those places?

Why does a hot piece of toast leave moisture on a plate?

388, pp. 428 ff; 389.

bubble nucleation

3.33
Salt water bubbles

Why are more bubbles produced when salt water is poured into salt water than when fresh water is poured into fresh water?

390.

condensation

adiabatic process

vortices

buoyancy

3.32
Contrails and distrails

Why do contrails often form behind airplanes? Why aren't they always produced? If you look closely you may see that a contrail actually consists of two or more streams that eventually diffuse and become indistinguishable. Why is there more than one stream at first? Why is there a clear gap between the airplane and the leading edge of the contrail? What's responsible for the bursting and blooming of contrails that makes them look like strung popcorn (Figure 3.32).

You may be fortunate enough to see both a contrail and its shadow on underlying clouds. But the distrail, a dark line left by an airplane flying through a cloud, is even more interesting. How does an airplane make that kind of trail?

362, pp. 120–129; 364, pp. 73-74; 380 through 387; 1537.

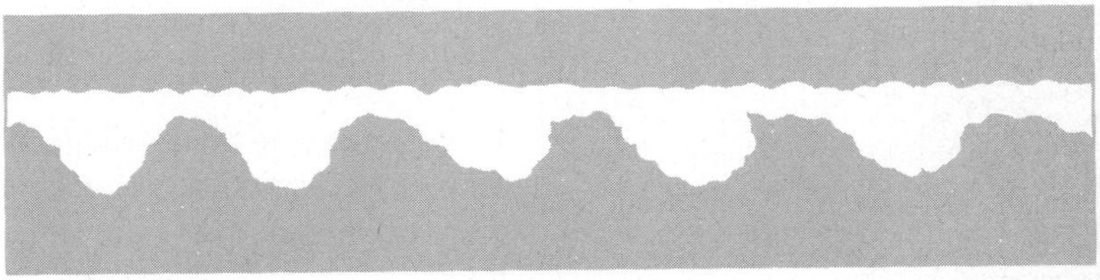

Figure 3.32
Side view of contrail that has burst to a popcorn appearance.

buoyancy

Bernoulli's principle

3.34

Fireplace draft

In a good fireplace the smoke goes up the chimney rather than out into the room, even if the fire is not directly beneath the hole. What causes this draft, and why is it better the taller the chimney? Why is the draft better on a windy day? Finally, why do some chimneys puff (Figure 3.34)?

44, p. 188; 318, pp. 225-230; 364, pp. 216-217; 391, pp. 111-112; 392, pp. 108-112; 393; 394.

Figure 3.34
Puffing chimney.

By permission of John Hart. Field Enterprises

buoyancy

3.35

Open-air fires

Many communities that still allow open-air fires forbid them during the daylight hours. Why would it matter whether the fires are during the day or evening?

buoyancy

turbulent eddies

3.36

Cigarette smoke stream

Why does cigarette smoke suddenly form swirls after rising smoothly for several centimeters (Figure 3.36)?

399, pp. 175-176; 400.

Figure 3.36

buoyancy

stability

lapse rate

3.37

Stack plumes

You would think an industrial stack plume would rise vertically or, if there is a wind, would rise at some angle. Yet the plume shapes shown in Figure 3.37*a* are often seen in a *uniform* horizontal wind. What causes these shapes? The last one with the peculiar periodicity is especially interesting. Why do some bent-over plumes split sideways downwind from the stack (Figure 3.37*b*)?

364, pp. 207–212; 395 through 398.

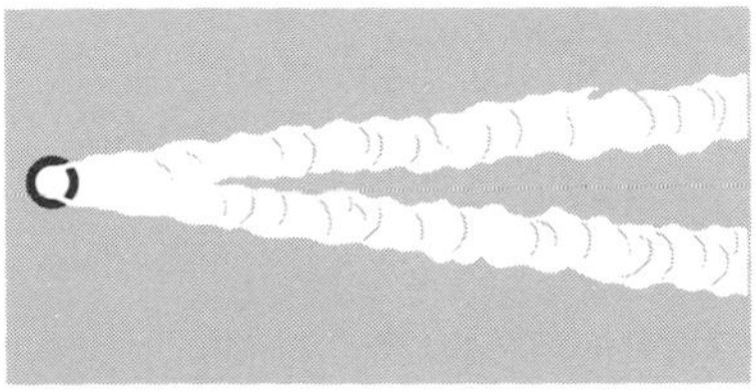

Figure 3.37b
Top view of a bent-over plume that has been split sideways.

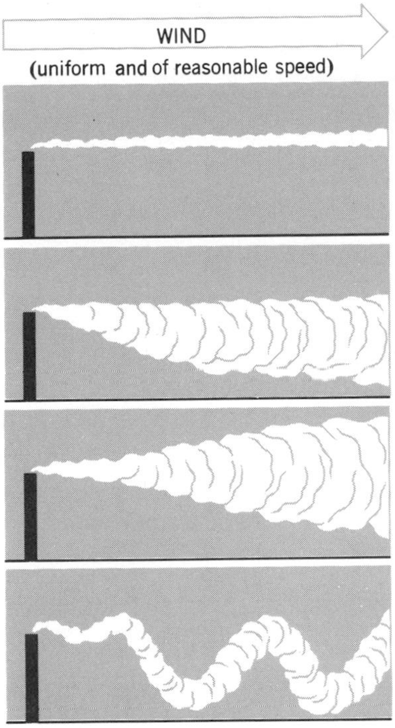

Figure 3.37a
[After Bierly and Hewson, J. Appl. Meteorology, 1, 383 (1962), permission granted by authors and the American Meteorological Society.]

ice crystal growth

capillarity

radiation absorption

3.38

Shades of ice coverings

If you observe a distant ice covering on a North Alaskan lake or river when it begins to melt in the late spring, large parts of the ice will look dark and other parts will look white. A walk across the ice can quickly (and painfully) teach you that the dark ice is weaker and should be avoided. Why is the ice light and dark, and why are the dark areas weaker?

338, pp. 120–126; 376.

supercooling

free energy

3.39

Freezing water

Why does water normally freeze at 0°C? What is so special about that particular temperature? Under some circumstances liquid water can exist at subzero temperatures; for example, water drops at temperatures as low as −30°C have been found in clouds. What must be done to make such supercooled water?

Can ice be heated above 0°C without melting?

338; 389; 402 through 404.

freezing

latent heat

evaporation

3.40

Freezing hot and cold water

In cold regions like Canada or Iceland, it is common knowledge that water left outside will freeze faster if it is originally hot. While this may seem completely wrong to you, it is not just an old wives' tale, for even Francis Bacon noticed it. Try putting warm and cool water in various containers either outside on a freezing night or in your freezer. If in any of your tests the warm water freezes first, then you'll have to explain why.

405 through 411.

thunderstorm thermodynamics

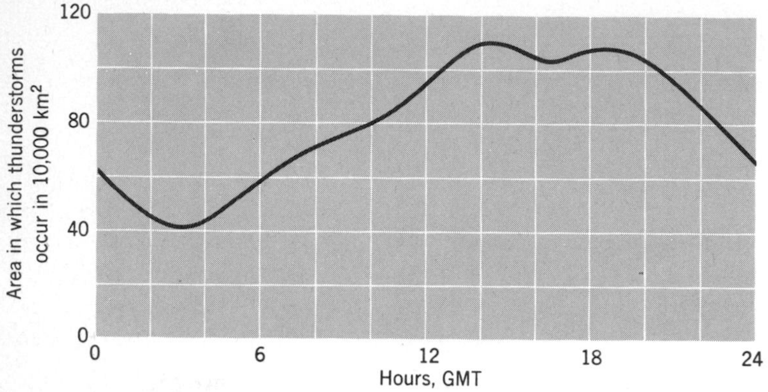

Figure 3.41
(After D. J. Malan, Physics of Lightning, English Universities Press, Ltd.)

3.41
Worldwide thunderstorm activity

If you plot the worldwide thunderstorm activity versus Greenwich Mean Time (GMT). you get a curve that has a maximum at 7 P.M. London time and a minimum at 4 A.M. London time (Figure 3.41). In other words, when it is 7 P.M. in London, the earth is experiencing the greatest amount of thunderstorm activity. Is any time dependence plausible? Is there any physical basis for this particular dependence?

219, pp. 123–124; 300, pp. 109–111; 332, Vol. II, Chapter 9; 388, p. 445; 401.

heat conduction

3.42
Getting stuck by the cold

If you touch a cold piece of metal such as a metal ice tray fresh from the freezer, your hand may stick to the metal. Be careful if you actually try this experiment, for you can lose the skin that sticks to the metal. Have water running in your sink and, immediately after touching the ice tray, dunk your finger and the tray under the water. Do *not* lick the tray, as some unknowing children do, for that may result in very painful injury.

Why does your finger stick to the tray? How cold must the metal be for this sticking to happen?

latent heat

3.43
Wrapping ice

Why does ice keep frozen longer if it is wrapped in a wet piece of paper?

160, p. 166.

density change with temperature

convection

insulation

3.44
Pond freeze

Why does the top of a pond freeze before the middle and long before the bottom? (There's more than one reason.) If this weren't so, there would be virtually no fresh-water fish outside the tropics.

In areas where water transportation is necessary, the formation of ice can be prevented by bubbling air up from pipes laid on the bottom of the lake or river. If ice is already present, the bubbles will even melt the ice although it may take four or five days to do it. How do the bubbles clear a river or lake in this way?

158, p. 288; 403; 412, pp. 495–496; 413, p. 139; 414, pp. 4–6, 58–61.

conduction

phase change

3.45
Skiing

What allows skis to glide over snow? Is it the same as the mechanism involved in ice skating? Could you ski on other frozen substances or is snow (water) unique? Can it get too cold to ski? Why are skis waxed? Finally, why do ebonite skis slide much better than metal ones?

421, p. 393; 422 through 424.

adiabatic compression

pressures and phase change

3.46
Ice skating

When you are ice skating, why do your skates slide along the ice surface? If you can, explain the physics involved with practical numbers. Obviously it can get too warm to skate. Can it get too cold? Is the ice that is found in very cold places, such as Greenland, slippery? Could you skate in a similar way on other frozen materials such as carbon dioxide (dry ice)? Suppose you had to walk across ice and you could choose between a patch of smooth ice and a patch of rough ice. Would you find one more slippery than the other?

166; 321, p. 274; 414, pp. 111-113; 418, p. 129; 419; 420.

conduction

phase change

3.48
Making a snowball

Why can't you make a snowball if the temperature is very low? What holds a snowball together, anyway? Approximately what *is* the lowest temperature at which you can still make a reasonably good snowball?

166.

adiabatic compression

pressure and phase change

3.47
Snow avalanche

How can sudden warmings and mechanical vibrations trigger snow avalanches? Why do many avalanches occur at sunset when there is a general cooling rather than a warming? There are even claims that a skier's shadow may be enough to set off an avalanche. Why would this happen?

In a dry snow avalanche a huge cloud of snow particles precedes the slide, crashing down the mountain side at speeds up to 200 miles per hour with enough force to destroy large trees and move steel bridges. According to one story about a skier caught in one of these snow slides (Figure 3.47), the skier and the slide reached the opposite slope with such speed that the trapped air was compressed and warmed and thus partially melted the snow. Within several minutes, however, the snow had refrozen, and when the rescue team reached the still-living skier, they had to saw him out.

415 through 417.

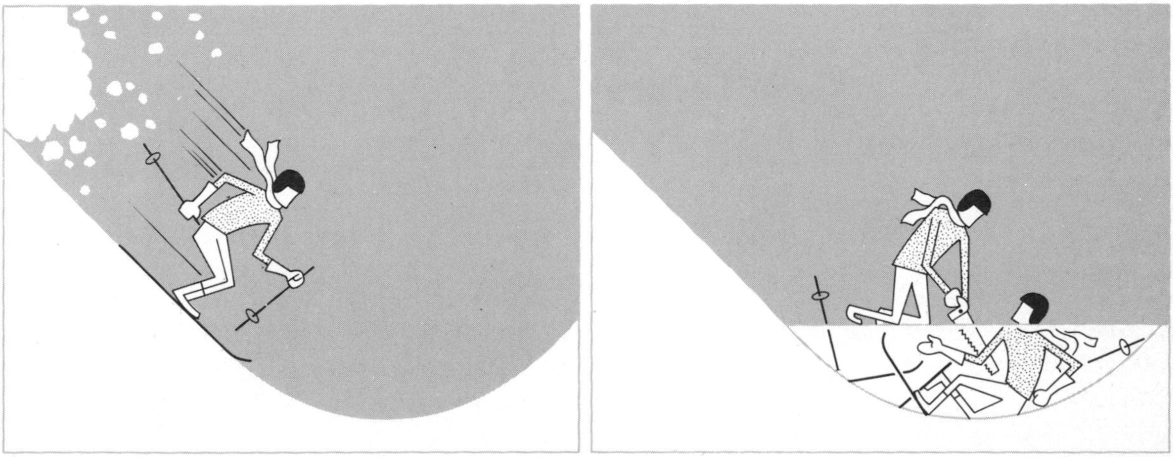

Figure 3.47

conduction

phase change

3.49
Snow tires and sand for ice

Sand and studded snow tires are both commonly used in winter driving on icy streets. Why is it that neither does you much good if the temperature is below zero? For that matter, why do they help for temperatures above zero?

166.

freezing point

3.50
Salting ice

When my grandmother makes homemade ice cream, she packs ice around the ice cream container, and then she salts the ice. Why does she add the salt? In a similar vein, why is salt put on icy roads? To both these questions you'll probably answer, "to lower the freezing point." Yes, but *how* does salt lower the freezing point? If the day is *very* cold, the salt won't improve the road contions. What is the lowest temperature at which it will still do some good?

How cold would it have to be for a body of salt water to freeze over?

413, pp. 187–188; 414, pp. 3–4, 12–15, 47–48.

freezing point

3.51
Antifreeze coolant

Why does a mixture of antifreeze and water freeze at a lower temperature than pure water? How does the antifreeze also provide protection against overheating in the summer? If antifreeze is so good in these respects, then why don't you completely fill the radiator with it and forget about the water? (Most antifreeze manufacturers suggest the mixture should not exceed about 50% antifreeze.)

330, pp. 227–228.

latent heat

3.52
Feeling cool while wet

Why do you feel cool when you first step out of a shower or a pool? Try to estimate your rate of heat loss. (One parameter now used to measure such a cooling effect in a wind is the windchill factor.)

Why are hosptial patients sometimes given methyl alchohol rubdowns to soothe them? Why not just use water?

When I was young and on vacation with my family, we kept a canvas water bag on our car's front fender. Though the day may have been hot, the water in the bag was always cool. Why was that? Can you calculate the temperature of the water for some given situation (air temperature, humidity, car speed)?

158, p. 324; 427, p. 64; 428; 429.

freezing point

3.53
Carburetor icing

On some days, even when the temperature outside is as high as 40° F,

my carburetor will ice up and cause my car to stall. Figure 3.53 shows the throttle plate being frozen in place, thereby stopping the air flow to the engine. What causes this icing? Is this more likely on a dry or on a humid day? Can it happen when the outside temperature is below freezing?

426.

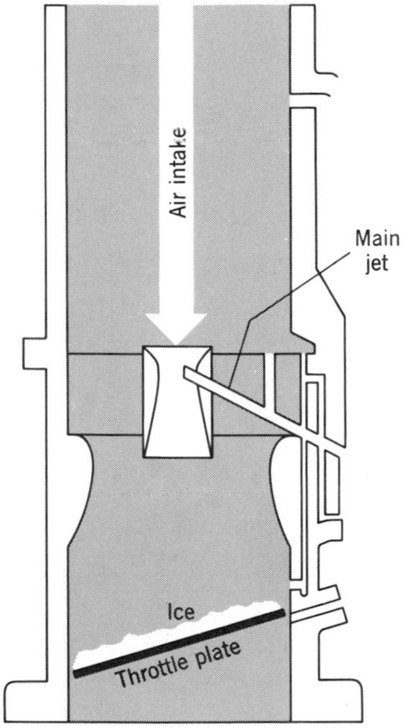

Figure 3.53
Carburetor icing.

latent heat

diffusion

3.54
Eating polar ice

Eskimos know that newly frozen sea ice is much too salty to eat or to melt for drinking but sea ice several years old is fine. They have also found that if the ice is pulled up onto shore and out of the water, the desalting is speeded up, especially if this is done during the warm spring and summer months. Why does the salinity decrease with time and, in particular, why does it decrease faster in the warm months when there should be more evaporation and a resulting *increase* in salinity?

338, pp. 95–97; 414, pp. 26-28; 425.

latent heat

3.55
A pan top for boiling water

If you boil a pan of water for spaghetti, why does the boil come much faster if the lid is left on? Well, there is less heat loss, right? But what does that really mean? Is there less convection or less infrared radiation? When the lid is on, isn't the lid itself nearly at the boiling temperature? Hence, won't there be nearly as much radiation and convection above it as above an open pan? If so, why does the water boil faster in a covered pan?

convection

latent heat

3.56
Briefly opening oven

My grandmother claims that on humid days her oven heats up faster if she opens the oven door wide and then closes it just before she turns on the heat. If this is true, then explain it.

160, p. 174.

latent heat

3.57
Water tub saving the vegetables

My grandmother puts a large pot of water in the cellar near her vegetables to protect them from frost. Why would the presence of the water help protect the vegetables?

160, p. 161; 329, p. 70; 438.

latent heat

3.58
Icehouse orientation

Before the refrigerator was invented, people in northern climates would store winter ice in icehouses for use in the summer. Among the features required of a good icehouse was proper orientation: its doorway had to face towards the east so the morning sun would eliminate the damp air. But this also meant the sun would warm the icehouse more than if it faced north or south, and so the dampness must have been far more undesirable than the extra warming. Why was that?

439.

heat conduction

heat pipes

latent heat

3.59

Heating meat with a "Sizzle Stik"

How can you get a roast to cook faster? Well, you can stick a metal rod into it as is commonly done in baking potatoes. Since heat is then conducted into the meat's interior quicker than directly through the meat, the meat cooks much faster. There is a device called the "Sizzle Stik", however, which abandons the metal rod in favor of a hollow metal tube containing a wick from end to end and some water (Figure 3.59). It is claimed that heat conduction is 1000 times better than with the solid tube, and indeed, cooking times may be cut in half. But how? Why would a hollow tube like this be better than a solid one? And why is there water and a wick in the hollow tube?*

430 through 432.

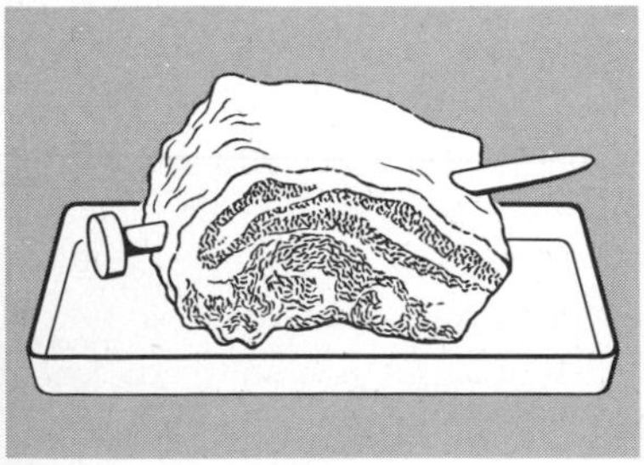

Figure 3.59
Sizzle Stick in roast. (After drawing by Horizon Industries.)*

*® Horizon Industries, Lancaster, Pennsylvania 17601, U.S.A.

pressure and phase change

latent heat

3.60

The highest mountain

On the earth why aren't there any mountains significantly higher than Mt. Everest, say, ten times higher? (Nix Olympica on Mars is over twice as high as Mt. Everest.) If there is some limit to mountain heights, then what determines it, and approximately what is the limit?

440.

conduction

3.61

The boiling water ordeal

One of the most fascinating examples of Oriental magic is the Yubana, or boiling water ordeal, of the Japanese Shinto following.

> In the Yubana, the performer approached a huge caldron filled with boiling water and suddenly thrust two clumps of bamboo twigs into the liquid, flinging it high and showering it all about his head, shoulders and arms. As the water reached the fire below the caldron, it produced great clouds of steam, which subsided only when the caldron was almost empty. The performer was then seen quite unharmed by the ordeal, proving the mighty power of Shinto.*

Boiling water would have burned the performer's skin, of course, so there must have been some trick. Hence, you should not try this experiment yourself. Would it have helped if the Shintoist timed his ritual so that his feat came soon after boiling commenced? What was the water temperature then?

449.

*From *Master Magicians* by Walter Gibson, Copyright © 1966 by Walter Gibson. Reprinted by permission of Doubleday & Company, Inc.

phase change

latent heat

bubble formation

3.62

Boiling point of water

What does it really mean to say that a pan of water is boiling? One hundred degrees centigrade is the commonly accepted boiling point of water at an atmospheric pressure of one atmosphere. How can any one temperature like this be called the boiling point? Why can water sometimes be heated above 100°C without boiling (still at a pressure of one atmosphere)? Finally, why is it claimed that once water has reached 100°C any additional heat input will not raise the water's temperature but will only increase the evaporation rate? Why can't the water beneath the surface get hotter than 100°C with an additional heat input?

441.

evaporation rate

3.63
A puddle's salt ring

When salt has been used to deice a sidewalk, why is it left behind in rings around the puddles as the puddles evaporate? The same thing can be seen on a larger scale in the white edges around lakes in dry areas. You can even see it in your own kitchen by saturating a glass of hot water with salt and then letting the solution set for a month. Afterwards, both the inside and outside surfaces of the glass will be coated with salt. Why is salt left on the outside of the glass?

360, pp. 21-23; 458.

ideal gas law

vapor pressure

latent heat

phase change

3.64
Dunking bird

The dunking bird, which is probably the most popular of all physics toys, is a glass bird that rocks back and forth and "drinks" from a glass of water (Figure 3.64*a*). You start the motion by wetting the head, after which the bird slowly begins to oscillate and eventually dunks its head into the water. The bird then rights itself and repeats the cycle without further assistance. As long as it keeps getting its head wet, it will continue to bob up and down. What makes it go?

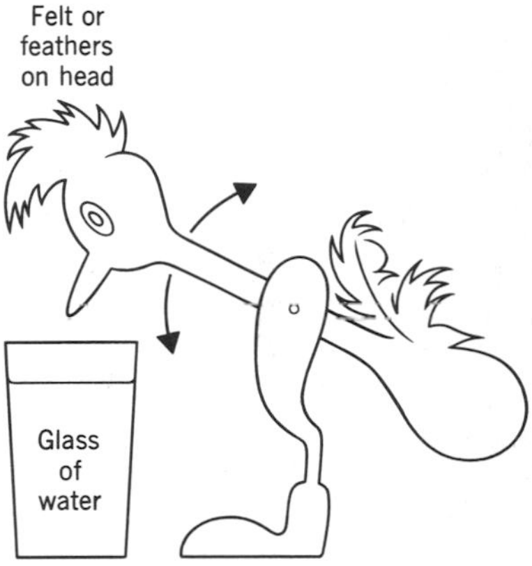

Figure 3.64a
Dunking bird.

Perhaps the dunking bird is a solution to next century's power needs. Just imagine—we erect a huge bird just off California, and as it continuously dunks its head into the ocean, it provides the entir- West Coast with energy. This might lead to a dunking-bird cult, however, and we would all end up paying tribute by dunking in unison three times to the west each morning (Figure 3.64*b*), so maybe we'd better just forget it.

433 through 437; 1457.

Figure 3.64b
The dunking-bird cult.

3.65

Dancing drops on hot skillet

If water drops are sprinkled onto a dry, hot skillet, the drops will dance and skim along the skillet's surface. Why don't the drops evaporate immediately? What makes them skim along? Surprisingly enough, the drops will disappear faster if the skillet is cooler. Why is that?

Examine a skimming drop closely and you will find it assumes a variety of odd shapes. The drop is actually vibrating but since your eye cannot follow the motion that quickly, you see a composite shape. To catch it in various vibrating states, use a stroboscope or a high-speed camera. In Figure 3.65 some of the fundamental shapes are sketched. Why do the drops vibrate?

155, p. 234; 160, pp. 171-172; 330, p. 254; 442 through 446.

Figure 3.65
[After N. J. Holter and W. R. Glasscock, J. Acoustical Soc. Amer., 24, 682 (1952).]

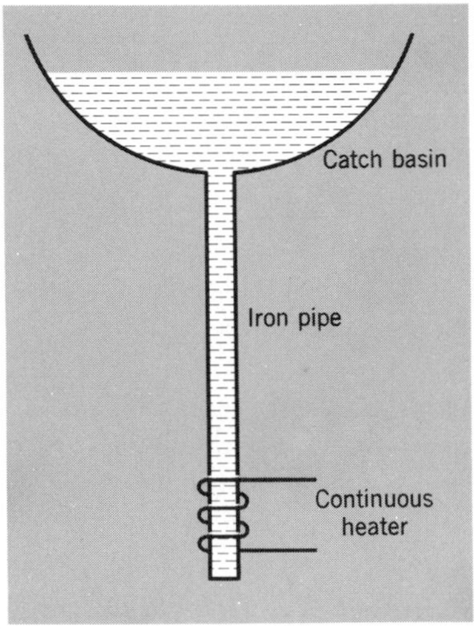

Figure 3.66
Artificial geyser. (E. Taylor after F. I. Boley.)

3.66

Geysers

What causes the eruptions of geysers and, in Old Faithful's case, what is responsible for the periodicity of the eruptions? Could their energy source be simple heat conduction through the surrounding rock, or is a faster heat supply needed?

Suppose you were to make an artificial geyser with a continuous heat source as shown in Figure 3.66. How deep should you make the tube, how much power should you provide for the heating, how often would it erupt, and how high would the water jump?

450 through 452.

vapor pressure

3.67
Percolator

How does my plain old, non-electric coffee percolator work? For example, must the central stem be relatively small? And, is all of the water at boiling temperature when the pot begins to perk?

253, pp. 110–111; 1533.

latent heat

3.68
Single-pipe radiators

While most steam radiators have two pipes (one inlet and one outlet), there is one system in which there is only a single pipe (Figure 3.68). As if that were not strange enough, it is said that the steam and returning water in that single pipe are at the same temperature. How could they be at the same temperature if the radiator is heating the room? Where does the radiated heat come from?

318, pp. 6–8; 418, p. 143.

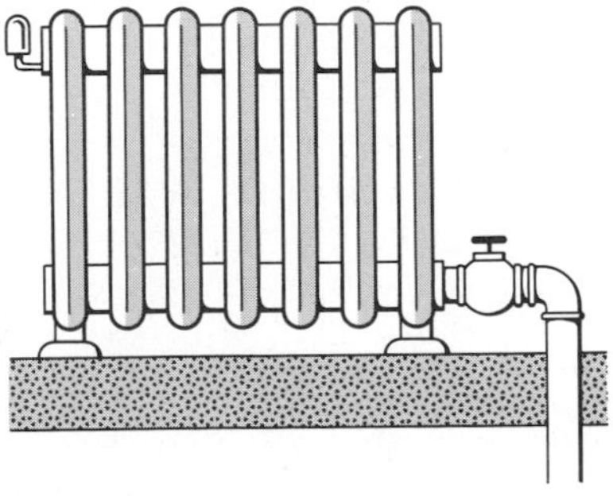

Figure 3.68

vapor, conduction, radiation

3.69
Licking a red-hot steel bar

Though fire walking has long been associated with Far East mysticism, there have recently been some scientific investigations into the feat and even a fire-walking display before thousands of people at a soccer match halftime. Even more amazing than the fire walkers, however, are those people who can briefly plunge their hands into molten metal and touch and lick (!) red-hot steel bars without the slightest injury. You may suspect deceit is involved, but the feat can actually be explained with good physics. Although I have dipped my fingers into molten lead without harm, you should not try these experiments yourself, for they are dangerous and can result in a very bad burn. Figure 3.69 shows that even good physics won't save an overconfident scientist.

Suppose a professional showman were to lick a red-hot steel bar. What might guard his tongue not only from a very serious burn, but indeed from any burn at all? Why should he use only extremely hot metal? Is there any danger in less hot metal? In fire walking is there any optimum speed with which to walk? In particular, can a fire walker walk too fast?

330, pp. 254–255; 447; 448.

steam flash

convection

3.70

Banging radiator pipes

What causes the hammerlike pounding of steam radiators?

253, p. 155; 318, pp. 9, 15; 453, p. 319.

thermal absorption

radiation

3.71

Wrapping food with aluminum foil

Ordinary kitchen aluminum foil has one shiny side and one dull side. Does it really matter which finish is on the outside when the foil is wrapped around something to be cooked, as a baked potato for example? Which finish should be outside when the wrapped material is to be frozen, and again does it really make a difference?

3.72

Old incandescent bulb

Why does an incandescent bulb become gray as it becomes old? Does it become uniformly gray, or is one side preferred?

3.73

How hot is red hot?

Probably you know that an object sufficiently heated will become incandescent. A red-hot poker in the fire is a common example. Can you estimate the temperature at which an object, let's say, the poker, first becomes visibly incandescent? Does it matter if the poker is black iron or shiny steel?

1583.

Figure 3.74 Refrigerator as an air conditioner.

3.74

Cool room with refrigerator

Once, on a very hot day, I tried to cool my dorm room by leaving my refrigerator door open (Figure 3.74). How much did I cool my room that way?

thermal absorption

and transmission

3.75

Black pie pans

Why are the bottoms of some frozen pie pans painted black?

If you make a pie yourself and you want the bottom crust browned, why should you use a thermal glass pan rather than a metal one? If you have to use a metal one, why should it have a dull finish, instead of a nice shiny one? You may very well already know why in principle, but does it really matter in fact? Try some simple experiments to see.

radiation

3.76

Archimedes's death ray

During the Roman attack of Syracuse about 214 B.C., the Greek scientist Archimedes supposedly saved his town by burning the Roman fleet with sunlight directed by mirrors located on the shore. Presumably, many soldiers simultaneously reflected the sun's image onto each ship in turn, and each ship was set on fire.

Considering that Archimedes did not have very large mirrors, would such a feat be possible? Can you estimate how many mirrors, let's say, one meter square, would be needed to set aflame dark wood 100 meters away within less than a minute? Should those mirrors be curved or flat if the target distance is variable? If they are flat, how large is the image of the sun on the wood? Finally, *could* Archimedes have destroyed the Roman fleet in this manner?

1574 through 1580; 1615; 1616.

pressure

nonlinear oscillation

3.77

Toy putt-putt boat

The putt-putt boat (Figure 3.77) has an unbelievable means of propulsion. Two pipes join a top section, the boiler, to the boat's rear. When the water-filled boiler is heated by a candle, the steam that is produced forces water out of the pipes and thus drives the boat forward. The boat should stop when the boiler runs out of water, but actually more water is sucked into the boiler through the pipes, and the process repeats itself. Thus, the boat putt-putts its way along. Why is water sucked up? When it is sucked up, why doesn't the boat move backward as far as it had previously moved forward?

454 through 457.

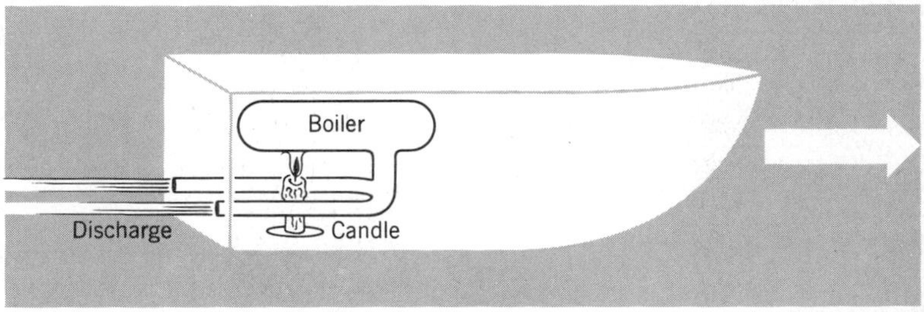

Figure 3.77
Cutaway view of putt-putt boat. [After I. Finnie and R. L. Curl, Amer. J. Phys., 31, 289 (1963).]

conduction

specific heat

3.78

Feeling cold objects

Shouldn't all objects at the same temperature feel like they *are* at the same temperature? You aren't reluctant to put your clothes on when they are at a room temperature of about 70° F, but how about sitting down naked in a dry bathtub at the same temperature? What's the difference?

462, p. 76.

Figure 3.79
"It's not the heat or the humidity, it's this damn 100% wool, fully lined burnoose."

radiation

convection

phase change

3.79
White clothes in the tropics

Why do people wear white clothes in the tropics (if, in fact, they do)? Supposedly it keeps them cooler. Is that a real and measurable effect? If they have light skin, does white clothing make any difference? Does the sun heat you primarily with ultraviolet, visible, or infrared light? How does white clothing respond to each of these frequency ranges? How much of the heating is from direct sunlight, and how much is from the environment? Finally, if you're traversing a desert, should you wear white clothing or go nude?

344, pp. 58-59; 459 through 461.

thermal conduction

and absorption

3.80
Cast-iron cookery

There is an ancient kitchen mystique about cooking in cast-iron pots and pans as opposed to steel ones. Cooks, from the gourmet to the occasional, swear there is less sticking and better, more uniform cooking with the cast iron pot. Is there any physical basis to that claim?

radiation

heating

flux

thermal conductivity

3.81
The season lag

Why exactly is it cold in winter and warm in summer? Is it because the earth is closest to the sun in summer and furthest away in winter? No, actually just the opposite is true (Figure 3.81).

Predict which months should be the coldest and which should be the warmest. You will probably pick, if your explanation is the common one, the months of November, December, and January for the coldest and May, June, and July for the warmest. However, the weather records and your own experience tell you that the coldest months are December, January, and February and the warmest are June, July, and August. As my grandmother says, "When the days get longer, the cold gets stronger." Why does the weather lag your prediction by one month?

388, p. 7.

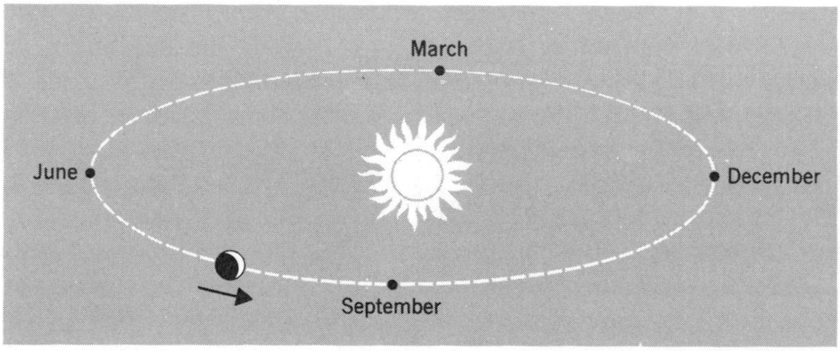

Figure 3.81
Earth's orbit around sun (not to scale).

temperature

kinetic theory

radiation

3.82

Temperature of space walk

What is the temperature of the space where an astronaut is space walking? If he held up a thermometer, what would it read?

radiation

3.83

Greenhouse

A greenhouse is somehow designed to keep plants in a warm environment. How does it do this? Does it have special glass or will any glass material do?

A controversial application of the greenhouse principle is in predicting the results of our atmospheric pollution. For example, a catastrophic warming of the earth might be caused by a high altitude, supersonic transport system. Why is this feared, and how could the more general pollution of the atmosphere lead to a runaway greenhouse effect? The subject is, of course, very complicated. In fact, some claim that the pollution will not bring a warming, but instead will lead to a cooling of the earth and possibly even another ice age. An intriguing account of the effects of clouds on the solar light input to the earth is in Fred Hoyle's excellent science fiction book, *The Black Cloud* (470).

219, pp. 153–154; 388, p. 22; 466, pp. 33–34; 467 through 469; 1544; 1545.

conduction

convection

radiation

3.84

Why do you feel cold?

If you stood naked out in a field on a cold winter day, why would you feel cold? For instance, is your body heat escaping to the air by heat conduction? Why would a fur coat make you feel warmer? Wouldn't it conduct heat too?

While indoors on a cold day, stand facing a large window and then turn the opposite way. Most likely your face will feel cooler in the first position. Why is that? After all, the air temperature doesn't change suddenly as you turn around.

In the movie *2001: A Space Odessey* an astronaut space walked without a spacesuit for a few seconds. (The author, Arthur C. Clarke, believes this could be done without harm to the astronaut.) During such a walk in deep space, would the man have a sensation of cold?

How is it that some people can adapt to very cold working conditions? Some people, in fact, court an adverse, cold environment for religious reasons or to prove their stoic nature. An extreme case of adaptation was discovered by Charles Darwin when he found the Yahgan Indians of South America living in temperatures near 0°C with little more than a fur cape draped over their shoulders. What physically changes in the body's

Figure 3.84
"He was streaking."

response to the cold to allow such adaptation?

Finally, when you do get very cold, why do you shiver?

253, pp. 140–142; 344, Chapter 4, 5; 412, p. 498; 428; 459; 460; 463 through 465.

heat loss

3.85

Wrapping steam pipes

Exposed steam pipes are often covered with asbestos to minimize heat loss, and so we might conclude that asbestos is a poorer conductor of heat than the room air. Otherwise why would anyone pay for asbestos insulation? But, as a matter of fact, asbestos is a *better* heat conductor than air. Why then is it used to cover pipes, if that seems precisely the wrong thing to do?

253, p. 74.

convection

3.86

Thunderstorm wind direction

"You don't need a weatherman
To know which way the wind blows"

---Bob Dylan,
*Subterranean Homesick Blues.**

When a thunderstorm is a few miles away and coming toward you, does the wind blow toward or away from the storm? Most likely you'll find that it changes direction as the storm gets closer. Why should it do that?

300, p. 4; 362, p. 47; 363, pp. 105–106.

convection

3.88

Insect plumes over trees

There have been many observations of dark plumes forming over tree tops near sunset (Figure 3.88). Though the plumes look like smoke, closer inspection re-

convection

3.87

Silvery waves from your finger

Sprinkle a small amount of aluminum powder into a squat jar of wood alcohol, screw on the top and put the jar in the refrigerator. Once it has cooled, remove it, and place your finger against the side of the jar. Silvery waves form and quickly spread away from your finger (Figure 3.87). What generates the waves? (The powder serves merely to make them visible.) What would happen if instead you pressed an ice cube to the jar's side while the jar and alcohol are at room temperature?

472.

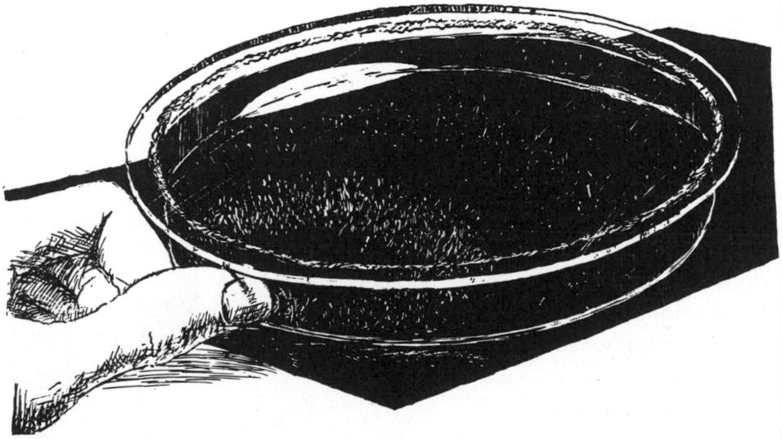

Figure 3.87
Waves spread from your finger across the alcohol. (From "The Amateur Scientist" by C. L. Stong. Copyright © 1967 by Scientific American, Inc. All rights reserved.)

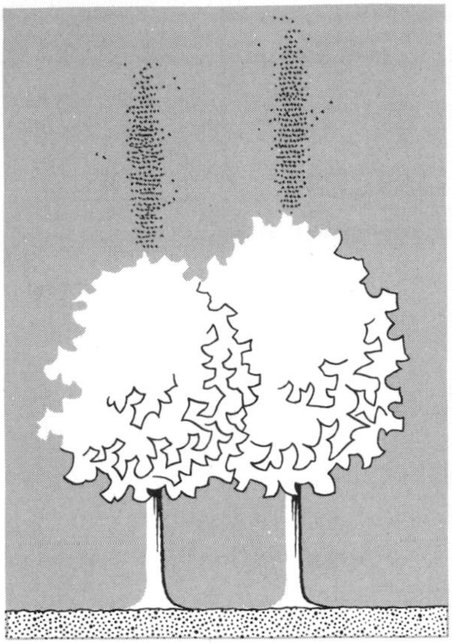

Figure 3.88
Insect plumes over trees.
[After J. H. Wiersma, Science, 152, 387 (April 15, 1966), Copyright 1966 by the American Association for the Advancement of Science).]

veals they are actually thick swarms of insects, usually mosquitos, that have gathered above the trees. The columns are vertical and well defined and may even suggest a small fire in the tree. They have also been seen over TV antennas and church steeples. In fact, there is even a story about a fire department rushing out to fight a church fire only to find that the plume above the steeple was insects and not smoke. Why are these insect plumes formed?

473 through 480.

convection

3.89

Shrimp plumes and Ferris wheel rides

Shallow water brine shrimp ascending in large numbers also take on the appearance of a plume (Figure 3.89). These plumes, which may be as large as several cubic meters, are always found over stones on the bottom. What's more, they are never found over shady stones, but only over those stones that enjoy some sunlight. In spite of this, however, the shrimp plumes frequently lean away from the sun. The questions to be asked about this are obvious. Why do the shrimp ascend in such large concentrations only over sunlit stones? If the sunlight is desirable, then why do the plumes frequently lean away from the sun?

A shrimp in the plume is carried up to the surface of the water, where it separates itself from the plume and swims back to the bottom. Why are the shrimp then drawn back to the plume to continue their Ferris wheel ride?

481.

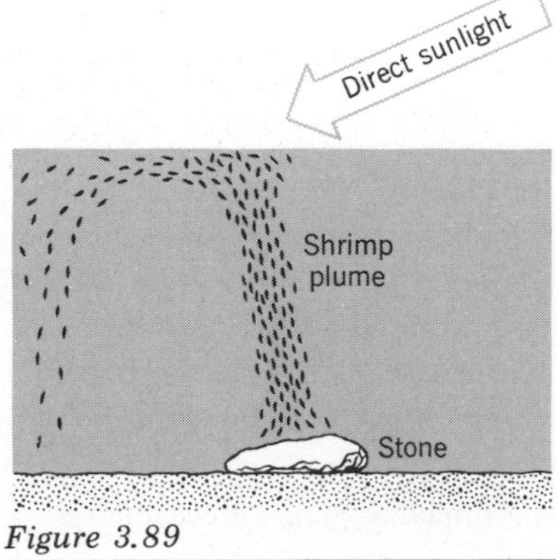

Figure 3.89

phase change

latent heat

human heat transfer

3.90

Heat stroke

If you ever mowed the lawn in the middle of summer as I used to do in Texas, you've probably wondered how your body stays as cool as it does. A significant amount of thermal energy is generated inside your body, up to 1400 kcal per hour during heavy physical exercise, and if that heat is not disposed of somehow, your body temperature could rise as much as 30°F per hour. Of course, that would soon be fatal. How is the heat dissipated? Can you trace the path by which it is lost?

Mowing the lawn in the midday on a once-a-week basis was miserable, for I always got heat exhaustion, yet there are people who do this daily without ill effects. Somehow the body becomes accustomed to working in the heat. What exactly happens? The heat is generated at the same rate internally, so the dissipation mechanism must somehow change.

High temperatures in Texas were usually bearable because the humidity was so low. Why is it so much more uncomfortable in places with high humidity?

344, pp. 57–59; 482.

cooling

conduction

thermal radiation

3.91
Cooling a coffee

Suppose you have just made a hot cup of coffee, but you've still got 5 minutes until class. If you want to bring your coffee to class as hot as possible, should you put the cream in now or just before class? When should you add the sugar? When should you stir it and for how long? If you don't want to stir it, should you leave the spoon in? Does it matter whether the spoon is plastic or metal? Would your answer be different if cream were black instead of white? Does your answer depend on the color of your cup? Make numbers for your arguments if you can.

transport and

temperature

3.92
Polaroid color development

If you take a color Polaroid picture on a cold day, you must develop it in a metal plate previously warmed by your body. If you don't, the colors will be off balance, because when the dyes are cold, they will not reach the positive in the proper amount of time. Why does the temperature affect the transit time of the dyes that way?

497.

convection

conduction

radiation

latent heat

3.93
Heat islands

Why is the temperature in a city higher than that in the surrounding countryside by 5 or 10 degrees (Figure 3.93)? In addition to there being more heat producers in the city, how is the temperature difference affected by a city's tall buildings, expanses of rock and concrete, quick rain drainage and snow removal, dust concentrations, frequency of smog and fog, etc.?

Meteorologists who map the temperature distribution of a city, whether large or small, find a "heat island" concentrated near the city's center. Temperatures are lower as one moves away from the heat island, toward the suburbs and countryside. One consequence of this is that springtime blooming of flowers should begin sooner near the city's center.

344, pp. 78–81; 483 through 493.

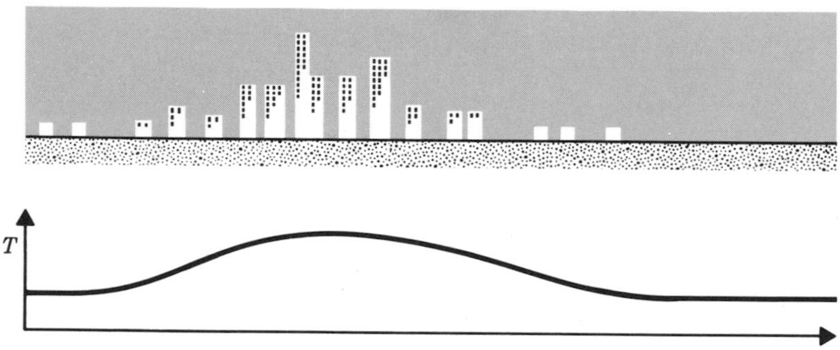

Figure 3.93
Heat island of a city.

kinetic theory

ideal gas law

3.94
Total kinetic energy in a heated room

A stove will warm the air in a room. Will it also increase the air's total thermal energy? (The thermal energy is kinetic energy of the air molecules.) Well, the air's thermal energy certainly depends on its temperature, and since the air is being warmed, the total thermal energy will be increased. That sounds correct, but one discussion of this claims that the total energy will not change. How can that be?

343, pp. 40–41; 494 through 496.

radiation

3.95

Smudge pots in the orchard

Why does a fruit grower put smudge pots in his orchard overnight when he fears a frost? Since the pots are placed so far apart, they surely can't provide much heat to warm the fruit. What's the point then? Does the grower ever use them during the day?

330, p. 398; 471, p. 130.

conduction

convection

3.96

A warm blanket of snow

Why is there less danger of crop damage on a sudden cold day if there is a good snow cover on the crops?

160, p. 183; 413, p. 205.

Wien's law

atmospheric transmission

3.97

Fires from A-bombs

> Of the multiple dangers to life which nuclear explosions present, . . . the resulting setting of innumerable fires is perhaps the worst. A single one-megaton bomb may ignite clothing, paper, dry wood, and other similarly combustible materials at distances up to 15 kilometers, and present capabilities make it necessary to scale this range upward by an order of magnitude. The resulting fire storm would in many populated areas "escalate" until destruction of life and property would be virtually total (219).

But if you are more than several kilometers from the blast site there is sufficient time (up to 3 seconds) to fall behind an obstacle for protection. First of all, how exactly does the blast cause fires several kilometers from ground zero? Second, why does this fire danger come at such a relatively long time after the explosion begins?

219, pp. 307-310.

crystal genesis

3.98

Growing crystals

Why does it take small particles, perhaps impurities, to start crystals growing in a supersaturated solution?

498.

3.99

Snowflake symmetry

Why are snowflakes six-sided (hexagons or six-pointed stars), and why are the six arms exactly alike? How does one arm know what its neighbors are doing as the snowflake forms?

388, pp. 449-453; 404; 499 through 506.

surface tension

wetting

3.100

Two attractive Cheerios

If two fresh Cheerios* are placed near each other while floating on milk, they will rapidly pull together. What force causes that attraction? Is it possible to get the Cheerios to repel each other for a suitably chosen liquid on which they are floated?

*An O-shaped breakfast cereal from General Mills, Inc.

capillarity

3.101

Cultivating farmland

Why are farmlands in semiarid regions frequently cultivated (the top soil is plowed and broken up into a loose texture)? If a footprint is left undisturbed in cultivated soil, the soil inside the footprint will become hard and dry. Why is that?

158, pp. 141-142.

surface tension

wetting

3.102

Wall curvatures of a liquid surface

Some liquids have surfaces that turn up near a glass wall; others turn down. Why do they do this? What force pulls them up or down? What is the fundamental difference (on a microscopic or atomic scale) between those that slope up and those that slope down? Can you calculate what surface shape is expected?

Some liquid drops will remain drops after being placed on a flat glass surface. What prevents them from spreading out? What is the fundamental difference between such a nonwetting liquid and a wetting one? Finally, what is the nonwetting drop's shape when it is sitting on the surface?

Suppose a nonwetting liquid is

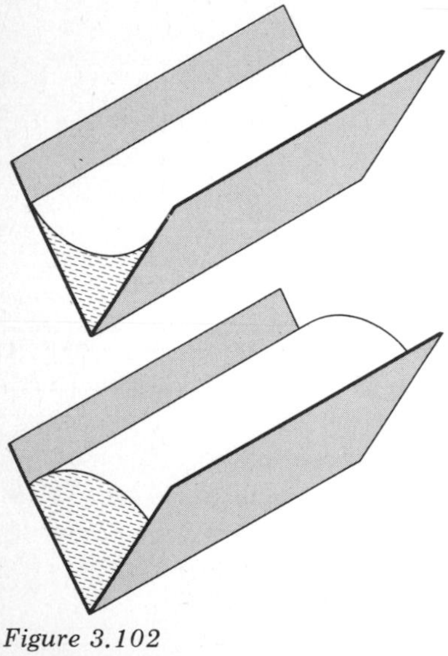

Figure 3.102
Which way does the nonwetting liquid curve?

in a small trough as shown in Figure 3.102. Which shape do you expect? Or is either possible, depending on the trough's angle? If the latter, at what angle is the liquid flat?

51, pp. 127–128; 321; 507 through 511.

osmotic pressure

atmospheric pressure

negative pressure

3.103

Rising sap in trees

How does sap rise in trees, especially in very tall ones (some redwoods are 360 feet high)? Certainly there is a pressure difference between crown and roots, but why? Does the tree act like a suction pump? If so, then shouldn't all tree heights be limited to 33 feet since that is supposedly the maximum height of a suction pump? Some other mechanism must be involved.

512 through 519.

osmotic pressure

capillarity

freezing

3.104

Ice columns growing in ground

Have you ever seen columns of ice growing out of the ground, perhaps about 1½ inches high? Upon close inspection you may find bits of soil and pebbles on top of the columns. Strangely enough, when the columns form, the ground itself is unfrozen and usually wet. What makes these columns grow? If the temperature is low enough to cause freezing, shouldn't there be ice on the ground? Finally, what will limit a column's height?

338, p. 133; 521.

capillarity

osmotic pressure

freezing water

3.105

Growing stones in the garden

If you have ever taken care of a garden, you may be aware of the annual crop of stones that must be cleared from the garden each spring. Though some regions don't have this problem, others, New England for example, have an abundant stone crop. Robert Frost's "Mending Wall" is about such a crop of stones.

The stones obviously migrate upward from the rock bed below the soil, but why? The stones, after all, are denser than soil and should gradually move down, not up. What's forcing those stones up? A simple simulation of this stone migration, suitable for the classroom lab, is given in Bowley and Burghardt (522).

403; 522 through 526.

osmotic pressure

capillarity

freezing

3.106
Winter buckling of roads

"Something there is that
doesn't love a wall,
That sends the frozen-ground-
swell under it
And spills the upper boulders
in the sun"
---Robert Frost, "Mending
Wall"*

If you have ever lived in the north you might have seen pavement develop bumps (on blacktops) or cracks (in concrete) or even become tilted during the winter (Figure 3.106). These bumps can sometimes be as high as a foot. What could cause this? My first guess would be that water underneath the pavement expanded in freezing, but it would require so much water to make these large bumps that such an explanation is hard to accept. So, what does cause the bumps?

338, pp. 131-133; 403; 520.

*From "Mending Wall" from ***The Poetry of Robert Frost*** edited by Edward Connery Lathem. Copyright 1930, 1939, © 1969 by Holt, Rinehart and Winston, Inc. Copyright © 1958 by Robert Frost. Copyright © 1967 by Lesley Frost Ballantine. Reprinted by permission of Holt, Rinehart and Winston, Inc., the Estate of Robert Frost, and Jonathan Cape Limited.

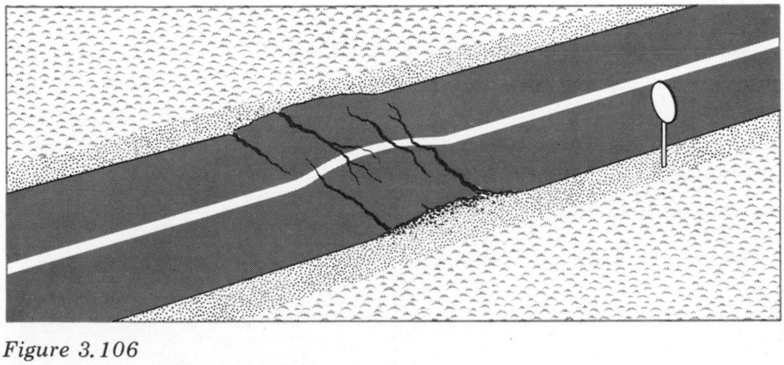

Figure 3.106
Buckling of road in the winter.

capillary and

osmotic forces

3.107
Shorting out a masonry wall

Masonry walls usually become damp, especially near the ground. One way to prevent this is to ground the wall electrically by running a wire from the wall to a metal stake in the ground (Figure 3.107). No batteries or other such power source are used, only a simple metal stake and wire. How would shorting out the wall in this way prevent moisture in the wall?

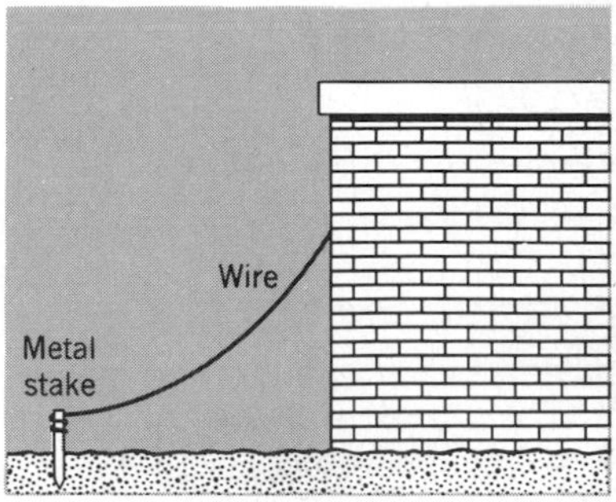

Figure 3.107
Drying a masonry wall by electrically grounding it.

527.

surface tension

3.108
Soap bubbles

What keeps a soap bubble together? Is it really spherical? What is the pressure inside the bubble? Does a bubble go up or down in air?

Is there any part of the surface that is most likely to burst first?

322; 528 through 532; 533, pp. 139 ff.

surface tension

buoyancy

3.109

Inverted soap bubbles

Inverted soap bubbles—where the water and air have traded places—can easily be made by carefully pouring soapy water into a dish of water from a height of a few millimeters. If you pour slowly, drops skim across the water surface. If you pour a bit faster, a drop may penetrate the water and remain there with a shell of air trapped around it, thus forming an inverse soap bubble (Figure 3.109).

Will these soap bubbles show colors as normal ones do? Will they have uniformly thick shells? Will the bubbles go up or down in the water dish? Finally, do you think there will be continuous evaporation from the inner drop into the air shell, eventually leading to a collapse?

534; 1608.

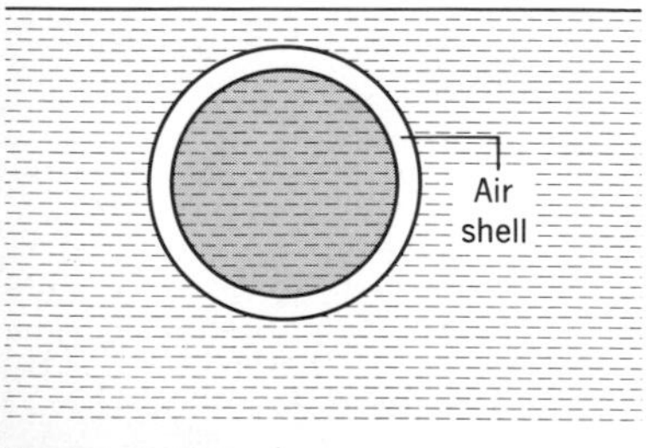

Figure 3.109

capillarity

3.110

A candle's flickering death

Why do many candles, especially small ones, flicker and pop in the last moments before burning out? What determines the frequency of the flickering?

535.

combustion

3.111

Dust explosion

One of my most delightful undergraduate tricks was to replace a friend's overhead light bulb with a short wire and a bag with some flour in it. The wire almost completed the circuit so that there was a spark when the light switch was thrown. Just before the victim appeared, I shook the bag to fill it with floating flour dust. Got the picture? My friend turned on his light, there was a spark, and the dust exploded, neatly covering his entire room with a layer of flour. Such dust explosions are very serious problems in some industries where static electricity builds up in a room full of dust. In either case, why does a spark cause an explosion of the floating dust?

536, pp. 383–384; 537 through 539.

combustion

thermal conduction

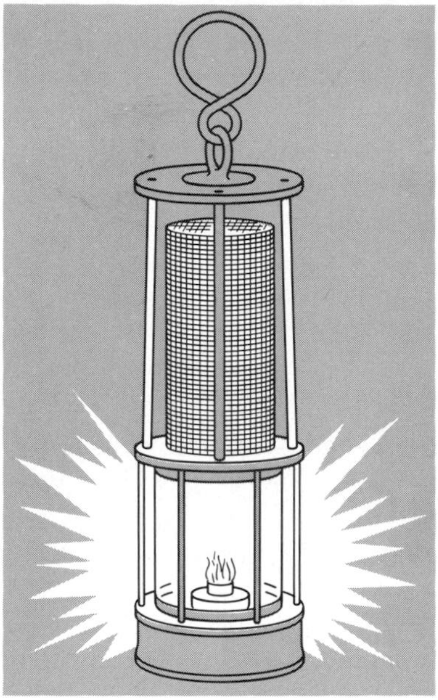

Figure 3.112

3.112

Davy mine lamps

The open flame miner's lamp is very dangerous if the miner encounters explosive gases. The danger can be avoided, however, if a fine mesh screen is placed over the flame holder as shown in Figure 3.112. The screen certainly can't prevent the explosive gas from reaching the flame, but it nevertheless prevents the explosion. How?

110, p. 171; 155, p. 232; 413, p. 205; 541, pp. 74–75; 542.

stress

desiccation

3.113
Mud polygons and drying cracks

You have frequently seen cracks in dried mud, but have you ever wondered why the cracks form or tried to explain their polygonal appearance? Sometimes the edges of the polygon will curl up, perhaps even so far that a tube develops, separates from the surface, and rolls away.

Ever since airplanes and aerial photography came into prominent use, giant polygons have been seen in the dry desert basin bottoms that have periodically had water. By giant I mean the widths of the polygons can be up to 300 meters and a fresh fissure may be as much as a meter wide and five meters deep.

Why do the cracks and tubes form? If the ground cracks into polygons, is there any reason to believe, as some authors have argued, that the polygons tend to be pentagons or hexagons? In other words, is there any preferential angle at which two cracks will intersect?

543 through 551.

stress

freezing

3.114
Thermal ground cracks

Mud cracks are not the only type of patterned ground you can find. For example, polygonal cracks are found in the permanently frozen ground of the arctic and subarctic regions. What causes the cracking in this case? Is there any preferred angle between cracks at intersections?

438; 552 through 556.

freezing

colloidal suspension

3.115
Stone nets

As a final example of patterns in the ground, stone nets—circles and polygons of sorted stones (Figure 3.115)—should be mentioned. What brings the stones from a random distribution into such geometric shapes?

556 through 558.

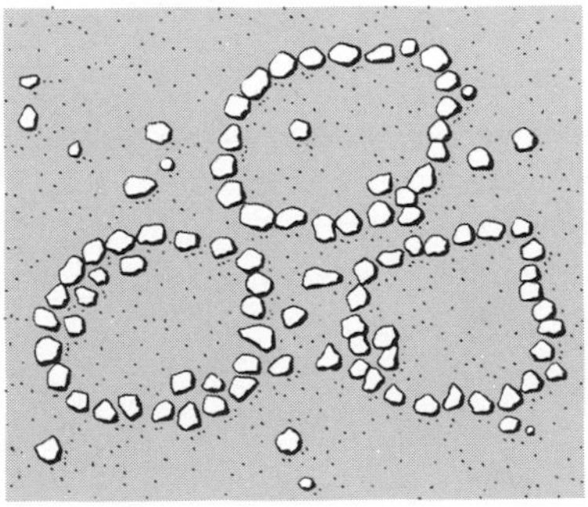

Figure 3.115
Naturally occurring circles of stones.

Figure 3.116
"Now, in the second law of thermodynamics.."

entropy

3.116
Life and the Second Law

"As you stay in a given place, things and people go to pieces around you."
---Celine

In thermodynamics one learns that entropy, which is a measure of disorder in a system, always increases in an irreversible process (the so-called Second Law of Thermodynamics). What about birth and life? Isn't the creation and growth of a human being a violation of this rule, for in that process, doesn't order increase? Isn't the rule also violated by the evolution of all animals over millions of years?*

559 through 562.

*A similar problem, whether or not quantum mechanics can explain life, is covered in Mehra (1569).

4

The madness of stirring tea

Hydrostatics

(4.1 through 4.14)

fluid pressure	buoyancy
Pascal's law	Archimedes' law

4.1

Holding back the North Sea

Remember the story of the Dutch boy who saved his town by thrusting his finger into a hole he discovered in the dike? How did he do it? How could one little boy hold against the pressure of the whole North Sea?

418, p. 68.

4.2

Breathing through air tube

To what depth can you breathe through a simple air tube while swimming under water? What determines the limiting depth?

563.

4.3

Measuring blood pressure

Why do doctors always measure blood pressure on your arm at a height about even with the heart? Couldn't they just as well measure it on the leg?

412, p. 191.

Exceptionally good references: Crawford's *Waves* (170) is the best example of real-world physics in a major textbook I have found. See also Tricker (399), Scorer (364), Lodge (923), and Schaefer (830).

4.4

Last lock in Panama

A ship is waiting patiently in the last lock of the Panama Canal as the water level is lowered. When enough drainage has taken place, the gate begins to swing open toward the ocean, and the lock director engages the machinery to finish opening it. The ship then begins to move out to sea without the aid of a tugboat and without using its own power. What forces it seaward?

564.

4.5

Panama Canal ocean levels

You may already know about the difference in ocean levels at the two ends of the Panama Canal. During the dry season the difference is small, but during the rainy season it can be as much as 30 centimeters. Why aren't the ocean levels the same?

565.

4.6

Hourglass's bouyancy

If an hourglass is floating in a narrow tube of water as shown in Figure 4.6, will it float again if the tube is inverted? The sand that was initially in the lower part of the hourglass is now pouring down from the upper part. The weight and volume of the hourglass are the same, however, so the hourglass should float back up to the top. Instead, it stays at the bottom of the tube until the sand has poured into the lower section. Why? Does the buoyancy of the hourglass really depend on whether the sand is in the lower or upper section?

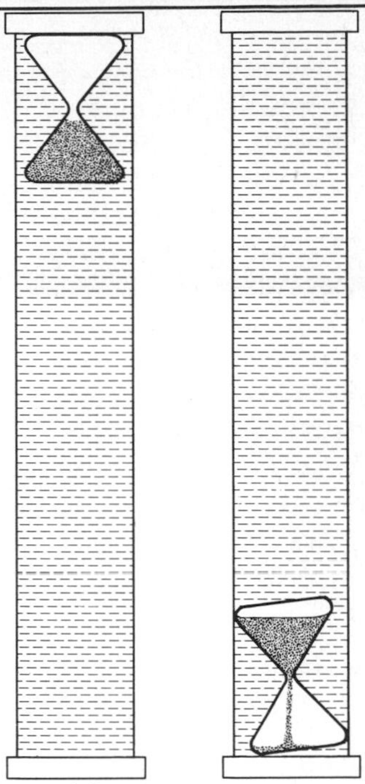

Figure 4.6
When the tube of water is turned over, why doesn't the hourglass float up? (From "Mathematical Games" by Martin Gardner. Copyright © 1966 by Scientific American, Inc. All rights reserved.)

566.

4.7

Boat sinking in pool

There is a famous problem about throwing a stone from a boat into the swimming pool where the boat is floating. When the stone is thrown from the boat, does the water level rise, fall, or remain unchanged? This problem was asked of George Gamow, Robert Oppenheimer, and Felix Bloch, all excellent physicists, and to their embarrassment, they all answered incorrectly.

What happens to the water level if a hole is made in the bottom of the boat and the boat sinks? If the water level changes, when does the change begin? In particular, does it begin to change when water first enters the boat?

567.

4.8

Coiled water hose

If you try to pour water into a coiled hose, as shown in Figure 4.8, no water will come out the other end. Indeed, surprisingly little water will even enter the hose. Why?

566.

Figure 4.8
(From "Mathematical Games" by Martin Gardner. Copyright © 1966 by Scientific American, Inc. All rights reserved.)

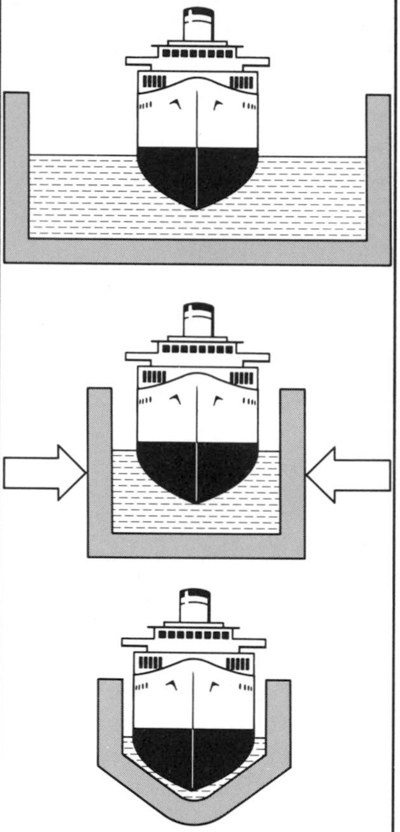

Figure 4.9
[After L. E. Dodd, Amer. J. Phys., 23, 113 (1955).]

4.9

Floating ship in dry dock

When a ship is put into dry dock, the water is removed as the dock is made smaller (Figure 4.9). What is the minimum depth of water under, say, a two-ton ship that will still support the ship?

567; 568.

4.10

Submarine stability

How does a submarine ascend and descend? How does it remain submerged at a fixed depth? Shouldn't changes in the water density at the submarine's depth make the submarine unstable? Sure, small corrections for the changes could be made, but such corrections are impractical. Besides, if quiet conditions are essential to avoid detection, then constant corrections are certainly forbidden.

Fortunately, there are many depths in the sea where a submarine is stable against the sea's perturbations. What is peculiar about those regions, which are called thermoclines?

570.

4.11

Floating bar orientation

Does a long, square bar float on a side or tilted over on an edge (Figure 4.11)? Even if you find the answer obvious, try floating several long square bars in a variety of liquids and then classify your results according to the relative density of the bar and liquid. Is your intuition correct?

569.

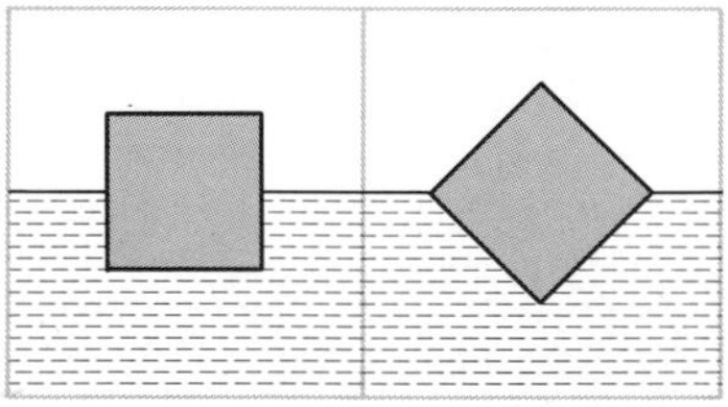

Figure 4.11

4.12

Fish ascent, descent

Do fish ascend and descend the same way as submarines? Do they compress and expand their swim bladder to change depth? This may be a common explanation, but it isn't correct because a fish has no muscular control over its swim bladder. So how do they do it?

Although fish can't survive rapid depth changes (in trawling, cod and hake are dead when pulled to the surface because of this), they can live at tremendous depths. For example, fish at 15,000 feet withstand a pressure of 7000 pounds per square inch. What provides the resistance to such pressure?

571.

air pressure

surface tension

4.13

Inverted water glass

Place a piece of cardboard over the mouth of a glass of water. (The glass does not have to be full.) Invert the glass, holding the cardboard in place. Now remove your hand from the cardboard—it stays in place and, therefore, the water stays in the glass. Why?

Try the same thing with a long glass tube (about 60 centimeters long and 3 or 4 centimeters in diameter) that is sealed at one end. Whether the inverted arrangement is stable or not depends on how much water is in the tube but probably not in any way you would have guessed. If the tube is nearly full or nearly empty, it is stable when inverted with the cardboard. But if it is about half full, the water falls out every time. Why?

572.

4.14

Floating bodies

Why do drowning victims first sink and then, after a few days, float to the surface?

gravity waves

Rayleigh–Taylor instability

4.15

Stability of an inverted glass of water

If the cardboard used in Problem 4.13 were to disappear suddenly with the water glass inverted, why would the water fall out? Yes, I know gravity will pull the water down, but how does the falling start? Isn't the water surface initially stable? Isn't it precisely the same forces holding it up against gravity? Once you decide why the falling begins, can you figure out how long it will take to empty the glass?

574 through 579.

buoyancy

stability

molecular and thermal diffusion

4.16

The perpetual salt fountain

Tropical and subtropical oceans have warm, salty water near the surface and cooler, less salty water below. A seemingly perpetual fountain may be made by dropping a tube to the bottom, and pumping water to the surface. The pump can then be removed, and the fountain will continue itself (Figure 4.16). What keeps the fountain going? Is it truly perpetual?

580, pp. 44–45; 581; 582; 1546.

Figure 4.16
Perpetual salt fountain in the ocean.

buoyancy
stability
molecular and thermal diffusion

4.17
Salt fingers

You can see a phenomenon related to the salt fountain in your kitchen by half filling an aquarium with cold, fresh water and then adding (carefully, without mixing) a solution of warm, dyed salt water on top. (The dye is only meant to be a tracer.) Immediately fingers of the upper solution extend into the underlying fresh water, making the boundary area translucent (Figure 4.17). You can see the fingers without the temperature difference if you pour a sugar water solution over a salt water solution. (Again, use a dye for a tracer.) What initiates the finger growth, and why are the fingers so stable?

582 through 590.

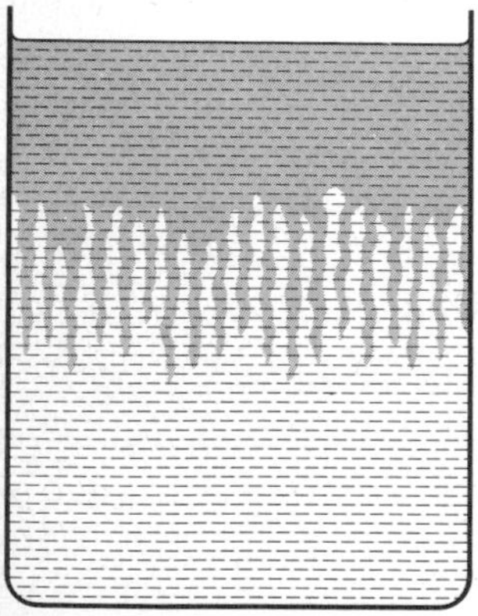

Figure 4.17
Salt fingers (exaggerated scale).

buoyancy
nonlinear system
Rayleigh instability

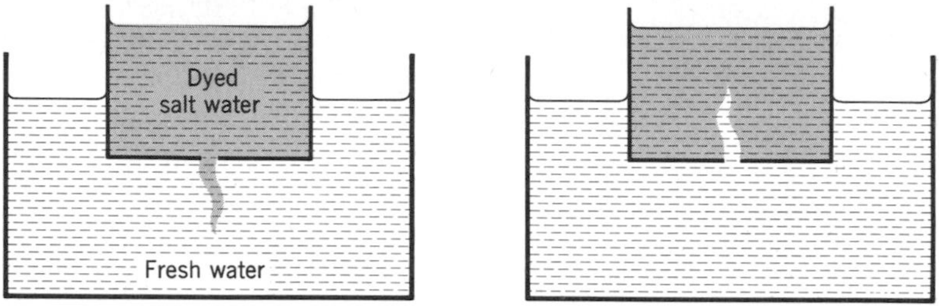

Figure 4.18

4.18
Salt oscillator

If you take an ordinary tin can, punch a pinhole in the bottom, fill it with saturated salt water, and partially immerse it in a container of fresh water, will the two solutions eventually mix? Well, yes, they will, but in a surprising way. (Color one of the solutions with a dye so you can see which is which.) There will be an alternating exchange of solutions, that is, salt water will flow down through the hole, then fresh water will flow up, and so on (Figure 4.18). This oscillation may continue for as long as four days, with an oscillation period of about four seconds. Why is there such an oscillatory exchange of fluid, and what determines the period?

591.

Bernoulli Effect
(4.19 through 4.40)

4.19
Narrowing of falling water stream

Why does a smoothly flowing stream of water from your faucet narrow as it falls? Is there some force squeezing it together? Can you calculate the change in the stream's diameter as a function of the distance from the faucet?

4.20
Beachball in an air stream

To catch the attention of customers, vacuum cleaner salesmen will sometimes reverse the air flow in a cleaner and then balance a beach ball in the exhaust jet (Figure 4.20). The ball is quite stable and can be held in place with the air jet at a considerable angle. Even a good slap will not be enough to release it from the jet. Why is it so stable? Will the

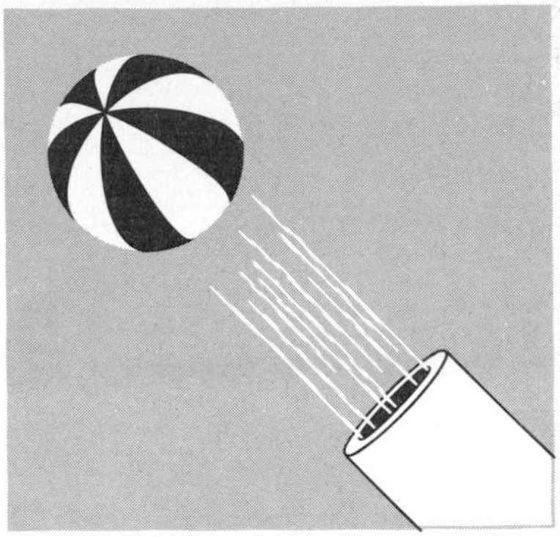

Figure 4.20

ball spin in any particular direction?

211, p. 155; 399, pp. 102–103; 592, p. 60; 593.

4.21

Toy with suspended ball

A toy, "á-Blow-Go"*, uses this suspension trick also. You balance a light ball by blowing through a small side tube, as shown in Figure 4.21. With a long, hard blow, the ball is lifted until it is pulled into the top of the tube and shot back to its original position. The point of the game is to circulate the ball this way as many times as possible within one breath. (My record is five

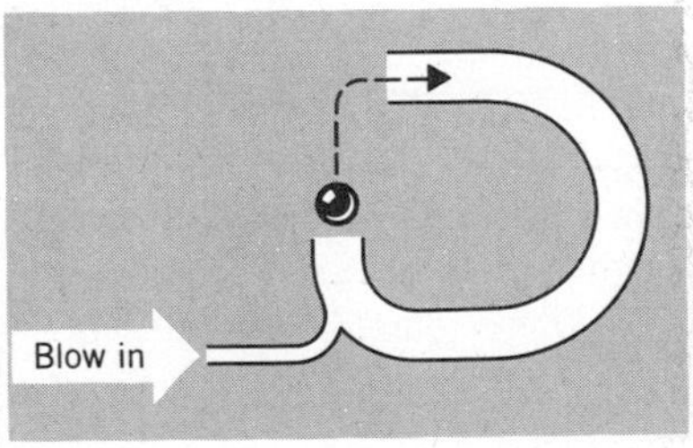

Figure 4.21
By blowing through the side tube, you make the ball circulate through the main tube.

complete circuits.) What makes the suspended ball stable, and what makes the ball enter the top tube?

Norstar Corp., Bronx, New York

momentum transfer

wetting

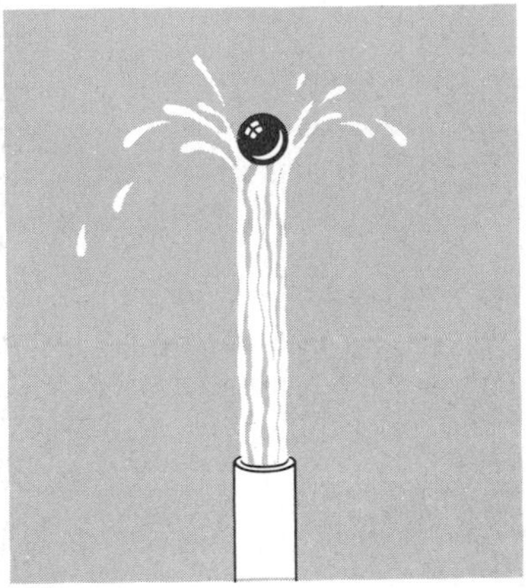

Figure 4.22
Ball suspended in water jet.

4.22

Ball balanced on a water jet

In another similar trick, a ball is balanced on a vertical water jet (Figure 4.22). Occasionally the ball may sit still for several seconds, but usually it wavers and bobs. Why doesn't the wavering cause it to fly out of the jet? What holds it in? Does this really involve the same physics as the beach ball problem?

To be honest, the ball does sometimes escape the jet, but in the course of its fall, it reenters the jet and is returned to its former position. It will even do this in a vacuum. What entices the ball back into the stream like this?*

595.

*For yet another suspension but with photons instead of air or water, see Prob. 5.104.

4.23

Egg pulled up by water

Let a faucet pour onto an egg floating in a glass of water (Figure 4.23). For flow rates above some critical value, the egg will rise as if it were attracted to the falling water. Why, and what determines the critical flow rate?

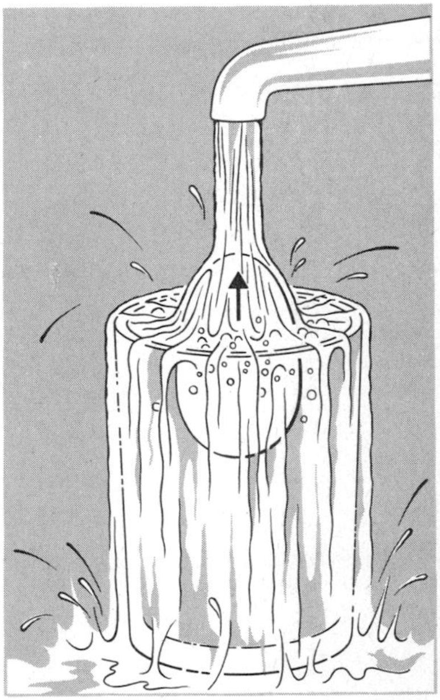

Figure 4.23
Egg pulled upward by water stream.

momentum transfer

wetting

4.24

Spoon in a faucet stream

If you hold a light spoon round side upward in a stream of water as shown in Figure 4.24, the spoon seems to be glued to the stream. You can move your fingers several inches away, putting the spoon at a considerable angle, and the spoon will still refuse to leave the stream. The falling water should, by all rights, push the spoon away, not attract it. What causes this?

592, p. 60; 595; 596.

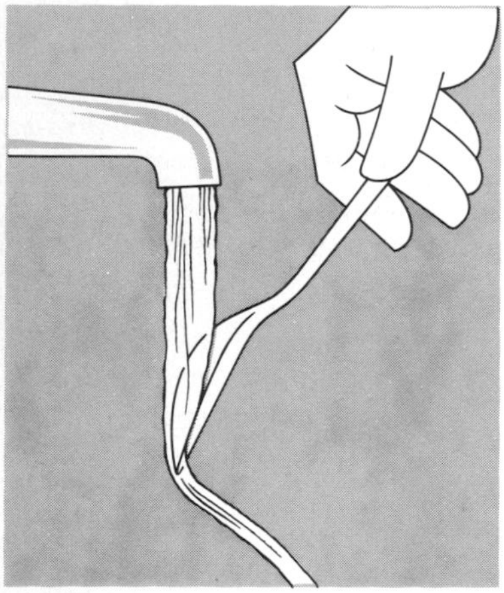

Figure 4.24
The spoon is kept in place by the water stream.

4.25

Water tube spray guns

If you put one end of a tube into water and blow across the open end (Figure 4.25), you can force water up the tube. With a strong blow

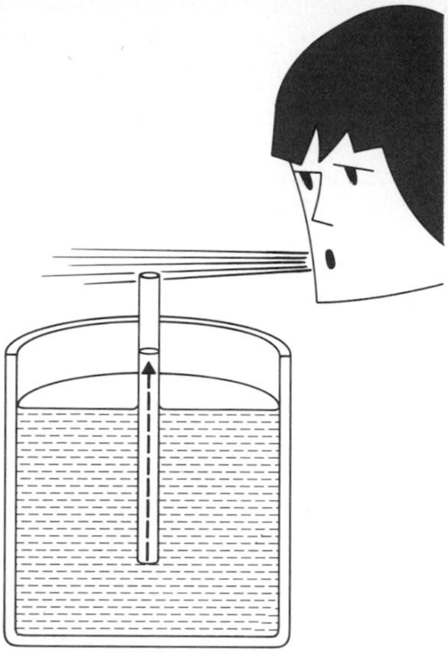

Figure 4.25
Water is lifted up the tube by the air blown across the tube.

across a short tube, you can soak your friends. The aerosol can is a more practical application: pressurized air blows across a narrow container of the material to be sprayed. How do such spray guns and cans work?

597.

4.26

Passing trains

When high-speed trains pass each other, they must slow down or their windows will be broken. Why? Will the windows be pushed into the train or sucked out? Will this happen if the trains are traveling in the same direction? If you stand near a high-speed train, will you be pulled toward or pushed away from the tracks . . . or both?

599 through 602.

4.27

Ventilator tops and prairie dog holes

Why is the draft through a ventilator pipe improved if the top of the pipe is surrounded with a cone (Figure 4.27a)? Similarly, why is the ventilation inside a prairie dog tunnel improved if the entrances are surrounded by high, conical mounds (Figure 4.27b)?

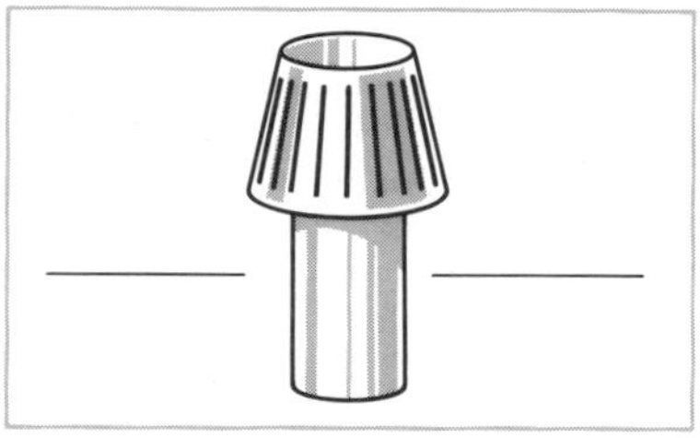

Figure 4.27a
Ventilator pipe with cone top.

139, pp. 179-180; 598.

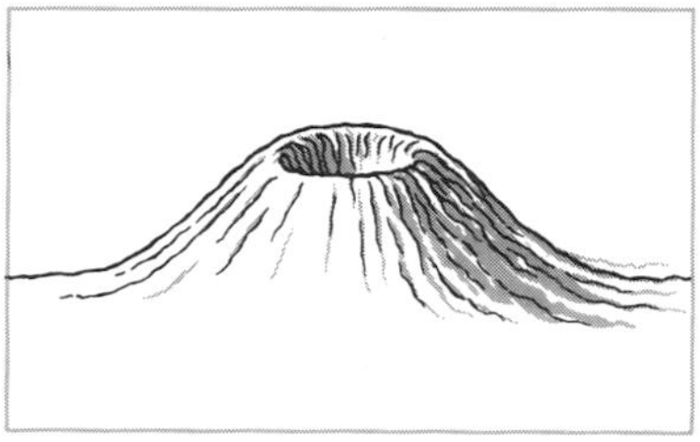

Figure 4.27b
Prairie dog hole with high mound.

pressure

4.28

Insects rupturing on windshields

Are insects squashed directly on the windshield of fast moving cars, or do they rupture in the air and then splatter on the windshield? If the latter is the case, then what causes the rupture? You may be tempted to blame the insect's fate on turbulence, but is there really that much turbulence? Why doesn't the strong, deflected wind stream carry the bugs safely over the car? (Figure 4.28 shows one way to avoid the bugs.)

364, pp. 12–13.

Figure 4.28
(By permission of John Hart. Field Enterprises.)

eddy formation

4.29

Flapping flags

Why does wind, even a uniform wind, make flags flap? What determines the frequency of the flapping?

124, p. 115; 453, p. 51.

Bernoulli effect

momentum transfer

4.30

Wings and fans on racing cars

Racing cars have gone through a great many changes over the years, some obvious, some subtle. One of the best developments was the addition of a horizontal wing above the rear of the car. When a car with such a wing entered a curve, the driver would tilt the wing forward. Upon leaving the curve, the wing was leveled again. This wing and its adjustments proved very useful in keeping a car on the road in turns, hence allowing much higher speeds there. Were it not for the danger of broken wings resulting in uncontrollable cars on the tracks, these movable wings would still be in use. But safety forced the racers to fix their wings in place. In either case, movable or fixed wing, how would a wing help in keeping the car on the road?

One of the strangest versions of a racing car has been the Chaparral 2J, which was built by Jim Hall who also pioneered the movable wing. The Chaparral 2J had two large fans in its rear designed to pull air beneath the car, through the fans, and out the rear. Skirts were built along the bottom sides of the car, hugging the road, so as to tunnel the air beneath the car. Again, Hall greatly increased the speed of his cars by increasing the traction. But how? Why would air tunneled beneath the car and out the rear increase traction? Can you estimate the resulting increase in traction and speed?

1581.

Bernoulli effect

momentum transfer

4.31

Lifting an airplane

"How does an airplane gain lift?" is a standard physics question, and the standard answer involves Bernoulli's principle, but is that the only, or even the major, factor? If the wings are shaped (as is in Figure 4.31) to produce a Bernoulli effect, then how do airplanes fly upside down?

The crucial point of the standard argument is that the air moves faster over the wing than under the wing, and this means, because of Bernoulli's principle, there is greater air pressure beneath the wing. Hence there is lift. Why does the air move faster over the top? Well, the two streams of air moving below and above the wing must cross the wing in the same amount of time. The air moving above has a greater distance to travel and thus moves faster. Here the standard argument stops. But why *must* the upper air traverse the wing in the same time as the lower air? This is rarely explained. As a matter of fact, the top and bottom streams have **unequal** traversal times. So, why does the wing have lift?

593; 603 through 605.

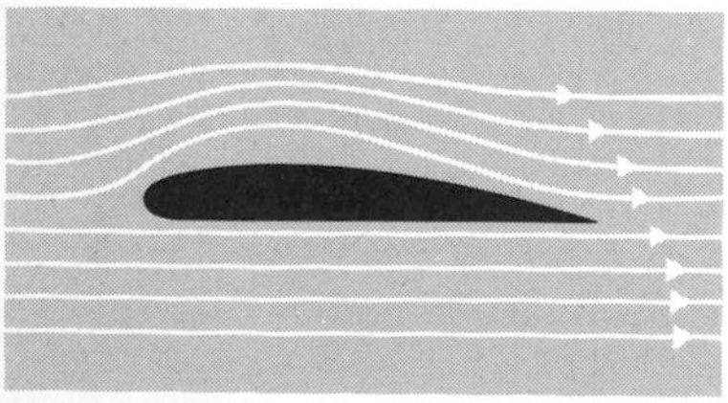

Figure 4.31
Cross section of airplane wing.

4.32

Pulling out of nose dive

Suppose a plane stalls and goes into a nose dive. Why must the pilot wait until he reaches a high speed, higher than his normal cruising speed, before he attempts to pull out of the dive?

603.

4.33

Sailing into the wind

It's not difficult to see how a sailing boat can be pushed along with the wind, or at some angle to it, as long as that angle is not too large. But not only can sail boats travel 90° to the wind, they can even sail into the wind at an angle of 45° or more. In this case the wind will obviously oppose the motion of the boat, right? So what does push the boat when it sails windward? Disregarding water currents, what angle will give the fastest boat speed?

611 through 613.

4.34

Frisbee

What keeps a Frisbee* aloft? Must it be spinning? It apparently doesn't have to be a disc, because Frisbee rings work almost as well.

*® Wham-O Manufacturing Company, San Gabriel, California.

4.35

Manpowered flight

Is it possible for a man to fly under his own power (Figure 4.35)? The question is an old one but far from dead. It now seems that present attempts to design manpowered aircraft will eventually lead to a working model.

Some of the problems in designing the aircraft are how much power can a man produce, and how much is needed for flight? How large should the wings be? Should they flap? Is the lift improved if you stay close to the ground?

606 through 610; 1518; 1519.

Figure 4.35
Man in glorious flight.

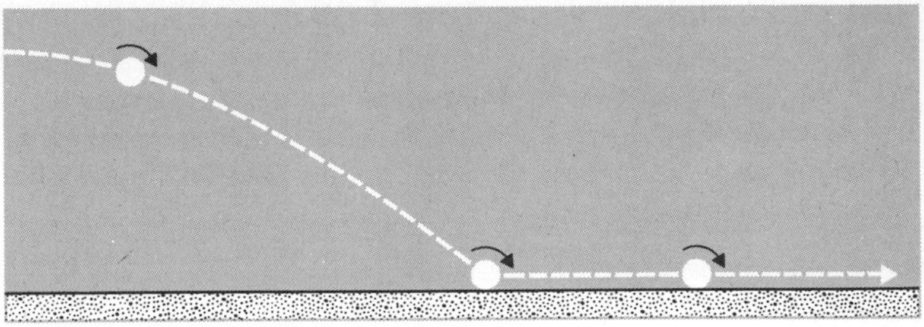

Figure 4.36
Top spin on golf ball causes it to roll forward.

4.36

Golf ball top spin

To gain distance, some golfers will give a top spin to their ball so that it will roll farther after it has hit the ground (Figure 4.36). Considering the ball's total trajectory, is this really a wise thing to do?

36, pp. 53, 138–139; 399, pp. 103–104; 593; 616 through 621; 1484.

4.37

Flettner's strange ship

In 1925 a most unusual ship crossed the Atlantic propelled by two large, vertical rotating cylinders (Figure 4.37). How did those rotating cylinders drive the ship forward?

In a more modern application, NASA has used the same principle by adding a horizontal rotating cylinder to an airplane's wing. How would such a cylinder provide lift for the airplane?

110, p. 22; 155, p. 117; 399, p. 105; 453, pp. 71–72; 615; 623.

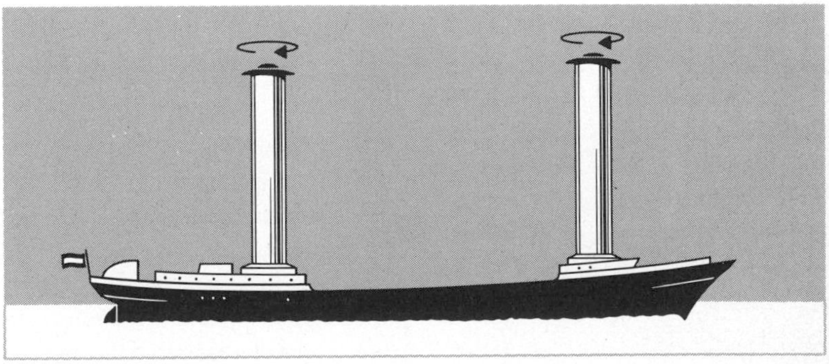

Figure 4.37
Flettner's ship propelled by two rotating cylinders.

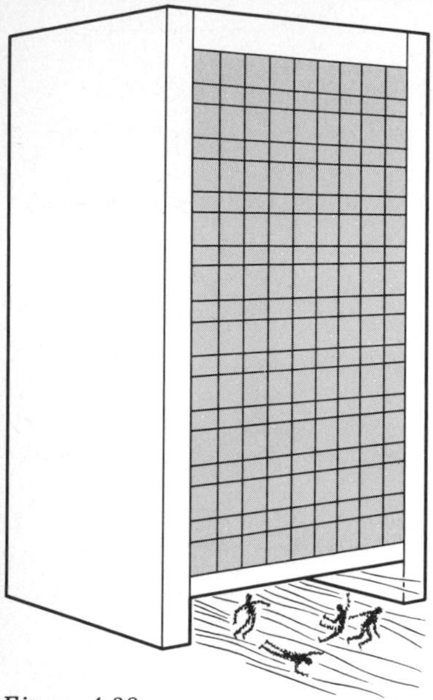

Figure 4.38
Strong winds through building.

4.38

Winds through a building

In one type of modern building design, the floors are hung like bridges between two solid walls and the ground level area is left open (Figure 4.38). This is an attractive design, but inconvenient in windy regions. For example, when the spring winds blew through one such building at MIT, wind speeds up to 100 miles per hour were measured, certainly much higher than elsewhere on the campus. (Students and junior faculty alike were bowled over by the wind; only full professors could withstand the gale.) What causes this enhancement of wind speed?

614.

4.39

Curve, drop, and knuckle balls

Can baseball pitchers really throw curve balls, drop balls, and knuckle balls? If they can, then explain *how* each is thrown. Does a curve ball break continuously or suddenly? Does a drop ball suddenly drop? And does a knuckle ball actually dance, as batters claim? How far will a major league pitcher's curve ball deviate from a straight line by the time it crosses home plate?

36, pp. 53, 138–139; 211, p. 156; 593; 615 through 622.

4.40

Curves with smooth balls

A smooth ball should not curve since, unlike a baseball, it has no rough surface with which to "grab" the air. You can nonetheless throw a curve with a smooth ball, but it will curve in precisely the opposite direction as will a baseball. Why?

593; 619 through 622.

Waves

(4.41 through 4.59)

wave speed (group and phase)

superposition refraction

interference dispersion

reflection

Bernoulli effect

flow around obstacle

driven oscillator

4.41

Building waves

How are periodic water waves built up by random gusts of wind that play along a water surface? Is the wind drag across the surface more important than vertical disturbances? Is there a minimum wind speed required to maintain the water waves? Do the waves provide a feedback to the wind flow to build up the waves even further?

399, pp. 141–147; 580, pp. 133–136; 624; 625.

wave interference

4.42

Monster ocean waves

There are many stories about ships at sea suddenly encountering incredibly large waves. For example, a wave 100 feet high was seen by a cargo vessel captain in 1956 off Cape Hatteras, and there were reports of 80 foot waves in the North Pacific in 1921. In 1933 a wave estimated at 112 feet high was seen by the U.S.S. Ramapo in the North Pacific. Imagine standing on the bridge beneath a wave 112 feet high!

Why do these waves suddenly appear and then disappear? If they are somehow caused by storms, then shouldn't there be more than one large wave? Could they be caused by a sudden underwater earthquake? (Can such earthquake waves be detected by a ship at sea?)

399, p. 138; 626, pp. 48–49; 627; 628; 629, pp. 53–60.

wave velocities

light scattering

4.43
Whitecaps

Why exactly do whitecaps form on the ocean and other bodies of water, and why are they white? In a moderate wind, why do they often appear in succession, each forming downwave of the previous one with a time interval of a few seconds between appearances?

390; 630; 631.

gravity and capillary waves

4.45
Whirligig beetle waves

When a whirligig beetle skims quickly along the surface of the water, why does it make pronounced waves in front of itself, but in back barely visible waves or none at all (Figure 4.45*a*)? If it skims slowly, there are no waves, front or back. Why? A boat doesn't do this; it always makes waves to the rear. What is so different about a skimming water beetle?

A similar asymmetry is present in the wave pattern around a narrow obstacle in a moving stream: the waves upstream have a much smaller wavelength than those downstream (Figure 4.45*b*). What causes the asymmetry, and what determines the wavelengths in the two cases?

633; 634.

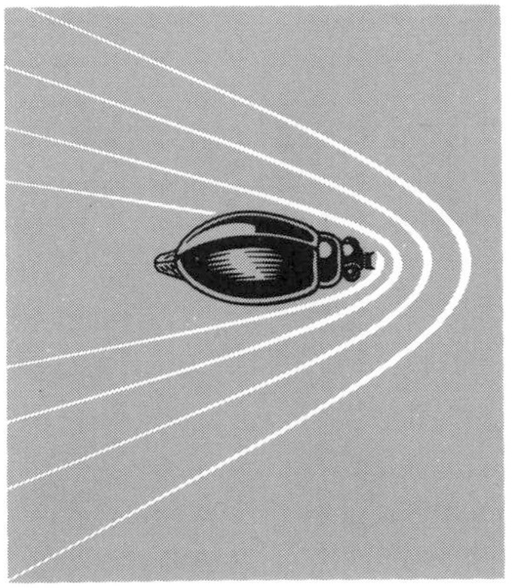

Figure 4.45a
Whirligig beetle waves.

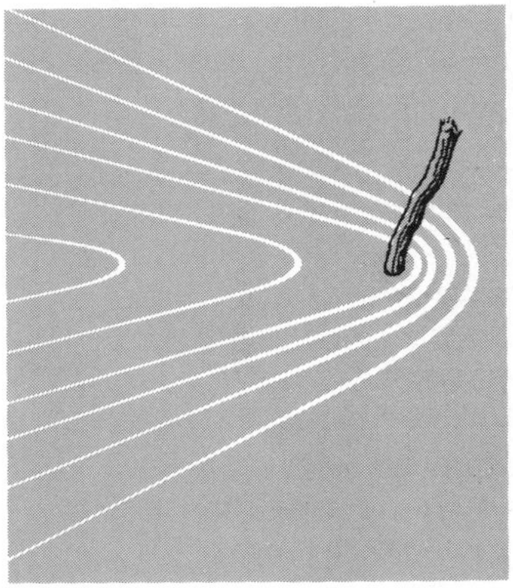

Figure 4.45b
Waves around stick in moving stream. [Both figures after V. A. Tucker, Physics Teacher, 9, 10 (1971).]

wakes

Bernoulli effect

4.44
Boat speed and hydroplaning

What determines the practical speed limit of boats, ducks, and other things larger than insects? If the limitation is friction from the water, then why does a longer boat generally have a higher maximum speed? Wouldn't a longer boat feel more friction and hence have a lower maximum speed?

Why can a hydroplane go much faster than a normal boat of similar length? It is, as you know, partially lifted out of the water. How is the lifting accomplished, and how does it permit such high speeds?

632; 633.

interference

dispersion

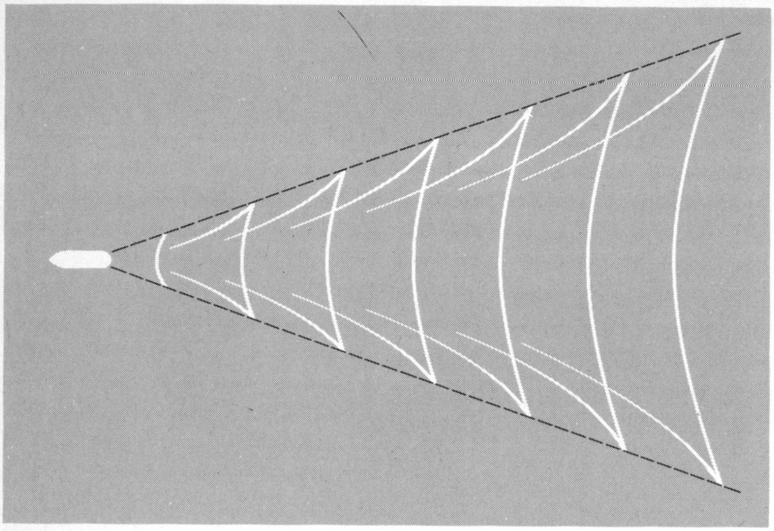

Figure 4.46
Ship waves as seen from above. [After H. D. Keith, Am. J. Phys., 25, 466 (1957)].

4.46
Ship waves

If you ever have a chance to fly over ships moving in deep water, examine their wave patterns. Notice the disturbed areas are always V-shaped with the same angle (38° 56′). As one writer put it, the V shape is present "whether the moving object is a duck or a battleship" (760). Why is that?

Inside the disturbed area, the pattern gets more complicated (Figure 4.46). Can you explain the origin of the two types of wave crests that are present? Are they also the same for a duck and a battleship?

How does the pattern change in shallow water? First, can you explain what "shallow" means? Shallow compared to what?

51, pp. 200-203; 399, Chapter 17; 635, Chapter 8; 636 through 640.

nonlinear wave

interference

4.47
Edge waves

While investigating water waves, Faraday discovered a very curious form of wave produced by a simple, horizontally oscillating plate slightly immersed in a water basin (Figure 4.47*a*). Ignoring wave reflections from the basin's sides, I would have guessed that only common, plane waves would be made. However, when the oscillating plate was immersed about 1/6 inch, he saw the following:

> Elevations, waves or crispations immediately formed but of a peculiar character. Those passing from the surface of the plate over the water to the sides of the basin were hardly [visible], but apparently permanent elevations formed, beginning at the plate and projecting directly out from it to the extent of 1/3 or

Figure 4.47a
Plate oscillating in water.

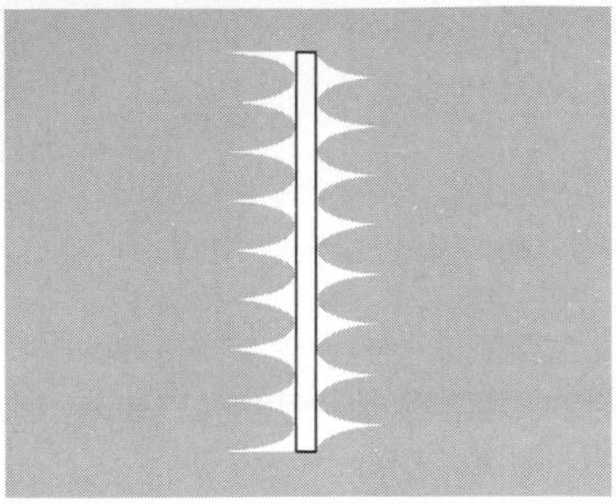

Figure 4.47b
Edge waves on the oscillating plate, as seen from above.

1/2 an inch or more, like the teeth of a very short coarse comb [Figure 4.47*b*] (643). Faraday also noticed these strange waves had half the frequency of the vibrating plate. Now how can a vibrating plate possibly set up standing waves whose crests are perpendicular to the plate?*

641 through 646.

*To see the edge-wave theory used to discuss rip currents on ocean beaches, see Refs. 647 through 651 and Ref. 1618.

refraction

4.48

Swing of waves to shore

When ocean waves reach the shore, why are they approximately parallel to the shoreline? Surely the waves originally come from a variety of directions.

360, p. 28; 399, pp. 95-96; 628; 635, pp. 133-136.

shallow water waves

Bernoulli effect

4.49

Surf skimmer

You can surf, in a sense, on water only one or two inches deep by riding a wooden disc skimming along the shallow surf (Figure 4.49). If you leap on it when it has sufficient speed, you may be carried 20 feet or more. What holds you up during such a ride, and why does this support disappear when the disc slows down? Why do longer boards travel farther? Shouldn't a longer board provide more friction and hence stop sooner?

626, pp. 152-156; 653.

Figure 4.49

shallow water waves

wave speed

4.50

Surfing

What rushes you to shore when you're surfing? Are you pushed by the wave, or are you continously falling downhill? Why are the best waves to ride those on the verge of breaking, and why is most surfing done in waters over gently sloping beaches? Why is the surfing position on the wave front relatively stable? Is a surfer more stable on a long board than on a short board?

626; 652, pp. 80-81.

buoyancy

wakes

4.51

Bow-riding porpoises

Porpoises are often seen riding motionlessly a few feet beneath the water surface near a ship bow. They make no swimming motions at all, so they somehow gain their propulsion from the ship itself. The technique must be well developed, for a porpoise can ride for more than an hour with little or no effort and can remain stationary, flip over on a side, or even slowly revolve around its body axis. There may even be two or three layers of the porpoises, all bow riding together. What actually carries the porpoises along?

A similar case is related by Jacques Cousteau in one of his underwater books (660). Sharks are often accompanied by small "pilot fish" that, according to legend, guide the shark. Cousteau saw one such pilot fish, a very small one, directly in front of the shark's head, somehow being propelled along by the shark itself. That was a precarious position indeed! How was the pilot fish pushed, and why was his position so stable?

654 through 660.

gravity

noninertial forces

static and harmonic theories of tides

4.52

Ocean tides

What causes the ocean tides? You may be satisfied in answering that the tides are driven by the gravitational attraction of the moon and sun, but let me ask a few more questions.

Does the water bulge on the moon side of the earth because the moon pulls the water vertically away from the earth? If it does, that seems strange because isn't the water's attraction for the earth much, much greater than its attraction for the moon?

If the earth's seas are pulled to the moon and the resulting bulge in the ocean is the high tide, then why are there two high tides a day? The earth turns once a day, and hence each point on the earth's surface should face the moon only once a day. Therefore, shouldn't there be just one high tide a day? However, since there are two high tides a day, the water on the earth should have two bulges, one of them being *away* from the moon (Figure 4.52). How do you explain the second bulge?

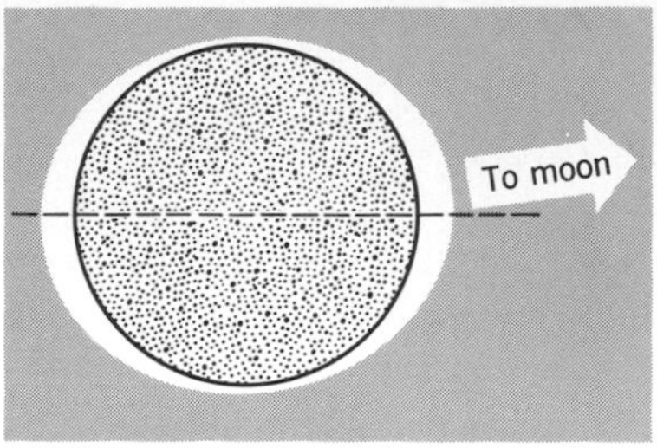

Figure 4.52
Two tides on the earth (exaggerated, of course).

Some seas, (the South China Sea, the Persian Gulf, the Gulf of Mexico, and the Gulf of Thailand, for example) have only one high tide a day. Why don't they have two? Still other places, such as the Indian Ocean, have alternating diurnal and semidiurnal tides. Again, why?

Finally, why isn't there a high tide when the moon is directly overhead? For some reason, there is always a lag.

111; 399, pp. 3-14; 661, Chapter 5, pp. 149-181; 662, pp. 26-32, 40 ff; 663, Chapter 4, pp. 11-55; 664, pp. 177, 179, 188 ff; 665, pp. 195 ff; 667 through 669; 1589.

4.53

Tides: sun versus moon

Which provides the stronger driving force on the tides, the moon or the sun? If you make a rough calculation to see, would you compare the direct gravitational pulls of the moon and sun on a piece of the earth's water? If you do, you'll find that the sun is the dominant body.

Why are there spring tides, which are the larger than average tides near the times of new and full moons, and neap tides, which are the lower than average tides near the first and third quarters of the moon?

399, pp. 15-16; 661, pp. 156-159; 662, pp. 32-33; 663, pp. 23-24, 35 ff; 664, pp. 189-192; 668.

angular momentum conservation

4.54

Tidal friction effects

As a tidal current flows across the ocean bottom, energy is lost to frictional heating. One consequence of this energy loss is that the earth's rotation slows, and the day gets longer.

Does the energy loss have any further effects? A system cannot have a change in its total angular momentum unless there's an outside torque. There is no such outside torque on the earth-moon

system, but we've got an earth with a decreasing spin. How then is the total angular momentum to be conserved?

Will this go on forever? Will the earth's day continue to get longer? Will there be any change in the apparent motion of the moon? One prediction is that some day the moon may travel backwards across the sky.

111; 661, Chapters 16, 17; 663, Chapter 11; 672; 673.

resonance

4.55

Seiches

Water in a lake often sloshes back and forth just as it does in a small rectangular trough. The residents around Lake Geneva long ago noticed this sloshing (called a seiche), which can reach three feet in amplitude, but they didn't understand what determined its periodicity or even what caused it. What does determine the sloshing frequency in a rectangular basin? What periodicity do you predict for Lake Geneva (average depth about 150 meters and length about 60 kilometers)? Finally, what makes the lake slosh?

170, pp. 45-46; 580, pp. 138-140; 635, pp. 423-426; 661, Chapter 2; 662, pp. 62-65; 663, pp. 7-8; 664, pp. 272-273.

shock fronts

water waves

wave speed

4.56

Tidal bores

In most rivers emptying into the sea, the tidal rise is calm, perhaps even imperceptible. But in others the rise becomes so rapid that an almost vertical wall of water, a bore, races up the river with great force (Figure 4.56). The English rivers Severn and Trent and the Canadian river Petitcodiac experience these water walls. The bore of the Amazon is an awesome sight, being a mile wide at places and up to 16 feet high, sweeping upstream at 12 knots. The most striking of them all, however, is the bore of the Chinese Tsien-Tang-Kiang, which has risen as high as 25 feet. The Chinese skillfully use the bore to float their junks upstream, ignoring the danger and the helter-skelter ride. Why do these bores form, and why don't all sea coast rivers have them? Does their speed depend on their height or the depth of the river?

399, pp. 33-66; 635 pp. 320, 326-333, 351 ff; 661, Chapter 3; 662, pp. 97-98; 663, pp. 8, 120-125; 664, pp. 320-321; 674 through 676.

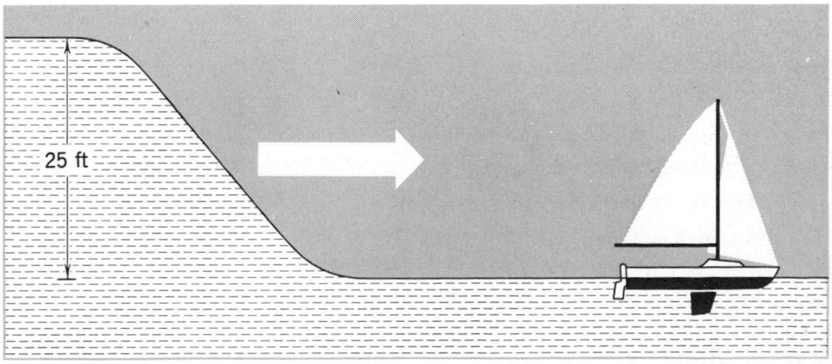

Figure 4.56
Tidal bore racing up river.

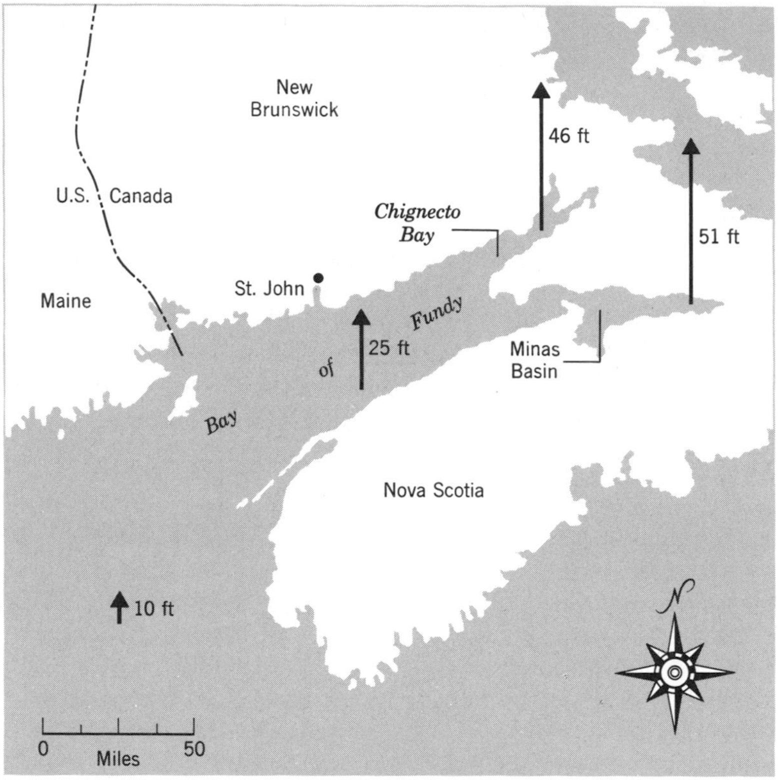

Figure 4.57
Tidal range in the Bay of Fundy.

4.57

Bay of Fundy tide

Why does the Bay of Fundy in Nova Scotia (Figure 4.57) have the world's largest tidal range (the change in water height due to the tides)? In some places the range is so large that men fish by erecting large nets during low tide and then during the next low tide, simply collecting the fish caught in the net during the high tide. At the mouth of the Bay, the range is not too large, about 10 feet during spring tides. Further up the Bay at St. John the range increases to 25 feet, and at the end of Chignecto Bay it is 46 feet. The largest range, 51 feet, is found at the end of the Minas Basin. (Winds can add as much as another 6 feet to these figures.)

Can a bay have an especially favorable length to enhance the tidal range? What would such a length be for a bay whose depth is like that of Fundy (75 meters)? How does that compare with Fundy's actual length?

399, pp. 27–29; 663, pp. 113-115; 664, pp. 235–236; 670; 671.

shock front

water waves

wave speed

4.58
Sink hydraulic jump

When a stream of water falls into my sink, the water spreads out in a relatively thin layer until it reaches a particular distance from the stream where the water suddenly increases in depth. Hence, a circular wall of water surrounds the stream (Figure 4.58). The same type of wall is made if the stream falls onto a flat plate, though the depth change is not as pronounced. What causes these jumps in water depth? What determines the radius at which a jump occurs? How high is the wall?

635, pp. 324 ff; 677 through 681.

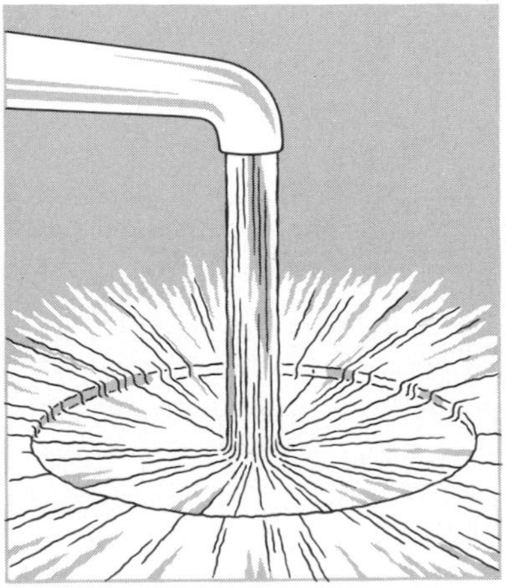

Figure 4.58
Hydraulic jump in the sink.

Figure 4.59
Standing waves in falling water stream.

4.59
Standing waves in falling stream

If you hold your finger or the flat of a knife in a thin water stream, a standing wave appears in the stream* (Figure 4.59). Why? What determines the spatial periodicity of this wave? Why does that periodicity depend on the distance between the flat surface and the faucet?

*Elizabeth Wood, personal communication.

4.60
Beach cusps

Why are cusplike formations, sometimes outlined on a side with small pebbles, very often found on sandy beaches (Figure 4.60)? Shouldn't the ocean waves striking smooth beaches be plane waves? Although some cusps are isolated and can be dismissed as flukes, there are many long beaches whose entire length is embroidered with periodically spaced cusps. What causes them?

629, pp. 386–389; 648, p. 5490; 650; 682 through 691.

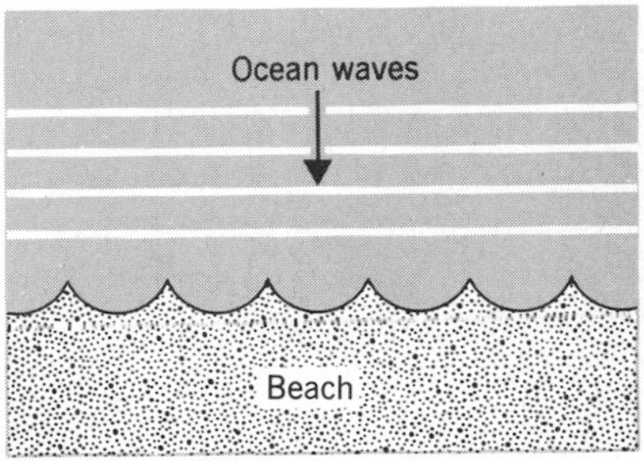

Figure 4.60
Beach cusps.

forces in rotating frame

friction

4.61
Ekman spiral

Suppose there is a steady wind blowing over the water somewhere in the middle of the ocean. In what direction is the net total mass transport of water by the resultant current? In the direction of the wind? Slightly to the left? Well, I understand that it is 90° to the right in the northern hemisphere and 90° to the left in the southern. Why 90°? The current off the California coast provides an example of this in shallower water. The winds there usually blow southward and parallel to the coast, but the top layer of the ocean moves toward the west.

580, pp. 76–79; 692.

vorticity

noninertial forces

friction

4.62

Stronger ocean currents in the west

Doesn't it strike you as odd that in both northern and southern hemispheres there are stronger ocean currents along the western sides of the oceans?

North Atlantic: Gulf Stream
South Atlantic: Brazil Current
North Pacific: Kuroshio
Indian Ocean: Agulhas Current

(The one exception is in the South Pacific, for there is no such large current off Australia.) Why is the west favored for strong currents?

666, p. 1025; 692 through 696.

secondary flow

centrifugal force

friction

4.63

Tea leaves

Why do leaves in a cup of tea collect in the center of the cup when you stir it? Since the tea is rotating, you may want to class this as just another centrifuge example, but wait—in a centrifuge don't the denser objects move *outward*? Hence, the centrifuge argument will only make the behavior of the tea leaves even more mysterious.

44, p. 189; 73; 700, pp. 84–85; 716.

secondary flow

centrifugal force

friction

4.64

River meander

Natural streams and rivers, especially the older ones, are rarely straight for any great length; they almost always meander back and forth (Figure 4.64). In some cases the weaving is so extreme as to cut off and abandon a loop, forming what is called an oxbow lake. Of course, the local terrain may force some sinuosity, but even still, shouldn't there be many more straight sections? What causes the meandering?

44, pp. 189–190; 73; 360, pp. 43–48; 364, pp. 78–79; 453, p. 146; 697, pp. 82–85, Chapter 9; 698, pp. 56–58; 699, pp. 144–145; 700, pp. 84–87; 701 through 715.

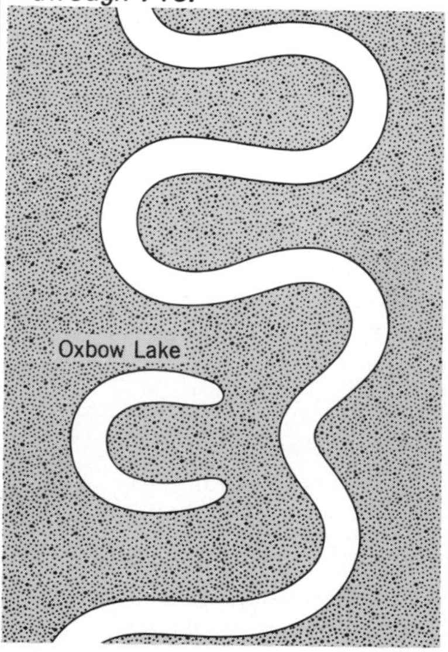

Figure 4.64

fluid flow around obstacle

pressure gradient

forces in rotating frame

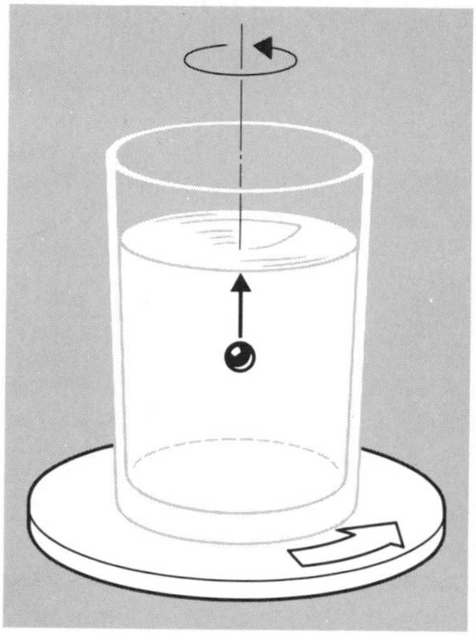

Figure 4.65
If the ball is released in the center of the rotating water, it takes longer to rise.

4.65

Rising ball in rotating water

Adjust a small ball's density (by partially filling it with water) so that it takes about 2 seconds to ascend through four inches of water. If the water is on a rotating turntable and the ball is on the center axis (Figure 4.65), the ascent time should be the same, shouldn't it? But as a matter of fact, if the rotational speed is 33 1/3 rpm, a four-inch ascent will now take about 30 seconds. Why is there such a big difference in rise time? Indeed, why is there any difference at all?

717 through 719; 1482.

pressure gradients

centrifugal force

4.66

Taylor's ink walls

If a drop of dyed water is placed in a glass of clear water, the dyed area will be about half a centimeter large. But if the drop is placed off center in a glass of water that is sitting on the center of a rotating turntable, the dyed area will be compressed into a thin vertical sheet that spirals around the center of the glass (Figure 4.66). What keeps the dye in such a sheet and prevents it from mixing with the clear water?

717; 720.

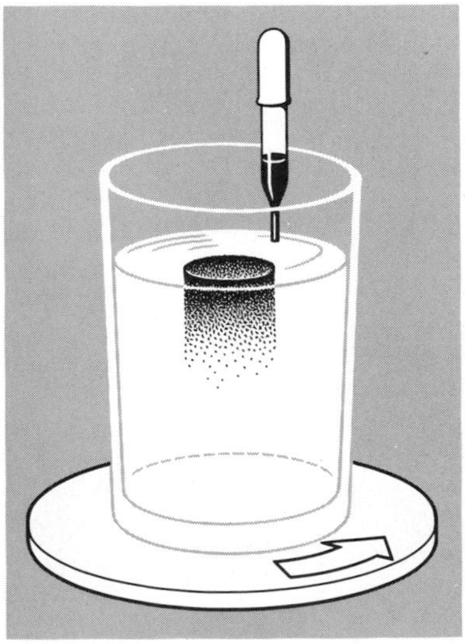

Figure 4.66
Taylor's ink wall in a rotating glass of water.

vortices

coriolis force

angular momentum

4.67

Bathtub vortex

Do northern hemisphere bathtubs really drain in a counterclockwise sense, as is commonly believed? If bathtubs do drain in opposite senses in the two hemispheres, does that mean the water doesn't rotate at all on the equator?

72; 721 through 736.

vorticity

4.68

Tornadoes and waterspouts

Do tornadoes and waterspouts turn in any particular direction, as do hurricanes? What makes them visible? Does water go up or down in waterspouts? Why do some tornado funnels hop along? Do adjacent funnels attract or repel each other? Finally, why do some funnels appear to be double layered, as if they consisted of two concentric funnels?*

226; 737 through 746; 1538.

*For more information on tornadoes, their cause and behavior, see Refs. 224, 225, and 747 through 750.

4.69

Soda water tornado

Place a recently opened bottle of soda water on a turntable's center and spin it at 78 rpm. Bubbles emerge from the soda water as you would expect, but when you add a small amount of sugar or some other granular substance, a tornado-like structure develops. What causes this vortex, and what provides its energy?

751 through 754.

buoyancy

4.70

Coffee cup vortex

Carefully stir a cup of hot coffee until you have a uniform swirl and then carefully pour a stream of cold milk into the center. A vortex will form in the center and a dimple may be noticeable. But if hot milk is used, the vortex will not develop. Why is there a vortex in the first case and not in the second.

755.

convection

vorticity

4.71

Dust devils

What drives dust devils, those whirlwind vortices that are often seen in deserts or other places with loose sand debris? Does their internal air move up or down, and is there a preferred sense of rotation as in hurricanes? How can

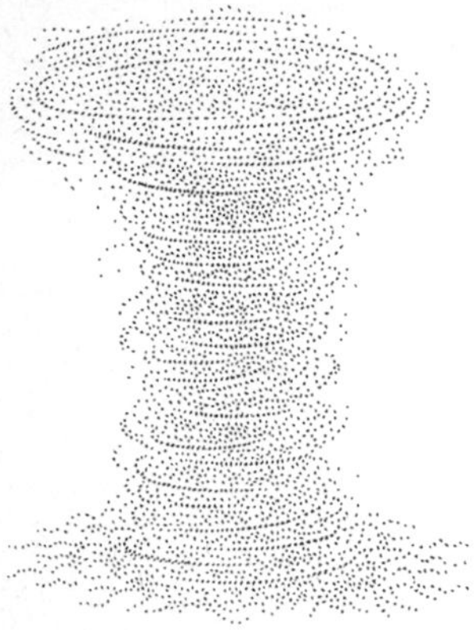

Figure 4.71
Dust devil.

seemingly small, local changes in the air trigger them? For instance, a jackrabbit tearing across the desert floor can leave a trail of dust devils. Why do nearly all dust devils die within only three or four minutes? Is it because of turbulence, or is the energy source removed? Finally, why are they shaped like an uneven hourglass (Figure 4.71) and not like a tornado funnel?

756 through 764; 1539; 1540.

4.72
Fire vortices

Why do tornadolike vortices frequently develop near volcanos, forest fires, and large bonfires?

765 through 772.

4.73
Steam devil

There is yet another natural vortex, but it is rarely seen. In the dense steam fog over some winter lakes, such as Lake Michigan, steam devils appear. You can simulate this by allowing cold air to blow over a bathtub full of warm water in a moist bathroom. What drives the steam devils?

773; 774.

4.74
Vortex rings from falling drops

If a drop of dyed water falls into a glass of clear water, you can see the vortex ring created by the splash and watch the ring as it expands and descends (Figure 4.74). Can you explain in simple terms why the ring is formed and why it expands? Which way does the fluid rotate in the ring? Finally, why are more (but less pronounced) rings also created by the same splash?

155, p. 103; 775; 776, pp. 522–526; 777.

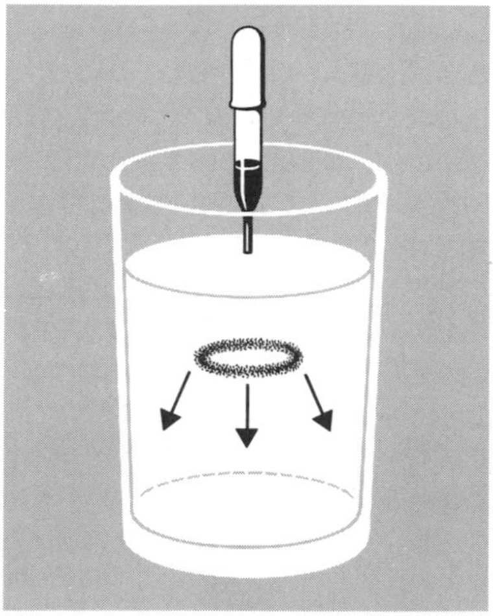

Figure 4.74
Falling and expanding vortex ring of dyed water.

4.75
Ghost wakes

If you quickly move a vertical piece of cardboard horizontally across a pool of water as shown in Figure 4.75a, two wakes will appear on the pool's surface. Why? If the cardboard is moved to the side as shown in Figure 4.75b, only one wake appears. Again, why?

1481.

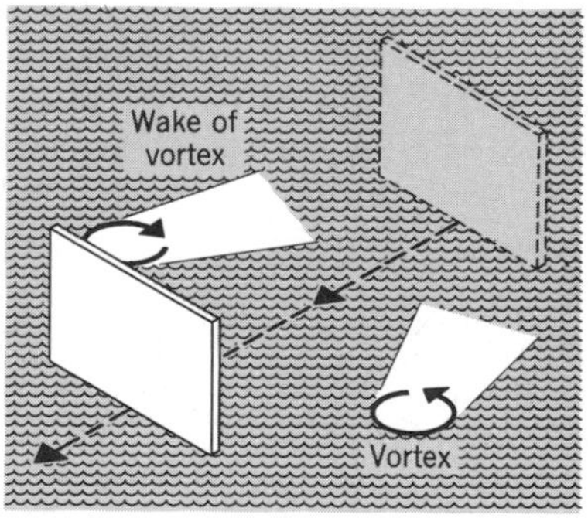

Figure 4.75a
Top view of moving cardboard and vortices.

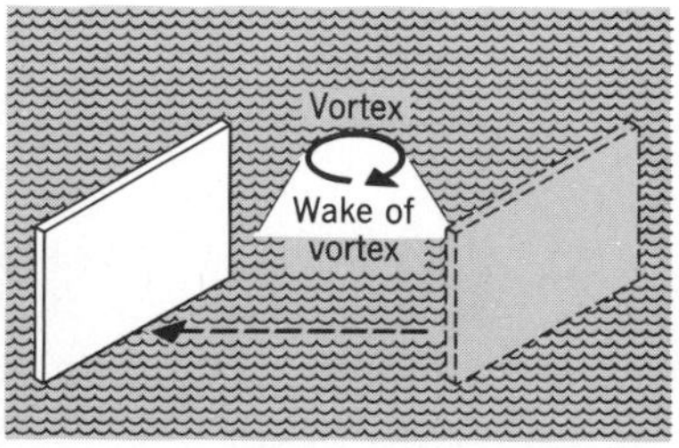

Figure 4.75b
Top view of moving cardboard and vortex. [Both figures after C. W. McCutchen, Weather, 27, 33 (1972).]

vorticity

adiabatic process

friction

4.76

Hot and cold air vortex tube

The Ranque-Hilsch vortex can mysteriously separate hot from cold air without any moving parts. If compressed air (at room temperature, say) is forced into the vortex tube through the side nozzle (see Figure 4.76), air as hot as 200°C will emerge from one arm of the vortex tube while air as cold as −50°C escapes from the opposite arm. There are no heating-cooling mechanical devices inside the tube, just a circular cavity with a center escape hole on one side and a valve at the end of the arm on the other side. How is the temperature difference created by this simple arrangement? Must we have a little man stationed in the tube, feverishly sorting out cold and hot air from the room-temperature air?

778 through 787.

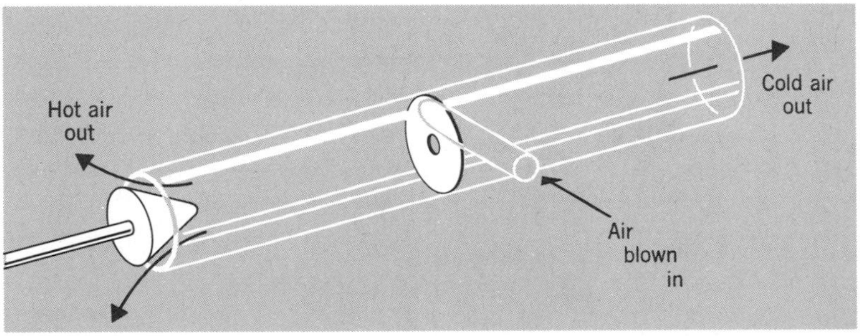

Figure 4.76
Compressed air blown into vortex tube separates into hot and cold air.

drag

eddies

4.78

Sinking coin

If a coin is dropped into a large container of water, will it sink with its edge or flat side downward? Will the same thing happen in a viscous fluid such as oil or a sugar solution? How will a cylinder sink?

Common sense probably tells you a sinking object will always assume the most streamlined orientation. However, for some parameters a coin and cylinder will sink in water with whatever orientation you initially give them. Making the disc larger or the fluid more viscous causes the disc to fall broadface. What forces the disc to present its broadest side? Why aren't smaller coins and cylinders also forced into the broadside orientation?

788 through 790.

eddies

aerodynamics

4.77

Birds flying in V formation

Do you think there is any physical reason for the V formation assumed by migrating birds? Or do you think it is simply an interesting behavioral response and serves no real purpose? If, perhaps, there is some aerodynamical basis for the formation, is it important that the formation be symmetric? Is it necessary that the birds synchronize the flapping of their wings? What advantage would the V formation have over any other formation (line abreast or zigzag, for example)? Why don't birds fly in schools like those of fish?

794.

wakes

4.79

Tailgating race cars

In stock car races what advantage is there for one car to tailgate another car (called drafting)? Is the lead car affected at all? When the trailing car suddenly pulls out to pass, why does it receive a whiplash acceleration around the lead car?

789.

wakes

eddies

4.80

Several sinking objects interacting

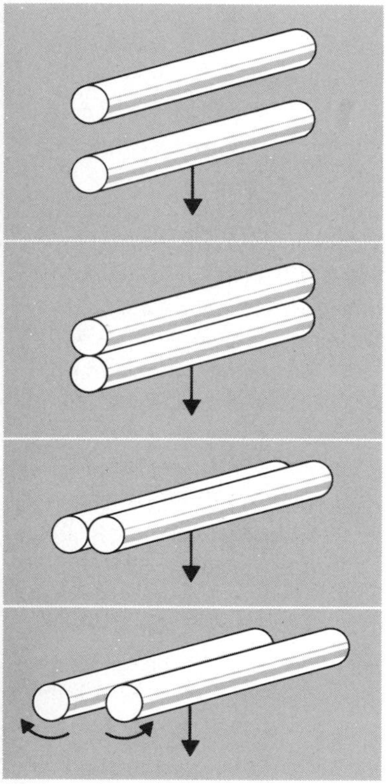

Figure 4.80a
Two views of two cylinders falling in a viscous fluid.

Several objects may interact in strange ways while sinking in viscous fluids such as oil or a sugary solution. Here are three examples.

Into a viscous fluid, drop two cylinders, one closely following the other. For certain ranges of viscosity and cylinder size and speed, the trailing cylinder may catch the leader and rotate about it until they are horizontally parallel, and then they will both rotate together and separate horizontally as they sink (Figure 4.80*a*).

In a simpler interaction, two discs dropped after a leader disc may catch the leader, and then the three will take on a stable butterfly configuration (Figure 4.80*b*).

Also, a compact cluster of three to six spheres will separate themselves into a horizontal, regular polygon, and this polygon will slowly expand as it falls.

Without getting into too much detail, can you roughly explain why each of these interactions take place?

789 through 793.

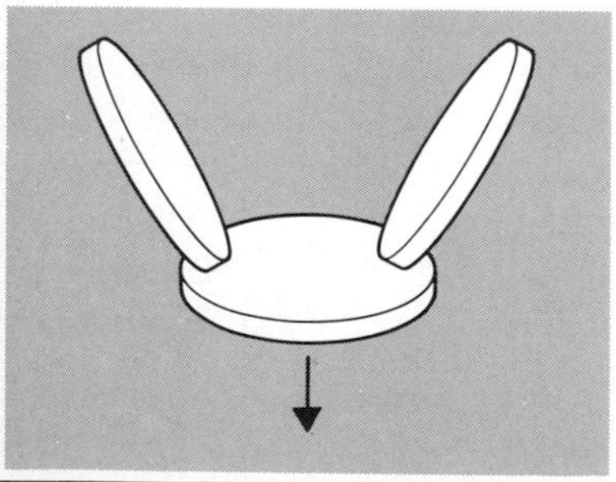

Figure 4.80b
Butterfly configuration of three discs falling in a fluid. [After K. O. L. F. Jayaweera and B. J. Mason, J. Fluid Mech., 22, 709 (1965).]

buoyancy

drag

wakes

vortices

4.81

Strange air bubbles in water

Closely examine bubbles rising through a glass of water. The very tiny ones (with radii less than about 0.7 millimeter) are spherical and rise to the surface in a straight line just as you would guess. Slightly larger bubbles (up to 3 millimeters in radius) are spherical but either zigzag or spiral upward. If the radius is even larger (more than 3 millimeters), the path is again straight, but for radii greater than 1 centimeter, the bubbles look like spherical caps and resemble umbrellas (Figure 4.81).

Why does a rising bubble's shape depend on its size? What forces the intermediate size bubble to zigzag and spiral, and what parameters fix the frequency of that motion?

776, pp. 367-370, 474-477; 796 through 801.

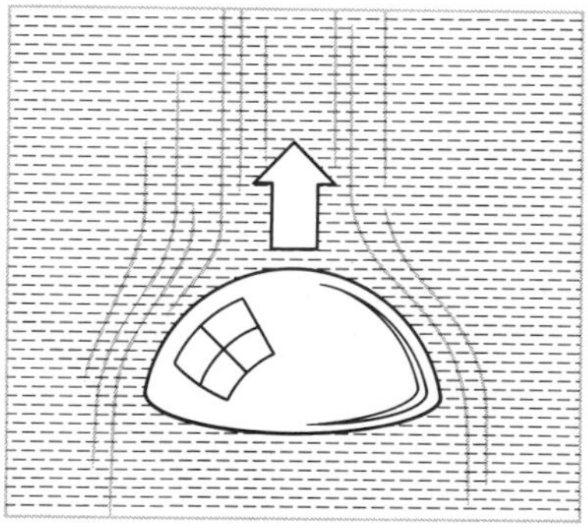

Figure 4.81
A large bubble rising in water resembles a spherical cap.

eddies

drag

4.82
Fish schooling

The schooling of fish certainly must have roots in social factors, but it must also offer a practical advantage to the fish, for when swimming in such a school, a fish's endurance is considerably increased, perhaps as much as sixfold. Why would there be an advantage for fish of similar size and shape to swim in regular arrays and in synchronous motion? In particular, what determines the distance between fish? Should one fish swim directly behind another? Why don't fish swim in the V formation that birds use?

1095.

eddies

4.83
Wind gusts on building

Why is the windward side of a building calmer than the rear in a strong and gusty wind? Shouldn't just the opposite be true?

453, pp. 138–139.

driven resonance

harmonic oscillations

4.84
Tacoma Narrows Bridge collapse

You may have heard of the failure of the Tacoma Narrows suspension bridge, because physics departments often have the spectacular film (1562) showing the bridge oscillating and eventually collapsing.

The bridge began its oscillations even when it was being built; in fact, the structure's rippling motion made the bridge workmen seasick. After it was opened to traffic, the motion was so pronounced that motorists came from miles away just for the thrill of being on the bridge. On days when the bridge oscillated as much as five feet, motorists on the bridge actually disappeared from each other's view.

Still, the bridge's collapse came as a complete surprise. Suddenly, on the morning of the collapse, the ripple ceased, and after a brief pause, the bridge went into a furious torsional oscillation. Two people on the bridge at the time crawled on all fours to escape. After trying to rescue a dog abandoned on the bridge, a professor could retreat only along the nodal line of the torsional oscillation. (His retreat is seen in the film.)

After 30 minutes of torsional motion a floor panel fell from the main deck. Another 30 minutes brought another 600 feet of deck down. Though the twisting then ceased briefly, it began again, and it took only several additional minutes to bring the remaining deck down.

The bridge designer (who died shortly after this tragic end to his career) could hardly be faulted, for at the time there was scant understanding of the aerodynamic behavior of suspension bridges. The repercussions in bridge building were enormous and long lasting.

The bridge failure is introduced in the physics classroom as an

Figure 4.82
"It all started with an innocent game of follow-the-leader!"

example of driven resonance. Although the wind was not blowing unusually hard that day, the bridge's oscillations grew in strength to catastrophic proportions. But why and how exactly did the wind do this? How would a fairly *steady* wind cause the rippling, which soon led to the torsional oscillations? Why would longitudinal oscillations be created? Since driven resonance implies a certain frequency match between the driving force and driven object, you must explain how the wind produced that frequency match.

How can a bridge's aerodynamic instability be minimized? One new feature resulting from the collapse was the placement of longitudinal gaps in the bridge's roadway, say, between the opposing lanes of traffic. Why would this help stabilize the structure?

802 through 812; 1556.

Kelvin-Helmholtz instability

convection

4.85

Air turbulence

What causes the bumps so frequently encountered by jet aircraft? Some disturbances are single jolts. Some force the airplane up and down as if it were a ship at sea. Others quickly heave the airplane to a different altitude, perhaps making the pilot lose control as a result. Often there are warning signs for these various types of disturbances, but some turbulence can occur in clear weather, with no clouds, and at altitudes of several kilometers. This turbulence was unknown until jet airplanes of World War II were first able to reach the relatively high altitudes at which it takes place. What is responsible for the clear air turbulence and the other types of disturbances? Why is it experienced primarily at higher altitudes?

819 through 822.

4.86

Watch speed on a mountain top

Why will a spring-driven watch run at a different speed on a mountain top than at a sea shore?

9, pp. 80–82.

turbulence

4.87

Wire mesh on faucet

Why is a wire mesh often placed over a faucet's outlet? It will, of course, catch small stones in the water supply, but people claim the water is also "smoother" or "softer" with the mesh in place. Why would that be?

turbulence

wave interference

4.88

Fast swimming pools

Why are some swimming pools said to be fast? Could different depths, different splash gutters, chemical additives, etc. noticeably influence a swimmer's speed?

edge oscillations

4.89

Nappe oscillations

When water is discharged over the spillway weirs of some dams, the falling water curtain may go into severe oscillations (Figure 4.89). The noise from the oscillations, in addition to the normal noise from water impact at the dam's foot, may even make the vicinity unbearable. What causes these oscillations, and why is there so much extra noise?

813 through 816.

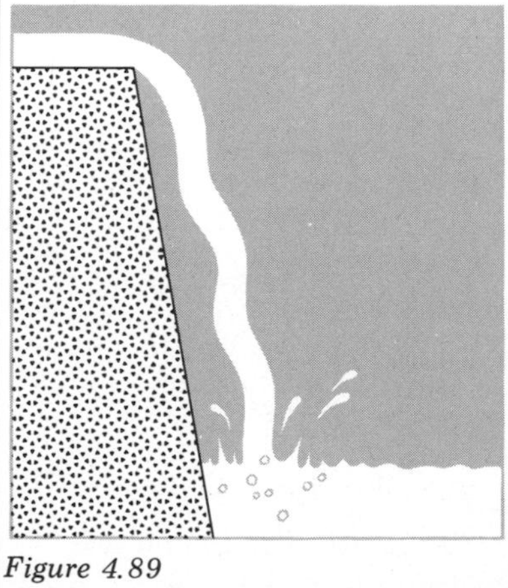

Figure 4.89

eddies

driven pendulum

4.90

Parachute holes

Why do parachutes often have central holes (Figure 4.90*a*), especially the conventional paratrooper parachutes? Isn't a hole a rather strange thing to have, for wouldn't you think it would be counter to the whole point of a parachute? If the hole is to reduce drag, why not just make the parachute smaller?

Some of the unconventional parachutes need even more explaining. For instance, some on stock car racers resemble two crossed-bandage strips (Figure 4.90*b*). Why would someone use such a drag chute? Wouldn't the drag be quite low?

Even in the absence of gusty winds, men using conventional parachutes swing to and fro during their descent. Since such swinging can be very dangerous during the landing, the men obviously are not doing it on purpose. What causes the swinging, and what determines its period?

817; 818.

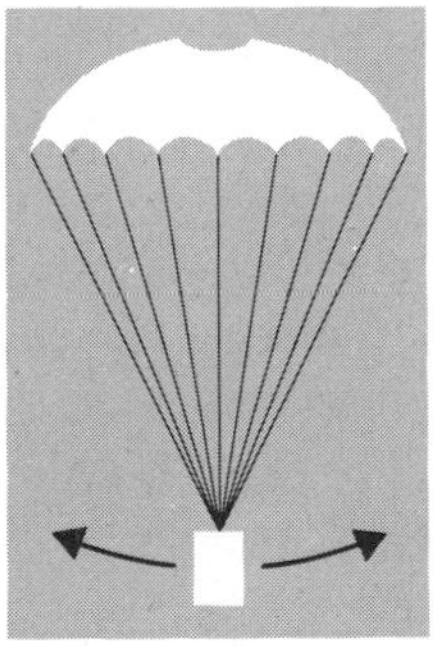

Figure 4.90a
Conventional parachute.

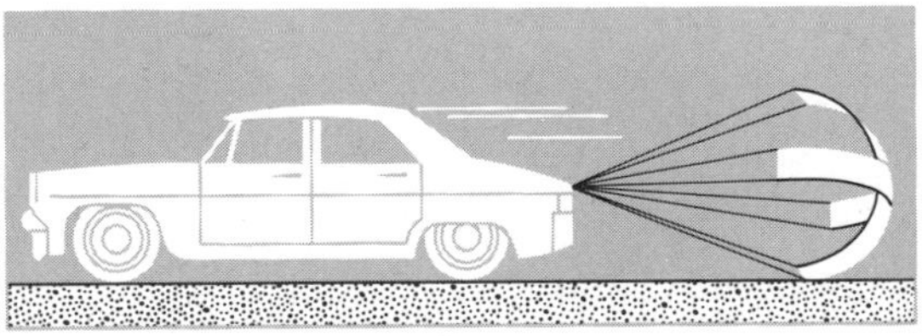

Figure 4.90b
Stock-car parachute.

turbulence

momentum transfer

hydrostatic forces

4.91

Speed of a drifting boat

A drifting boat is commonly thought to travel faster than the stream. Indeed, since a drifting boat can be steered, doesn't it have to? But how can the boat, which supposedly is just being pushed along by the stream, be moving faster?

453, p. 179; 824; 825.

flow around obstacle

particle transport

eddies

4.92

The gaps in snow fences

If you want to stop snow drifts near a roadway, railroad track, or walkway, why do you put up a snow *fence*. . .why not a snow *wall*? Granted a fence may be less expensive, but wouldn't a wall do a better job than a fence with all its gaps?

453, p. 334; 600; 826.

flow around obstacle

particle transport

4.93

Snow drifts

Snow drifts are much more pronounced around posts and trees than on the wind-facing sides of houses. Why is there such preferential unloading of drifting snow around the narrower obstacles?

364, pp. 12-13; 453, p. 333; 826.

drag

eddy formation

4.94

Streamlined airplane wings

Why are the trailing edges of airplane wings sharp? (To say that it's just for streamlining is not enough.) Why do some planes have swept-back wings and others not?

603; 605.

air drag

4.95

Skiing aerodynamics

Aerodynamically, what is the best position a skier can assume in a downhill race? Winners in the olympics and other world meets are often determined by time differences between skiers of as little as 0.01 second. Because of the crucial need for sound knowledge about the stance as well as the equipment of a skier, the French conducted wind tunnel experiments and developed the "egg position" (Figure 4.95a). Although this in not the best position for drag reduction, it is a practical one to assume in a strenuous race.

How about the other two positions shown? Before the testing a good many of the skiers had instinctively adopted the lowest possible position, dropping the arms alongside the legs (Figure 4.95b). As it turned out, the high crouch (Figure 4.95c) gives remarkably less drag than the lower crouch with lowered arms—but still not as little as the French egg position. Why?

823.

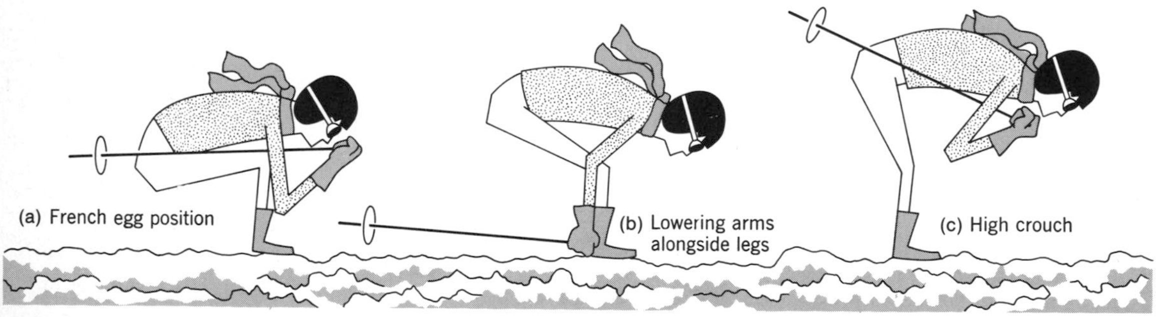

Figure 4.95
Three skiing positions.

air drag

4.96

Dimpled golf balls

Why are golf balls dimpled? In the very early days of golf, the balls were smooth, and it was only accidentally discovered that scarred balls traveled further than the smooth, unscarred ones. If today's dimpled ball is driven, say, 230 yards, a smooth ball similarly struck would travel only 50 yards. Does this make sense? Shouldn't the smoother ball go further because it will have less air drag?*

593; 827; 828.

*In the last few years a newer golf ball design—one with randomly spaced, hexagonal dimples rather than the old, regularly spaced, circular dimples—has been sold with the claim of an additional six yards in average flight distance.

air pressure

momentum transfer

4.97

Flight of the plucked bird

How do birds fly? Yes, I know they flap their wings up and down, but how does that keep them aloft and moving forward? Well, maybe the bird flaps backwards on the downstroke, thereby propelling itself forward. No, slow motion movies show the wing moving forward not backward, on the downstroke. Perhaps the best clue to the bird's flight lies in the ancient Greek myth of Icarus who flew too close to the sun, lost the feathers glued to his arms, and then plunged to his death. Must a bird have feathers to gain lift and forward drive? Can a plucked bird fly?

604.

pressure

stability

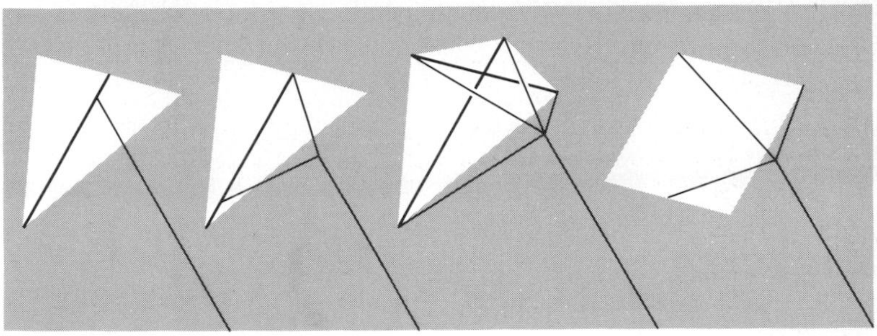

Figure 4.99
Several bridling techniques for kites.

4.99

Kites

What keeps triangular and box kites aloft, and which type is more stable? Why do some kites have tails? Finally, what advantages do the various bridling techniques shown in Figure 4.99 give?

829.

convection

vortices

lift and drag

4.98

Bird soaring

What allows birds to soar so effortlessly and so continuously? If they are riding on winds deflected upward by trees and hills, for instance, then why can they soar just as well over flat land and water? If they gain lift by gliding into a wind whose strength increases with height, then why do they seem to soar so much better on wind-free days? Finally, if they ride thermal currents upward, then why can you sometimes see one group of birds soaring while another group, either below or above the first group, must flap their wings to remain aloft? Besides, if the lift is produced by thermals originating on the ground, shouldn't larger birds have an easier time soaring near the ground? Actually, they can rarely soar there.

Some birds stalk ocean liners across long stretches of open water, somehow gaining their propulsion by gliding near the ship waves. How do they do this?

364, pp. 13–15, 120–121; 604; 852, pp. 127–131; 853 through 862.

roll vortices

convection

condensation

4.100

Cloud streets

Sometimes the sky is covered with long, straight rows of cumulus clouds called cloud streets. What orders the clouds this way, and in particular what determines the spacing between rows? Why aren't cloud streets made more often?

361, pp. 4–13, 39, 43; 362, pp. 28–30; 364, pp. 154–155, 175; 1456.

convection
surface tension
nonlinear fluid flow
stability
condensation

4.101

Coffee laced with polygons

If you examine a hot cup of coffee under a strong light that is incident nearly parallel to the surface of the coffee, you will find the surface laced with polygonal cells (Figure 4.101*a*). They disappear, however,

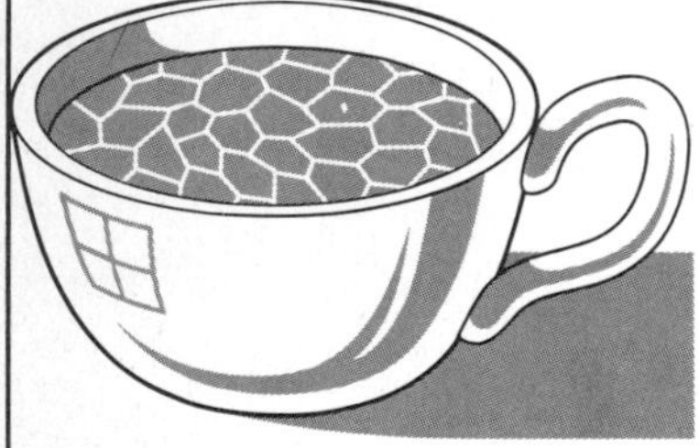

Figure 4.101a
Polygons on coffee surface. (After V. J. Schaefer, American Scientist, 59 (Sept.-Oct. 1971).)

as the coffee cools. You can also destroy the cellular appearance by putting a charged rubber comb (charge it by running it through your hair) near the coffee.

Other liquids show surface designs too. James Thomson, a famous Nineteenth–Century physicist, noticed the rapidly varying surface designs in a pail of hot soapy water and in strong wines. Later, the Frenchman Bérnard was able to make regular patterns in oil surfaces when the oil was heated from below. His regular polygons would slowly evolve into a beautiful hexagonal, honeycomb structure (Figure 4.101*b*). Still other

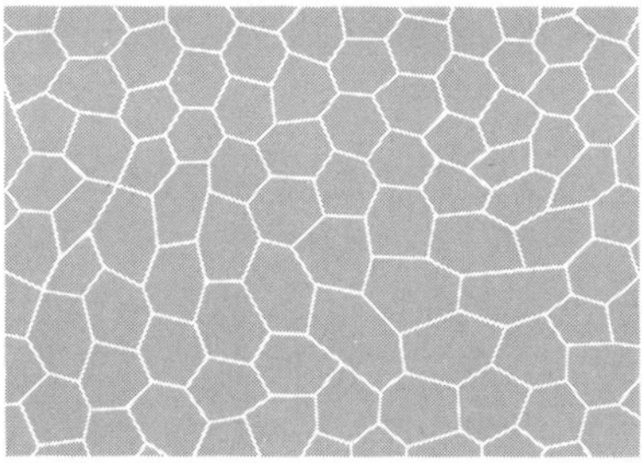

Figure 4.101b
Hexagonal Bernard cells.

fluids gave a roll-like appearance (Figure 4.101*c*). Recently, cellular surface designs were attempted on board spacecraft while under zero gravity.

Figure 4.101c
Surface with roll-like structure.

In these examples, why do rolls and polygons (especially honeycombs) form on the fluid surfaces? Is the same physics actually responsible for all of the examples? Why do the coffee cells disappear when there is a charged body nearby? Finally, do these several types of surface designs depend on gravity?

360, pp. 93-94; 453, pp. 418-421; 580, pp. 113-115; 830 through 849.

row vortices
gravity waves

4.102

Longitudinal sand dune streets

Looking down on desert sand dunes from a high altitude airplane, one sees "curious long, narrow dune belts running across the desert, roughly from north to south, in almost straight lines [Figure 4.102]," (863) as if one were viewing well-designed parallel streets. The dune belts are characteristic of virtually every major desert in the world, and they all run roughly north to south and have spacing of about 1 to 3 kilometers.

Leaves scattered over lake surfaces and surface seaweed also collect into rows, though the scale is smaller, with the rows being only

Figure 4.102
Sand-dune streets as seen from a high altitude.

100–200 meters apart and up to 500 meters long.

In these examples what determines the direction the rows and belts run? If it is the wind, then do the rows and belts run parallel or perpendicular to it? Moreover, what determines the spacing between them?

580, pp. 18–19, 119–120; 862 through 868.

eddies

saltation

4.104
Sand ripples

Why are the sides of a sand dune covered with sand ripples? What exactly determines the spacing of those ripples?

The sandy bottoms of streams are also often covered with sand ripples or waves. What causes those, and again, what determines the periodicity of the waves? If you watch them closely for a long time, you may find them traveling upstream. Why do they do that?

144; 453, p. 334; 629, pp. 381-386; 687; 688; 697, pp. 55–59; 698, pp. 134–136; 869 through 874.

vorticity

4.103
Smoke ring tricks

To amuse me during the long summer days of a small country town, my grandfather would blow smoke rings for hours on end. In one of his simpler tricks he would send a ring toward a wall, and the ring would expand as it approached the wall.

His best trick, however, was blowing one smoke ring through another, larger one. After the speedier trailing ring passed through the leading one, the former leading ring contracted and speeded up while the former trailing one expanded and slowed down (see Figure 4.103). Their roles were exchanged, and the new trailing ring then passed through the new leading ring. This game of leapfrog continued until the smoke rings became too dispersed for further play.

You can see the same thing by dropping a colored drop into a beaker of water. Upon hitting the surface, the drop forms a ring that both expands and descends.* A second, closely following drop will produce another ring that will pass through the first, and the game of chase begins.

Exactly how are smoke rings formed, and how do they retain their shape for so long? Why does a smoke ring expand as it approaches a wall? Finally, what causes the chasing game of two smoke or water rings?

36, p. 1; 51, pp. 161–167; 109, p. 7; 453, p. 75; 721; 850; 851.

*See Prob. 4.74.

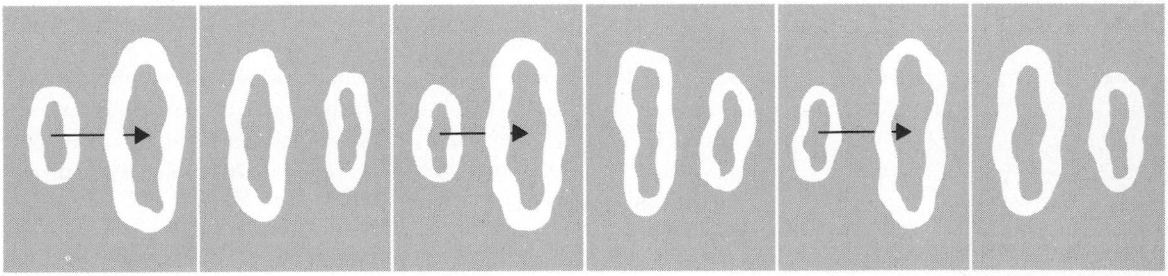

Figure 4.103
My grandfather's smoke-ring trick.

forces in liquids

cavitation

vapor pressure

4.105

Siphons*

How do siphons work? In particular, if they depend on atmospheric pressure, then why can some liquids be siphoned in a vacuum? Do they depend on gravity? When the siphon tube is first lowered into the liquid, why doesn't the siphon start itself? What force pulls the liquid up the first arm (denoted *A–B* in Figure 4.105) against gravity? Finally, how is the height of a siphon limited, especially when the siphon works in a vacuum?

875; 876.

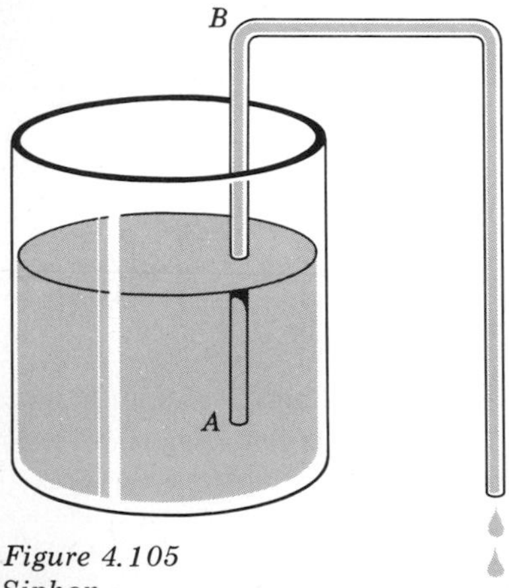

Figure 4.105
Siphon.

*For several curious types of siphons devised by Hero of ancient Greece, see Ref. 877.

saltation

flow around obstacle

friction

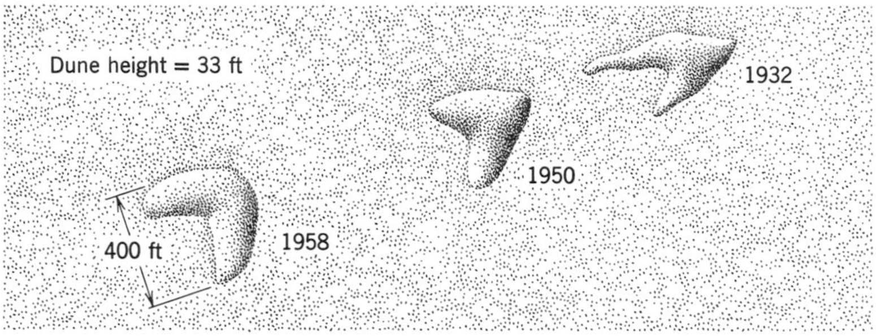

Figure 4.106
The march of a sand dune over 26 years. (Adapted from Geology Illustrated by John S. Shelton. W. H. Freeman and Company. Copyright © 1966.)

4.106

Marching sand dunes

I would have thought winds would tend to disperse a sand dune, but Figure 4.106 shows a typical case in which they marched a dune across a desert floor. The dune's character and identity remain intact even after 26 years of travel. How, exactly, are dunes moved by the wind?

144; 453, pp. 333–334; 698, pp. 141–142; 699, p. 198.

siphon

entrainment

4.107

The Crapper

How does a flush toilet work? What forces the water, etc. (especially the etc.) to enter the pipes? When the water from the tank comes into the bowl, is it merely falling from a water container above? Why do most toilets have a second, smaller hole in the bowl?

One of the most interesting books I have come across in writing *The Flying Circus* is *Flushed with Pride: The Story of Thomas Crapper* (878). It was Crapper who developed the flushing toilet. (Obviously he also contributed to the American language.)

Now you may not appreciate this, but it was tough work developing the flush toilet, and serious research was conducted by Crapper and others. Of course in their experiments these researchers had to simulate the actual material toilets normally handle. Toilet testing must have reached its zenith when in 1884 they achieved

"a super-flush which had
completely cleared away:
10 apples averaging 1 3/4 ins. in diameter,
1 flat sponge 4½ ins. in diameter,
3 air vessels,
Plumber's "Smudge" coated over the pan,
4 pieces of paper adhering closely to the soiled surface."*

A truly remarkable feat of technology!

253, pp. 334–335; 317, pp. 95–97; 318, pp. 108–113; 878; 879, pp. 260–261.

*From the book, *Flushed with Pride* by Wallace Reyburn. Copyright © 1969 by Wallace Reyburn. Published by Prentice-Hall, Inc., Englewood Cliffs, N. J. Permission also granted by Wallace Reyburn and Macdonald & Co.

drop aerodynamics

4.108
Street oil stains

On some roads on which the traffic speed is sufficiently high, oil stains are annular with an unstained section of road in the center of each stain (Figure 4.108). With slower traffic the stains are just splotches. Why do the annular stains appear, and how fast must the cars be traveling for them to be formed?

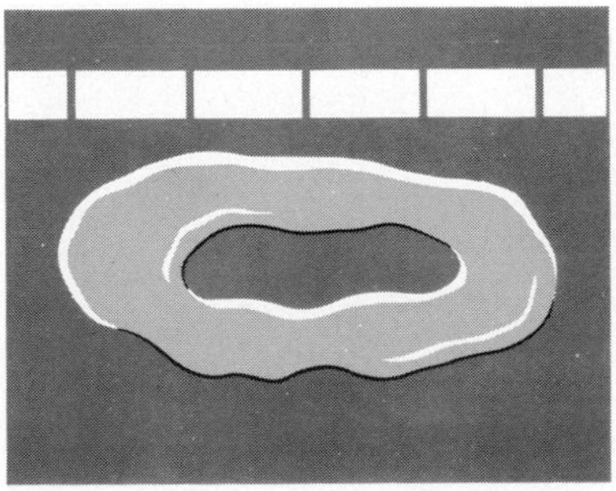

Figure 4.108
Street oil stain.

887, p. 187; 888.

surface films

boundary layers

4.109
Lake surface lines

Here and there on the surfaces of lakes and streams you can see thin, almost invisible lines. They are more noticeable if the water is flowing because then a small ridge of water builds up on one side of a line (Figure 4.109). What do you think these lines are, and why are the ridges formed?

Powder sprinkled on the ridge will reveal a two dimensional flow pattern of streetlike channels on the opposite side of the line (Figure 4.109). What causes such a pattern?

893; 894.

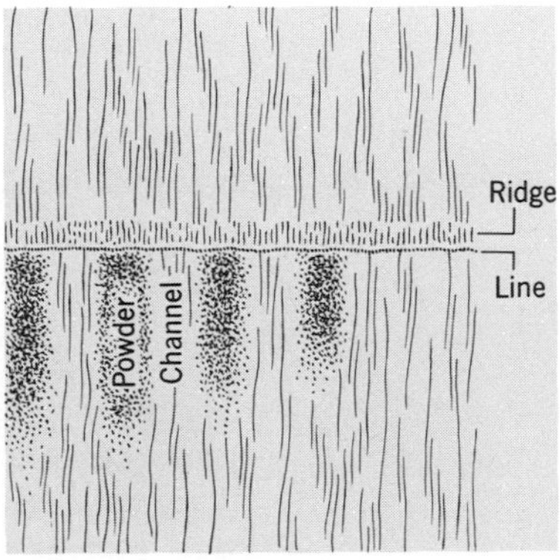

Figure 4.109
The line and ride on a stream or lake water (overhead view).

surface film

4.110
Milk's clear band

The next time you're mulling over a glass of milk, examine the milk film at the edge of the milk as you tip the glass. Between the film left on the bottom of the glass and the milk there is a clear area a few millimeters wide (Figure 4.110). Why is the clear band present?

458.

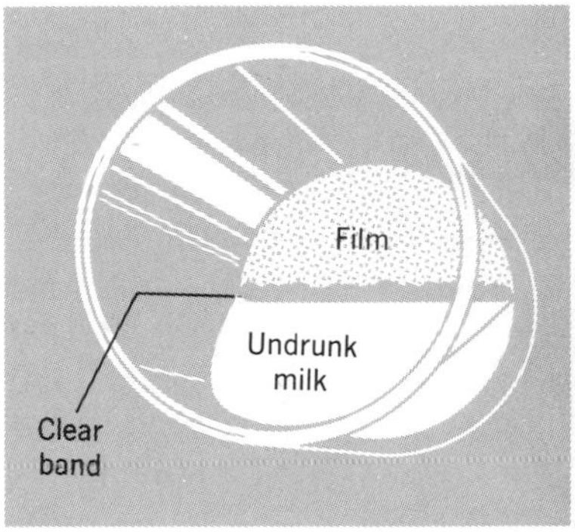

Figure 4.110
The clear band in a tilted glass of milk.

water waves

surface tension

4.111
Spreading olive oil on water

In *Prospero's Cell* (889) Lawrence Durrell describes the nighttime spear fishing in the lagoons beneath the Albanian hills. For spear fishing the water must be clear and calm because even a slight breeze severely distorts the image of the fish and ruins the aim. The fisherman can cope with a small breeze, however, by sprinkling a few drops of olive oil onto the water. Why do these few drops calm the water?

890, pp. 631-632; 891; 892.

internal waves

wave damping

4.112

Marine organic streaks

Biologically active regions of the oceans are often covered with long, wide streaks where the rippling of the water is suppressed by a film of organic material. When the illumination is just right, the sight can be beautiful. The organic streaks apparently do not depend on the wind like the seaweed streets discussed in Problem 4.102, for they are best seen in a light breeze, not a strong one. (The breeze really does nothing but heighten the contrast between the size of the ripples on the free water and on the streak.) What cause the film to form into streaks this way?

895 through 898.

4.113

Splashing milk drops

When a milk drop splashes on a liquid surface, a crater is thrown up, eventually breaking into a crownlike structure (Figure 4.113). As this crown subsides, a liquid jet (the "Rayleigh jet") leaps up

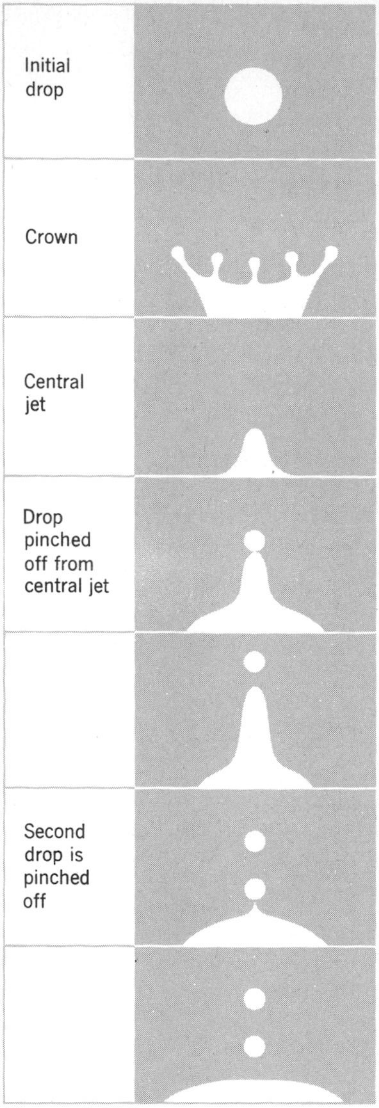

Figure 4.113
After P. V. Hobbs and A. J. Kezweeny, Science, 155 (3766), 1112-1114 (1967). Copyright 1967 by the American Association for the Advancement of Science.

from the center of the former crater, which then pinches off and ejects one or more small drops. Why does the crater rim break into the crownlike formation, and why does the central jet form and then pinch off drops? The pinching itself especially needs explaining.

Suppose the milk drop experiment is done in outer space (with no gravity). Will the same type of splash occur? In fact, will there be any splash at all?

880 through 886.

surface tension

pressure

centrifugal force

4.114

Water bells

If a water stream falls onto the center of a disc, the water will spread over the disc and form a transparent sheet as it flows off. The sheet may even close back onto the center support of the disc, forming a beautiful bell shape (Figure 4.114). What forces the sheet back in like this, and what determines the actual shape of the bell?

899 through 904.

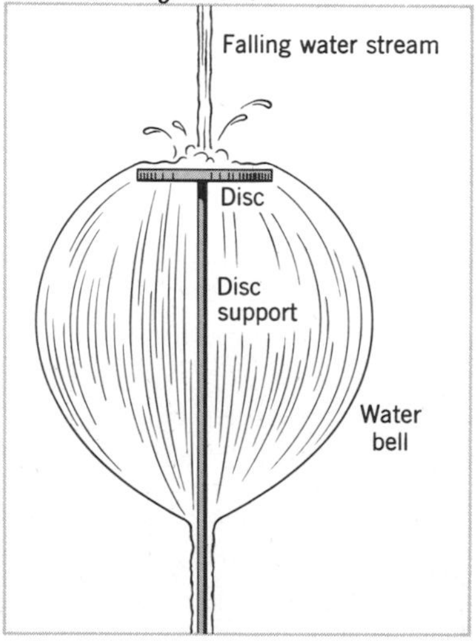

Figure 4.114

surface tension

momentum conservation

water waves

4.115

Water sheets

If two identical water jets are directed toward each other, beautiful thin sheets of water can be produced (Figure 4.115). Why do the streams form sheets rather than just break up? Why do the sheets eventually disintegrate at some particular distance from the impact point?

The shape of the edge and the stability of the sheet fall into three main types which depend on, among other things, the rate of water flow. For low speeds, the sheet is stable with circular edges. For the next type, at higher speeds, two things can happen: the edge may be cusp shaped or waves may be set up on the sheet. In the third type, for even higher speeds, the sheet will flap like a flag in the wind. Roughly, what causes these differences?

904 through 910.

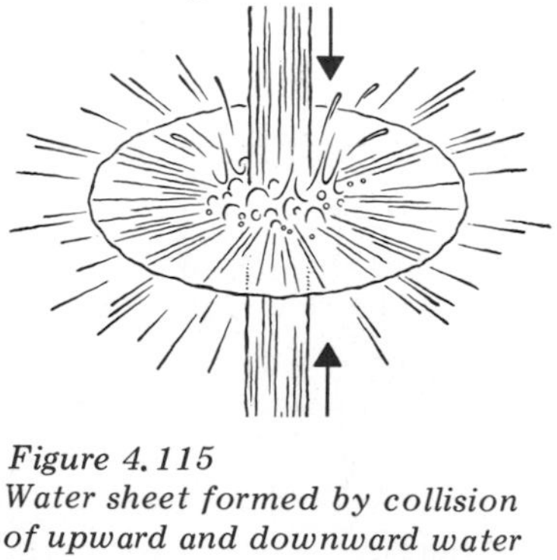

Figure 4.115
Water sheet formed by collision of upward and downward water jets. [After G. I. Taylor, Proc. Royal Soc., A 209, 1 (1960).]

4.116

Gluing water streams

Punch several adjacent holes in the side of a can, parallel to the bottom. Fill the can with water and run your finger through the leaking streams. For some reason, the streams now converge and remain together even after you've removed your finger (Figure 4.116). What keeps them together?

Figure 4.116
Three water streams seemingly glued together.

surface tension

4.117

Pepper and soap

If you dip a small piece of soap into a bowl of water sprinkled with pepper, the pepper will immediately race away from the soap. Why? How fast do you think the pepper grains are moving?

321; 592, p. 40.

Bernoulli effect

boundary layer

pressure differential

4.118

Pouring from a can

When I pour my beer, why does it insist on running down the side of the can instead of falling straight down from the lip (Figure 4.118)? What determines how far down it adheres to the can? How fast must I pour the beer to prevent any such "sticking"?

Your first impulse will most likely be to attribute the phenomenon to surface tension or adherence of the liquid to the container. However, neither is responsible for the spilt beer. What is, then?

911; 912.

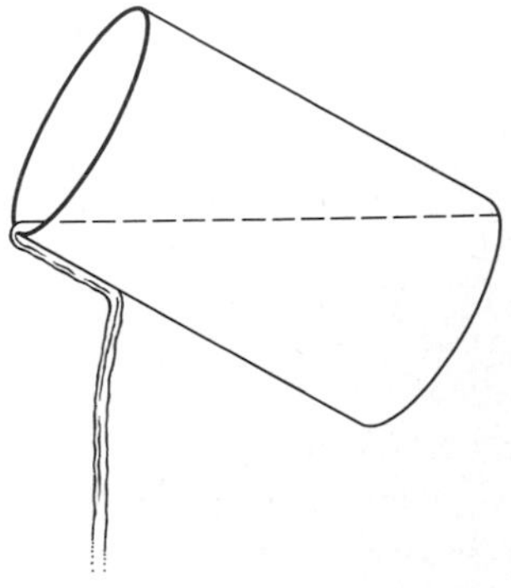

Figure 4.118
Fluid stream is forced back along the can.

surface tension

film creep

4.119
Tears of whiskey

After pouring a shot of whiskey into an open glass, you will see a fluid sheet that first creeps up the side of the glass and then forms tear drops around the side. What causes that upward creeping to such surprising heights?

832; 848; 849; 1530.

4.120
Aquaplaning cars

If you lock the brakes on your car while moving at high speed on a wet road, the car will act like an aquaplane. That is, the tires will skim along on a thin sheet of water and will not actually touch the road. Why does this happen, and why doesn't it always happen on wet roads even when the brakes are not applied? Is there any tread design that will minimize this effect?

913.

surface tension

4.121
Floating water drops

Water drops can often be seen skimming across a water surface, almost miraculously escaping consumption by the surface. What delays the death of these drops?

534; 914 through 916; 1608; 1609.

Non-Newtonian Fluids

(4.122 through 4.131)

4.122
Soup swirl reversal

The next time you fix tomato soup, give the soup a good swirl in the pan and then lift your spoon out. The swirl in the soup dies out, as you would expect, but during the last few seconds the soup turns in the opposite direction. What makes it do that?

917.

elastic fluids

4.123
A leaping liquid

Some hair shampoos (and several other liquids) display a curious leaping tendency when being poured into a partially filled dish. If the falling stream is thin enough, the liquid will form a small hump near the stream's entrance point. Then the stream will seemingly leap back up from the surface as shown in Figure 4.123). Each time this happens the hump disappears and must be rebuilt before another leap occurs. What causes the hump and the leaping stream, and what is so unique about the liquids displaying this property?*

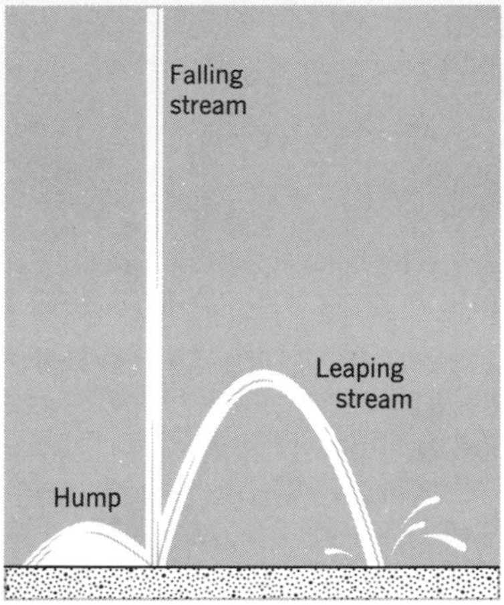

Figure 4.123
[After A. Kaye, Nature, 197, 1001 (March 9, (1963).]

917; 921; 922; 923, pp. 249-251.

*A. A. Collyer, personal communication.

Weissenberg effect

viscosity

stress

4.124
Rod-climbing egg whites

When a glass of water is placed on the center of a rotating turntable, the surface of the water curves up towards the outside of the glass because of centrifugal force. The same shape is also obtained if the glass is held fixed and a rotating rod is inserted along the central axis of the glass.

Not all fluids, however, behave

in this common-sense way. Egg whites, for instance, will have the proper curved surface on a rotating turntable but will behave strangely with the rotating rod. Rather than curve up toward the outside, the egg white will climb the rod (Figure 4.124). Gelatin dissolved in hot water will show normal behavior at first, but as the mixture cools, it will begin to display this strange urge to climb the rod. Since the centrifugal force is certainly still present because of the rotation there must be an even stronger force pulling the fluid inward and up the rod. What is this force?

917; 921; 923, pp. 231-236; 924, p. 375; 925, p. 121 ff; 926, pp. 52-53; 927, p. 671; 928, pp. 522-524; 929 through 934; 1620.

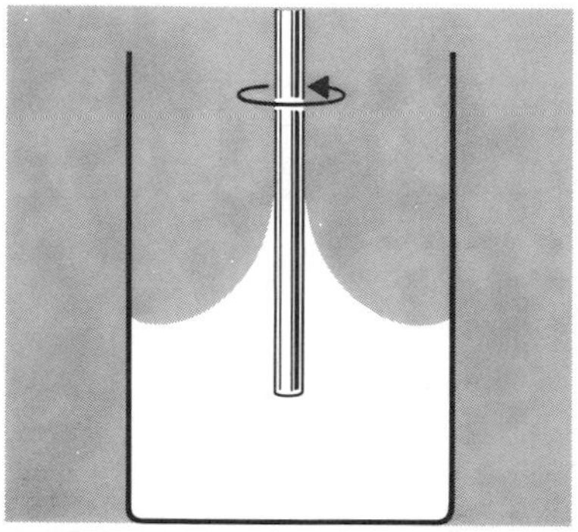

Figure 4.124
Egg white climbing the stirring rod.

viscous fluid flow

4.125

Liquid rope coils

If you pour thick oil, honey, or chocolate syrup onto a plate from a reasonable height, the stream will begin to wind itself up a short distance above the plate (Figure 4.125). Why does this coiling occur, and what affects the diameter and height of the coil and the rate at which if forms?

918 through 920.

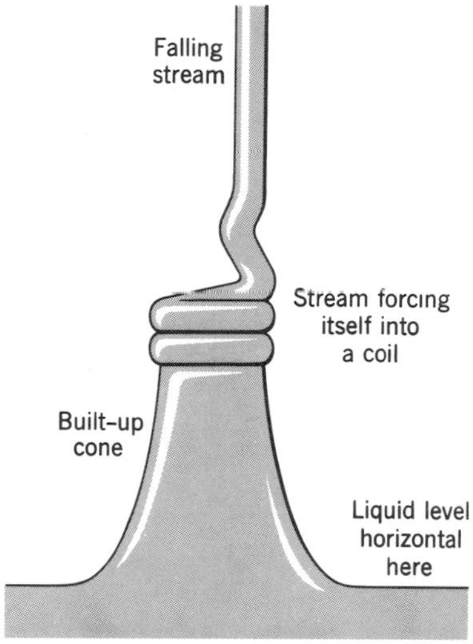

Figure 4.125
Liquid rope coil. [After G. Barnes and R. Woodcock, Amer. J. Phys., 26 (4), 205 (1958).]

viscosity

shearing

sol-gel change

4.126

Thixotropic margarine

Many common, household fluids would be useless if they were not thixotropic, that is, if their viscosity did not decrease when the fluids were subjected to shearing forces. For example, margarine would not spread very well at room temperature were it not for its decrease in viscosity when being sheared by the knife. Thixotropy is just as important in painting with one-coat paints. The paint must be viscous enough to give a smooth coat without running, so the viscosity must be low when the paint is sheared by the brush. But it must increase quickly enough after brushing to prevent running. There are many other thixotropic fluids: ketchup, gelatin solutions, mayonnaise, mustard, honey, and shaving cream. What effect must shearing forces have on the structure of these liquids to cause the decrease in viscosity?

110, pp. 185-186; 921; 923, pp. 246-248; 924, p. 374; 925, pp. 144-149; 930; 934; 935, pp. 405-407; 936 through 938; 1547.

dilatancy

stress

4.127

Die-swelling Silly Putty*

Do you expect a fluid to change its volume as it emerges from a pipe through which it is being pushed? Most fluids don't, their diameters upon emerging being the same as the pipe's inside diameter. An exception, however, is Silly Putty, a silicone putty sold in toy stores. Pack a small tube full of Silly Putty, let it stand for a while to settle, and then push it through the tube. As soon as it emerges, it expands noticeably (Figure 4.127). Such an effect, called die swell, ob-

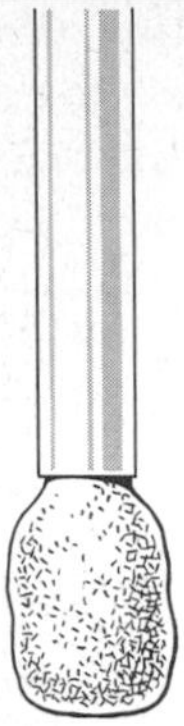

Figure 4.127
The Silly Putty expands when pushed from the tube (die swell).

viously stems from a peculiar property of this fluid, the Silly Putty, but what exactly causes it to swell, what other fluids respond similarly, and why don't all fluids behave this way?

917; 921; 923, pp. 242–244; 935, pp. 405–407; 1620.

*® Silly Putty Marketing, Box 741, New Haven, Conn., U.S.A.

4.128

Bouncing putty

Silicone putty also displays several seemingly incompatible properties. Hit it with a hammer, and it shatters. Bounce a ball of it, and it bounces better than a rubber ball. Leave a ball of it undisturbed, and it will gradually flatten. Apparently it behaves like a liquid but demands certain response times to external forces. Accordingly it will shatter if struck quickly or will bounce elastically if hit a bit slower. Gravity acting for a long time will cause it to flow. What is it in the putty's structure that determines such response times?

917; 923, pp. 236 ff; 939.

siphoning

elasticity

4.129

Self-siphoning fluids

Some fluids, such as polyethylene in water,* can siphon themselves out of containers (Figure 4.129) if you will only initiate the siphoning by pouring out some of the fluid. What pulls such a liquid over the wall of the container, and what holds the stream together?

917; 940.

Figure 4.129
A fluid that can siphon itself out of the glass.

*Collyer (917) gives directions for making such a fluid. Also see Edmund Scientific Company, 430 Edscorp Bldg., Barrington, New Jersey 08007, U.S.A.

hydrostatic pressure

viscosity

4.130

Quicksand

If you discover yourself stuck in quicksand, why is lying down on your back the best thing to do? (Once you lie down and free your legs, you can then roll toward shore.) If you should have to pull yourself, someone else, or an animal out of quicksand, why is it best to pull slowly? Does the viscosity of the sand change when you pull more quickly? If so, why? Why do deeply entrenched people and animals have bulging eyes?

941.

fluid flow

diffusion

4.131

Unmixing a dye solution

If a drop of dye is mixed into a solution by rotation, is there any way to unmix it?

Between two coaxial glass cylinders of nearly the same diameter, pour some glycerol and then carefully add several drops of dye (Figure 4.131). Rotating the inner cylinder 10 times or so apparently mixes the dye pretty well. However, if you turn the cylinder back the same number of turns, the

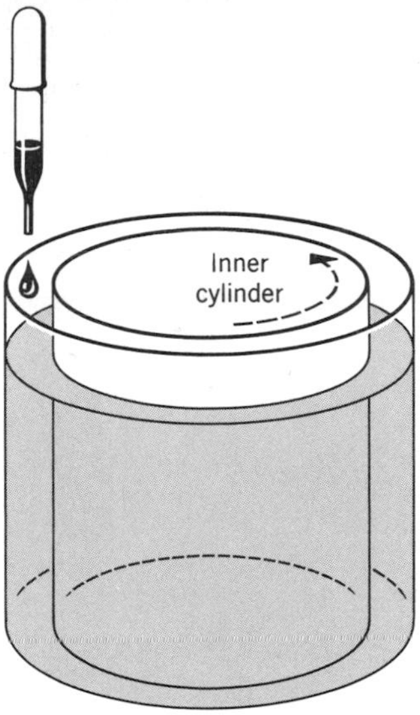

After putting drop of dye into the glycerol between the cylinders, carefully rotate the inner cylinder 10 times one way and then 10 times back

Figure 4.131
The two cylinders for unmixing a dye solution. (The space between the cylinders has been drawn large for clarity.)

dye unmixes itself back to approximately its initial distribution. Why? If you wait too long to make the reverse turns, this won't happen. Again, why?

942; 943.

5

She comes in colors everywhere

Ray Optics

(5.1 through 5.47)

reflection

refraction

dispersion

5.1

Swimming goggles

Why is it that when you are swimming underwater you can see much better if you wear goggles?

A particular Central American fish, the *Anableps*, seems to have the best, or the worst, of both media, for it swims just beneath the water surface with its large eyeballs protruding above the surface. Each eyeball is thus half in and half out of the water. Considering your need of goggles to see underwater, how can the fish see in both air and water this way?

170, p. 534; 332, Vol. 1, p. 36-3; 462, p. 116; 944; 945; 1570.

5.2

The invisible man

The invisible man in H. G. Wells' famous novel was invisible because he changed his body's index of refraction to an appropriately chosen value (Figure 5.2). What do you think the value was? No one was able to see the invisible man. Could such an invisible man, with that value of the index of refraction, see anything at all himself?

Figure 5.2
The invisible man.

5.3

Playing with a pencil in the tub

If your bathtimes have become dull and uneventful and you need something to spice them up, bring a pencil along and examine its shadow on the bottom of the tub. If you dangle the pencil half-submerged, you will find that the shadow does not entirely resemble a pencil. Rather, it looks like two rounded rods separated by a white gap as shown in Figure 5.3. Why is there a gap, and what determines its width?

951; 952.

Figure 5.3
Shadow of a partially submerged pencil. [After C. Adler, Am. J. Phys., 35, 774 (1967)].

5.4

Coin's image in water

If you place a coin in a transparent, open jar filled with water and look down through the water surface from the appropriate angle, you can see the coin's image on the surface of the water (Figure 5.4). Putting your hand on the far side of the jar usually has no effect on the image, but if your hand is wet, the image will disappear. Why?

950.

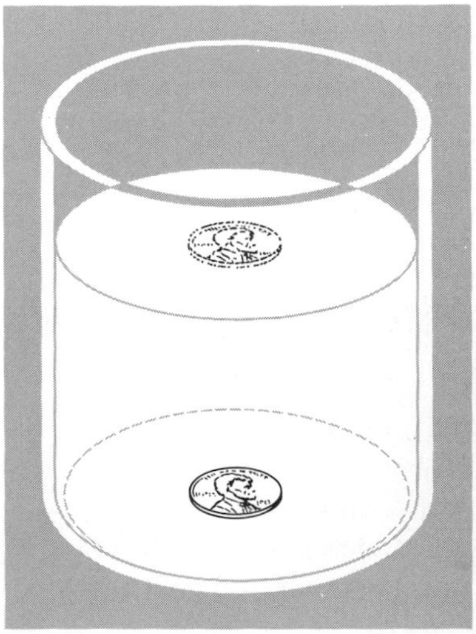

Figure 5.4
Reflection of coin in side of glass of water.

Exceptionally good references: Minnaert's book (954) is absolutely first class; his paper (991) updates the book. O'Connell's book (966) on the green flash is fascinating. Also, Wood (360), Larmore and Hall (983), and *Weather* (a journal).

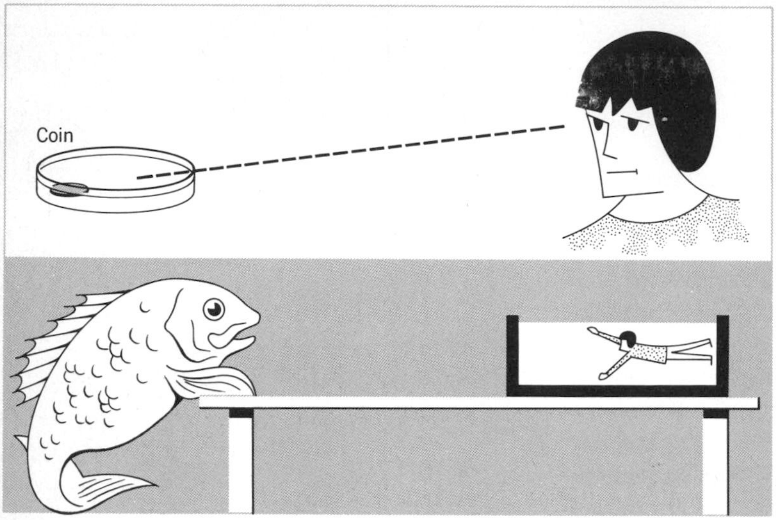

Figure 5.5

5.5

Distance of a fish

If you look down at a fish in a tank of water, you will see it at an apparent depth which is not as great as its actual depth. Is the apparent horizontal distance to the fish also distorted? The horizontal distortion may depend on whether you use one eye or two. Try it out by placing some object in a shallow dish of water and looking at it from a distance with your eyes nearly level with the surface of the water (Figure 5.5). First judge the object's distance with your head upright and then with your head tilted 90°. If the distance seems different depending on the position of your head, can you explain why?

946 through 949.

5.6

Ghosting in double-walled windows

What causes the double image, ghosting, of distant objects in double-walled windows? Under critical conditions, such as in air traffic control at airports, the ghosting can be not only annoying but also dangerous. Assuming some realistic situation, can you calculate the angular separation of a true image and its ghost? How will the separation vary with time of day and weather conditions?

953.

5.7

Mountain looming

There are places in the world where, in late afternoon and early evening, mountains can be seen rising out of the horizon on the ocean. The mountains are real, but they are too distant to be seen normally. First, in the early afternoon, a hazy patch peaks above the horizon. Then, as the afternoon wears on, the patch grows, quickly sharpening into obvious mountains near sunset. The individual peaks can even be recognized. How is this type of mirage created?

164, p. 469; 165, p. 164; 954, p. 41; 957 through 960.

5.8

Fata Morgana

Fata Morgana is the most beautiful of all mirages, and though it is very rare in some areas, it is common in the Straits of Messina between Italy and Sicily. When there is a layer of cold air over warmer water, one may see fairy castles rising out of the sea, constantly changing, growing, collapsing. According to legend, the castles were the crystal home of Morgana the fairy. This mirage is the most difficult of the mirages to explain because there are several competing effects involved, but can you unravel the effects?

164, pp. 474–475; 954, pp. 52–53; 955; 957; 958; 961.

5.9

Oasis mirage

What causes the water mirage commonly seen on hot streets? What features partially convince you that there is water in the street? Also, why do there seem to be palm trees around oasis mirages (even in areas where such trees cannot grow)? To a thirsty man, of course, the palm trees are more than enough to convince him of a water supply (Figure 5.9).

A pelican discovered on a hot asphalt highway in the midwest might have almost met its end because of the water mirage.

> The miserable bird had obviously been flying, maybe for hours, across dry wheat stubble and had suddenly spotted what he thought was a long black river, thin but wet, right in the midst of the prairie. He had put down for a cooling swim and knocked himself unconscious.*

Did the bird see a mirage?

165, pp. 164-165; 170, pp. 391-392; 219, pp. 295-296; 533, pp. 75-76; 954, pp. 45-46; 955 through 957.

*C. A. Goodrum, *The New Yorker*, **38**(8), 115 (1962).

Figure 5.9
Mirage. (By permission of John Hart. Field Enterprises.)

Figure 5.10

5.10

Wall mirage

Minnaert (954) describes a multiple image mirage that can be seen along a reasonably long wall facing the sun. (He suggests a length of 10 yards or more.) Place your hand against the wall and watch a bright metal object a friend brings near the other end of the wall (Figure 5.10). When the object is a few inches from the wall, it will appear distorted and you will see a reflected image in the wall as though the wall were a mirror. On a very hot day you may even see a second image as well. Why is there an image of the object inside the wall?

954, pp. 43-44.

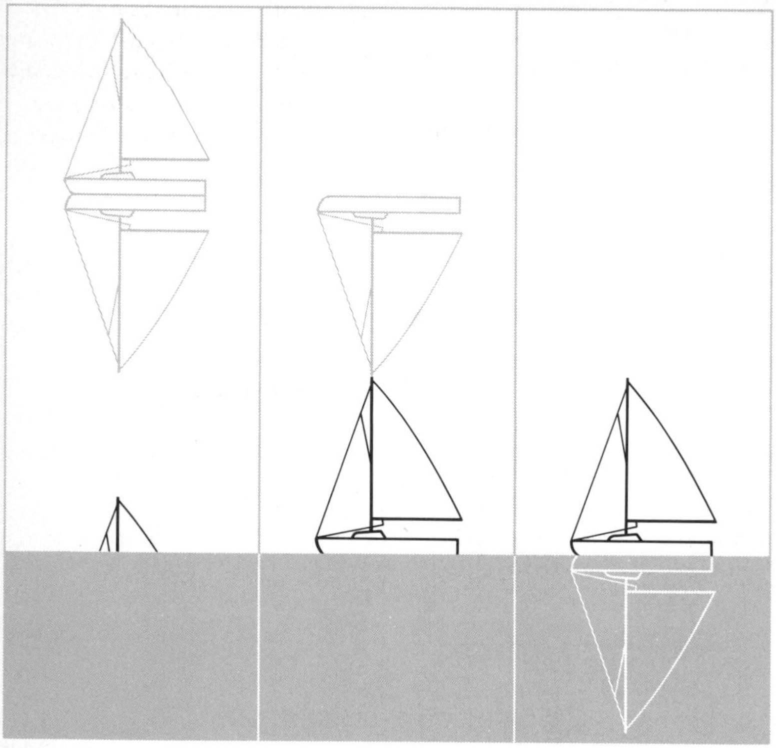

Figure 5.11

5.11

Paper doll mirage

A different type of mirage, the "superior mirage," involves one or more images of an object as shown in Figure 5.11. What's responsible for those images?

164, pp. 470–473; 165, p. 164; 954, p. 51; 955; 958.

5.12

One-way mirrors

One way mirrors are used a lot in spy movies, but are they really *one-way*? Try to devise a glass or a glass coating so that room scenes will pass in only one direction. If this is impossible, then how do the so-called one-way mirrors work?

984.

5.13

Red moon during lunar eclipse

Why is the moon red during a lunar eclipse—that is, when the moon is in the earth's shadow?

954, pp. 295–296; 983, pp. 21–22.

5.14

Ghost mirage

There are, or course, many curious stories of strange mirages. Can you explain the following one?

One hot August afternoon a woman was picking flowers from the wet ground when . . .she suddenly perceived a figure at the distance of a few yards from her. It was standing on a wet spot where there was a little thin mist (possibly steam) rising, and wavered a little, never remaining still, though she says, 'it had a great deal of bulk.' It was on a level with herself and formed a species of triangle, with herself and the sun. She was looking *towards* the sun, but not directly to it.

She thought at first that the figure might be a delusion: it stood exactly facing her and she first discovered it to be her own image by perceiving that like herself, it held. . . a bunch of flowers. She moved her hand with its nosegay and the figure did the same. The dress and flowers were precisely similar to her own and the colours as vivid as the reality. She could see the colouring and the flesh: it was

like looking at herself in
a looking-glass (962).

Needless to say, this soon unnerved the woman, and "she fled down the steep hillside, often stumbling, to rejoin her friends, both of whom had seen the figure" (962).

962.

Figure 5.15
How many "you's" do you see?

5.15

Number of images in two mirrors

How many images of yourself do you see while standing in front of two adjacent plane mirrors such as you find at a clothing store (Figure 5.15)? How does the number of images depend on the angle between the mirrors? Does it matter where you stand? If it does, where do you stand to see the most images? Are your answers the same for the number of images you will see of a package lying next to you?

985 through 989; 1524.

5.16

The green flash

Just as the top of the setting sun disappears beneath a clear, flat horizon, you may be able to see, for 10 seconds or so, a distinct green flash from the sun. Why does this happen? Could it be an optical illusion (say, an afterimage of the sun)? This was the common opinion for a long time, until photographs were made of the flash.*

In higher latitudes it can be seen for longer times "Members of Byrd's expedition to the South Pole are reported to have seen it for 35 minutes while the sun, rising at the close of the long winter night, was seen to be moving almost exactly along the horizon" (978).

Clear horizons, such as over the Pacific, are a definite asset. "According to Rear Admiral Kindell, strong and brilliant flashes were seen by him and other members of the U.S. Navy during the Okinawa campaign of 1945 at almost every sunset on clear days" (978).

A similar effect, although very rare, is the red flash that may appear when the sun peaks out beneath a cloud.

164, pp. 58-63; 165, p. 160; 362, pp. 152-153; 966 through 981; 1614.

*O'Connell's book (966) is full of green flash photographs.

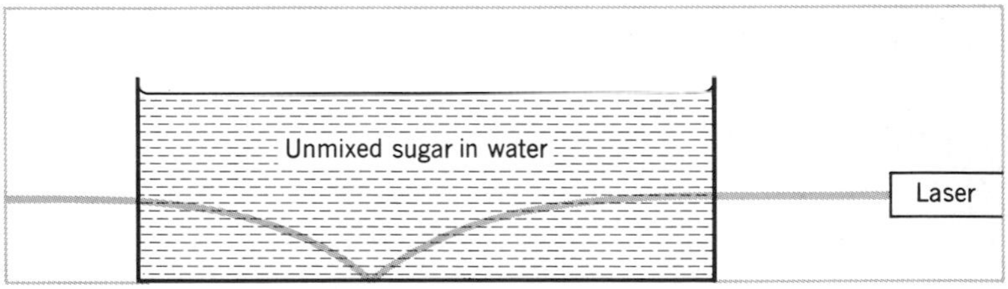

Figure 5.17
Laser beam bouncing in a sugar solution. [After W. M. Strouse, Am. J. Phys., 40, 913 (1972).]

5.17

Bouncing a light beam

If a narrow light beam (such as a laser beam) enters a container of water in which several lumps of sugar have been added without stirring, the light beam will bend and then bounce off the bottom as shown in Figure 5.17. What makes the beam bend down? What makes it bounce? And finally, once it is going up, what makes it bend down again?

963; 1551.

5.18

Flattened sun and moon

What causes the apparent flattening of the sun and moon when they are near the horizon? Can you roughly calculate the amount of distortion?

164, p. 470; 219, pp. 297-298; 954, pp. 39-40; 964; 965.

reflection

polarization

Brewster angle

5.19

Blue ribbon on sea horizon

The horizon on the sea often appears to be a much darker blue or gray than the sky or the rest of the sea. In fact, if you're standing on a beach, it almost appears that someone has stretched out a bright blue ribbon to mark the horizon. The ribbon disappears, however, if you lie on the beach or if you climb to a greater height. One clue about what causes the ribbon might be that the light from it is almost completely linearly polarized. Can you explain the ribbon and the polarization?

1500.

5.20

30° reflection off the sea

If you look at the sea just below the horizon, you will see reflections of objects that are more than 30° above the horizon. Objects less than 30° above the horizon are not reflected. Why? Is the minimum reflection angle determined by the average wave slope that, because of your observations, must be about 15°? Actually, it is not. Can you think of any other reason why the reflection is restricted in this way?

954, pp. 23-25; 990.

5.21

Lunar light triangles

When the moon is reflected in the sea or a lake, why is there a luminous triangle on the surface of the water (Figure 5.21)? What determines the shape and width of the luminous area? Why is there a corresponding dark triangle in the sky above the water?

399, pp. 243-246; 954, pp. 23-27, 138-139; 991; 992, pp. 74-80.

Figure 5.21
Lunar light triangle in the water and a dark triangle in the sky.

By permission of John Hart. Field Enterprises.

5.22

Shiny black cloth

Why do some types of cloth glisten while others do not? Black felt has a shiny side and a dull side. Some wall paints are glossy black; others are flat black. Since black absorbs visible light, how can a black surface be shiny?

253, pp. 278-279; 533, pp. 33-35.

pinhole optics

5.23
Inverted shadows

Punch a pinhole in an opaque sheet of paper, hold the paper a few inches from one eye, close the other eye, and then carefully hold a thin nail between the pinhole and you (Figure 5.23a). Move the nail around until a shadowy figure appears in the circle of light from the pinhole (Figure 5.23b). What causes that figure, and why is it inverted from the nail? Also, why does the figure appear to be on the far side of the pinhole?

533, pp. 49–51; 993; 1582.

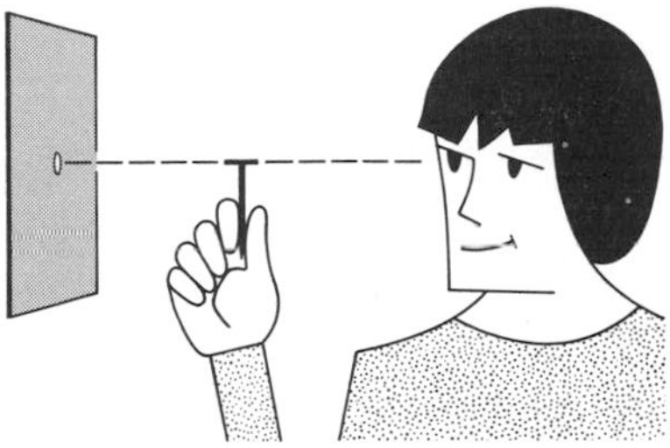

Figure 5.23a
Hold a thin nail between your eye and a pinhole.

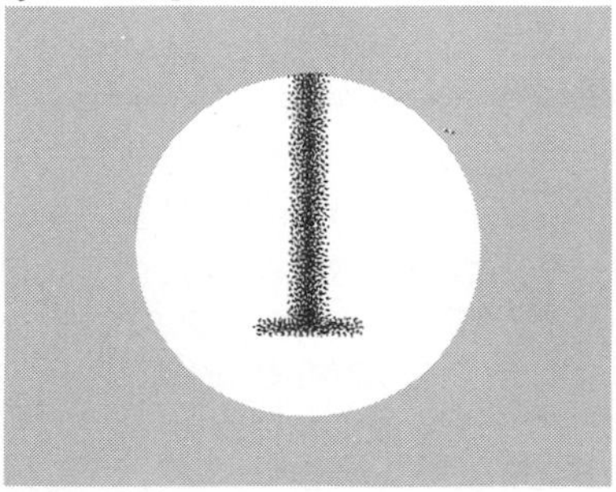

Figure 5.23b
Shadowy image of the nail in the pinhole image.

ray optics

diffraction

abervation

resolution

photometry

5.24
Pinhole camera

The simplest type of camera, and the easiest to build, is the pinhole camera. Moreover, there are some definite advantages to using a pinhole instead of a lens. For example, there is no linear distortion, and there is tremendous depth of field. Are there aberrations of any significance? In particular, is there any chromatic distortion in a simple pinhole camera? Finally, what is the best hole size, and what happens to your pictures if the hole is larger or smaller than the best size?

994 through 998; 1501 through 1503; 1586.

5.25
Eclipse leaf shadows

If you look at the shadows of leaves during a solar eclipse, you will see images of the eclipsing sun projected onto the ground. Why are these images made? Are they present all the time or just during an eclipse?

360, pp. 66–67; 533, pp. 29–31; 999.

5.26
Heiligenschein

Some morning when the grass is sparkling with dew, look at the shadow of your head on the grass. Around the shadow will be a bright light called the heiligenschein. How, exactly, does the dew cause this brightening, and why isn't there heiligenschein around your entire shadow? Do the blades of grass play any part in the effect besides holding up the dewdrops? Can you also explain the very bright heiligenschein that astronauts see when walking on the moon? (It certainly isn't due to dew-covered grass.)

164, p. 556; 165, p. 180; 360, p. 68; 362, pp. 136–137; 380; 954, pp. 230–234; 983, Chapter 2; 1000 through 1008.

5.27
Bike reflectors

If you shine light on a bike reflector at virtually any angle, the light will be reflected back to the source. Why is the reflector so good at this? An ordinary mirror will reflect well, of course, but it will not return the light to the source unless the incident light is perpendicular to the surface. What, then, is different about the bike reflector? If a narrow beam of light is reflected by a bike reflector, how wide will the return beam be?

170, p. 158; 1011.

5.28

Brown spots on leaves

It is a bad idea to sprinkle water on tree leaves during the day, because the water drops leave brown spots on the leaves. What causes the spots?

5.29

Rays around your head's shadow

I looked at the fine centrifugal spokes round the shape of my head in the sunlit water. . . Diverge, fine spokes of light from the shape of my head, or anyone's head, in the sunlit water! ---Walt Whitman, "Crossing Brooklyn Ferry", Leaves of Grass

These rays of light surround the shadow of your head if the shadow is cast upon slightly turbulent water. If the water is calm or has regular waves, the rays do not appear. Why?

954, pp. 333–334; 1009; 1010.

5.30

Cats' eyes in the dark

Why do a cat's eyes shine so brightly in the dark when you illuminate them with a flashlight? Why aren't they so noticeably bright during the day? Does the amount of reflection depend on the angle between your line of sight and the incident light beam? Why don't our eyes shine as much when illuminated at night or with a flashbulb?

954, p. 350; 983, p. 36; 1012.

5.31

Brightness of falling rain

Occasionally you can see distant rain falling, and in some case you may notice that "when these regions of falling precipitation are illuminated by direct sunlight, a distinct horizontal line can be seen, above which the precipitation appears much lighter than below" (1013). What is responsible for the change in brightness?

1013.

5.32

Rainbow colors

The color separation in the primary rainbow is usually explained as simple refraction and reflection of the light rays within raindrops. However, since the light rays are incident on a drop's surface within a wide range of angles to that surface (Figure 5.32), shouldn't the emerging light rays, even those of a particular color, also leave the drop in a wide range of angles? Why, then, do you see a particular color subtending a particular angle from the rainbow?

As a matter of fact, are rainbow colors actually as pure as a prism's? If the simple refraction explanation is correct, shouldn't the rainbow have pure colors?

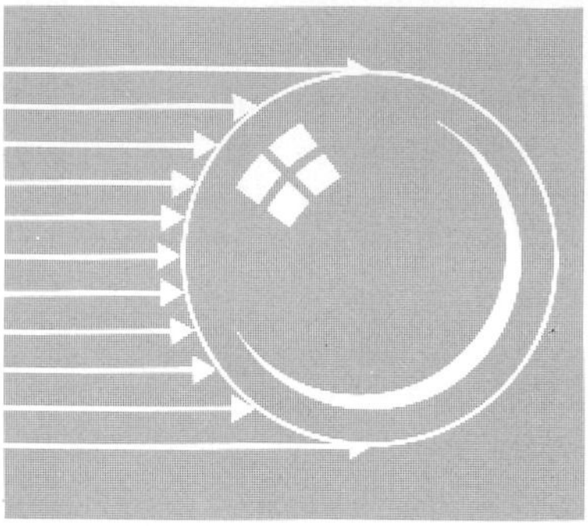

Figure 5.32
Light rays from sun incident on a water drop.

Why is the color sequence in the secondary rainbow opposite that in the primary rainbow, and why is the secondary rainbow so seldom seen? As a matter of fact, why can only two rainbows be seen in the sky? If the primary rainbow is due to a single reflection of light rays inside the raindrops, and the secondary rainbow is due to a double reflection, should not there be more rainbows resulting from further internal reflections?

A double rainbow can also be seen in the beam of a searchlight during a light rain at night. As the beam sweeps through the sky, the rainbows slide up and down the beam and may even disappear briefly. Can you account for such motion of the rainbows?

164, Chapter 3; 165, p. 177; 380; 954, pp. 174–179; 983, Chapter 3; 1014, Chapter 13; 1015 through 1021; 1499; 1627 through 1631.

5.33

Pure reds in rainbows

Why can pure reds be found only in the vertical portions of rainbows when the sun is relatively low?* (The sun must be low so that the vertical portions of the rainbows can be seen; if you are viewing the rainbow from a high point, the sun will not have to be so low.)

1022.

*Even the most commonplace features of the outside world still afford fresh understandings and surprises. Fraser (1022) points out that this simple feature of pure reds being restricted to the vertical portions of the rainbow somehow escaped notice until his paper of 1972. As another example of modern work, it has only been recently that photographs of the infrared rainbow have been taken (1023, 1024), thus allowing man to see for the first time what has periodically hung in the sky for millions of years.

5.34

Supernumerary bows

Sometimes several pink and green bows can be seen below and adjacent to the primary bow. Very rarely they can also be found above the secondary bow. What causes these additional bows? Don't they come as a surprise if you draw too simple a picture of the rainbow? Why aren't they found between the primary and secondary bows?

164, pp. 477, 483; 954, p. 178; 983, Chapter 6; 1014, pp. 241–242; 1019 through 1021; 1025.

5.35

Dark sky between bows

Why is the region of sky between the primary and secondary rainbows darker than the rest of the sky?

164, pp. 482–483; 954, pp. 179-180; 983, p. 56; 1020.

5.36

Rainbow polarization

Is the rainbow polarized? If it is, can you explain its polarization?

361, pp. 8-9; 954, pp. 181-182; 983, pp. 59 ff; 1014; 1020; 1630.

5.37

Lunar rainbows

Lunar rainbows are very rare. Is this only because moonlight is so much dimmer than sunlight, or is there some other reason?

164, p. 476; 954, p. 189; 1020; 1026; 1027.

5.38

Rainbow distance

How far away from you are rainbows formed? That is, how distant are the water drops? Is it possible to have a rainbow a few yards away from you?

If you look at a rainbow in your garden sprinkler, you may very well see two bows crossing over each other (Figure 5.38). Why?

164, p. 496; 954, pp. 169, 174; 1020.

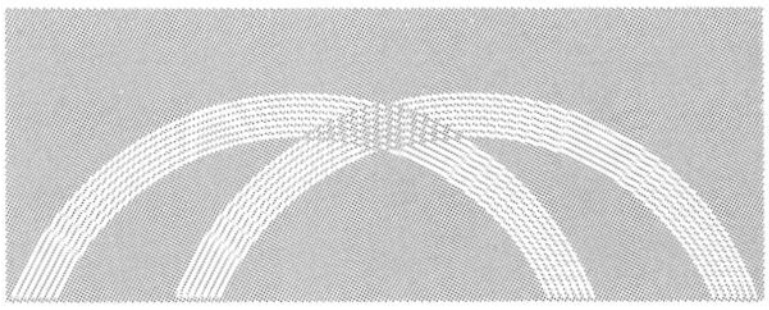

Figure 5.38
Rainbows seen in water-sprinkler spray.

5.39

Rainbow pillar

What causes the very rare pillar of light that has been seen at the foot of some rainbows (Figure 5.39)? [Minnaert (991) gives a photograph of such a pillar, along with the comment that these pillars have not yet been explained.]

991; 1028, plate 24; 1029.

Figure 5.39
Rainbow pillar at the foot of a rainbow.

5.40

Reflected rainbows

If you ever get a chance to see both a rainbow and its reflection in water, you'll notice they are different in shape and position. If a cloud is present, for example, you may see something resembling Figure 5.40. Why is there a difference in the cloud's position relative to the rainbow?

164, pp. 497–499; 165, p. 175; 954, pp. 186–187; 983, p. 68; 1020, pp. 272–275.

Figure 5.40

Figure 5.41a
Dewbow in grassy field.

Figure 5.41b
Dewbows as seen by someone under a street light at night. (After J. O. Mattsson, S. Nordbeck, and B. Rystedt, Ref. 1030.)

5.41

Dewbows

What causes the rainbows seen on dew-covered grass fields (dewbows) and on ponds with oily surfaces? In particular, can you explain their shape (Figure 5.41*a*)? Why do dewbows formed by streetlights have yet another shape (Figure 5.41*b*)?

164, pp. 499–500; 165, pp. 175–176; 954, pp. 184–186; 1020; 1030; 1031.

5.42

Sun dogs

Sun dogs (mock suns or parhelia) are bright images of the sun that sit to one or both sides of the sun. They are normally outside the 22° halo (if the halo is visible), as shown in Figure 5.43, being further away the higher the sun is in the sky. When the sun is higher than 60°, however, the sun dogs disappear. Can you explain what produces the sun dogs and why their position and existence depend on the sun's height? Also, why are they so much more colorful than the 22° halo?

164, pp. 510 ff; 165, pp. 169–171; 361, pp. 24–25; 362, pp. 140 ff; 380; 954, pp. 196–197; 983, pp. 70–73, 84 ff; 991; 1044 through 1051.

5.43

The 22° halo

Halos around the moon and sun are fairly common in most areas. The primary halo is 22° from the sun or moon (Figure 5.43) and is colored red on the inside and white or blue on the outside. Except for the corona immediately surrounding the sun or moon, the sky inside the 22° halo is dark.

Certainly the halo is caused by scattering of the light somewhere in the atmosphere, but what kind of scattering could give such a uniform design? For example, would you expect to get a 22° halo from sunlight scattered by high altitude dust? Also, why is the area within the halo dark?

Almost universally the halo has been thought to be a sign of imminent rain. Is there any truth to that belief?

164, pp. 512-513; 165, pp. 169-174; 219, pp. 298-299; 360, pp. 78-79; 361, pp. 24-25; 362, pp. 140-143; 954, pp. 190-195; 983, Chapter 4; 991; 1033 through 1050; 1610.

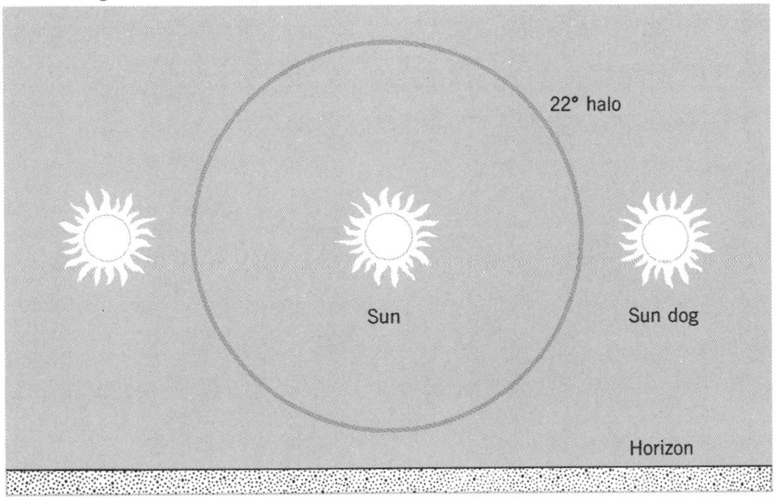

Figure 5.43
The 22° halo and sun dogs around the sun.

5.44

Fogbows

Why are fogbows—rainbows formed in the fog—whitish bands with orange on the outside and blue on the inside? Why are they about twice as wide as normal rainbows?

Can fogbows be produced by streetlights? If so, what difference do you expect from the fogbows formed in sunlight?

165, p. 175; 380; 954, p. 183; 1020; 1030; 1032; 1628.

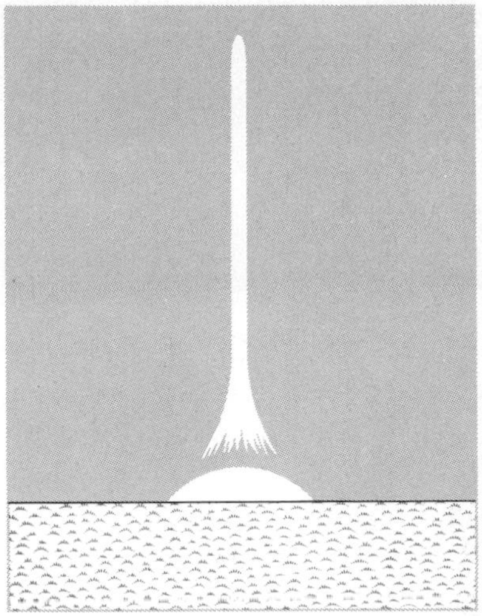

Figure 5.45

5.45

Sun pillars

Pillars of sunlight above and below the sun (Figure 5.45) can be seen fairly often near sunset or after sunrise. The columns may be white, pale yellow, orange, or pink, so they are quite pretty. Under some conditions they can even be seen above and below outdoor artificial lights such as streetlights. What causes these pillars?

164, pp. 543-544; 165, pp. 169, 172; 361, pp. 32-33; 362, pp. 148-149; 954, pp. 201-202; 983, pp. 135 ff; 1028, plate 23, p. 245; 1033; 1035; 1065; 1066; 1504.

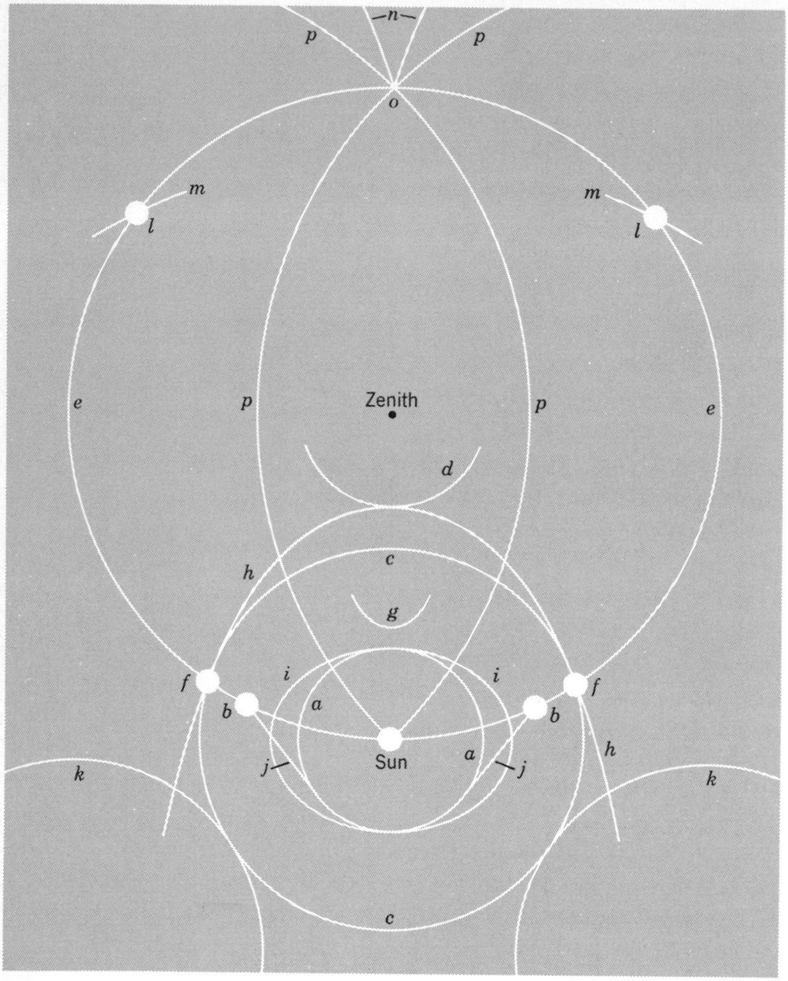

(a) 22° halo. (b) Sun dogs to 22° halo. (c) 46° halo. (d) Circumzenith arc. (e) Parhelic circle. (f) Sun dogs to 46° halo. (g) Parry arc. (h) Supralateral tangent arcs to 46° halo. (i) Tangent arcs to 22° halo. (j) Lowitz arcs. (k) Infralateral tangent arcs to 46° halo. (l) Paranthelia. (m) Paranthelic arcs. (n) Narrow-angle oblique arcs to anthelion. (o) Anthelion. (p) Wide-angle oblique arcs of anthelion.

Figure 5.46
Some of the possible arcs, halos, and sun dogs around the sun. (Not all can be visible simultaneously.)

5.46

Other arcs and halos

The full array of possible arcs and halos could be awesome if all of them were visible at once. (See Figure 5.46.) Usually, however, you will see only a few arcs or halos. Some are so rare, in fact, that their existence is still controversial. The Lowitz arc, for example, has apparently only recently been explained (1058). Several of the arcs can change shapes tremendously as the sun changes height, so it pays to watch as long as possible, making occasional sketches. See if you can explain the ones you do find.

164, Chapters 4, 5; 165, pp. 169-174; 361, pp. 28-29; 362, pp. 140-149; 380; 954, pp. 190-206; 983, Chapter 4; 991; 1034 through 1038; 1044 through 1064; 1514; 1515; 1622.

electric field

reflection

5.47
Crown flash

Concurrent with a lightning stroke in the main body of a storm cloud, there may be a brightening that ripples upward and outward through the top of the cloud. Is this brightening (called "crown flash" and "flachenblitz") an unusual type of discharge, or is it a peculiar reflection of light from the initial lightning stroke?

301, pp. 50–51; 1067 through 1069.

Polarization
(5.48 through 5.57)

5.48
Polarization for car lights

Polarized plastic sheets were first developed to cover car headlights so as to reduce the glare of an approaching car at night. How could this be accomplished, and what would be the best orientation of the polarized sheets? Don't forget that you still want to see the oncoming car, so the light shouldn't be entirely blocked out. Will the tilt of the windshield matter? Could you obtain similar results with polarized sunglasses?

1070, pp. 111–114; 1071 through 1074.

5.49
Polarized glasses and glare

Why do polarized sunglasses reduce glare? (Unpolarized sunglasses just decrease the total amount of light entering your eyes and do not preferentially block the glare.) When will polarized sunglasses improve a fisherman's ability to see beneath the water?

1070, pp. 100–102.

5.50
Sky polarization

Why is the light coming from a clear sky polarized? Where should the region of maximum polarization be? Can you verify your prediction by using a pair of polarized sunglasses? Is light from clouds polarized? Why are some areas of the sky unpolarized? Why is the polarization in some parts of the sky perpendicular to that predicted by conventional theory? Can you also find these neutral points and areas of perpendicular polarization with your sunglasses?

164, pp. 571–575; 165, pp. 194–204; 170, pp. 413–414; 360, pp. 62–63; 362, pp. 152–153; 446, pp. 43–45; 533, pp. 193–196; 954, pp. 251–254; 1070, pp. 98–99; 1075, pp. 12–17; 1076 through 1079.

5.51
Colored frost flowers

Some morning after a cold night, examine the thin, transparent frost flowers on a window facing the sun. If the flowers have started to melt and have formed a pool of water at the bottom of the window pane, look for reflections of the flowers in the pool (Figure 5.51). They will appear as patterns of colored fringes. What causes the color in these reflections?

1080.

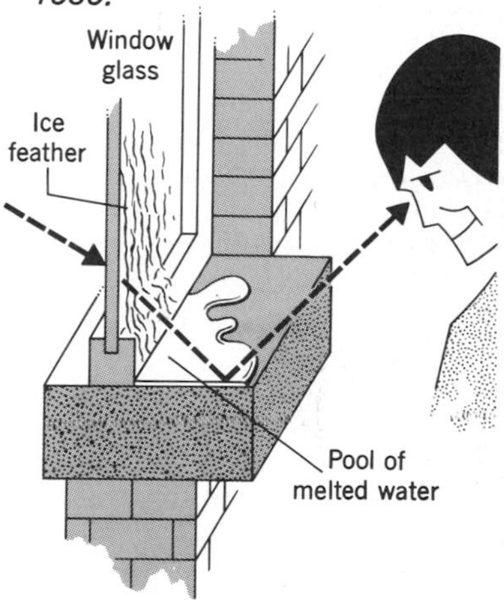

Figure 5.51
Optics for seeing frost flowers. [After S. G. Cornford, Weather, 23, 39 (1968).

5.52
Cellophane between two polarizing filters

Light will not pass through two polarized sheets whose polarization directions are perpendicular. But if clear cellophane is inserted between them, light is transmitted, the amount of transmission depending on the cellophane's orientation.

If you replace the cellophane with a piece of plastic food wrap you will find that very little light is transmitted. By stretching the food wrap, however, you can once again get a large transmission. What is the fundamental difference between cellophane and unstretched food wrap that accounts for the difference in transmission? How are the optical properties of the food wrap changed by stretching?*

170, pp. 420 ff; 360, pp. 14-16; 1077; 1078; 1081; 1082, pp. 79-93.

*For a whole bagful of optical devices and tricks that can be made with cellophane, tape, etc., see Chapters 8 and 9 of Crawford's excellent book *Waves* (170). Also see Refs. 1096 and 1097.

5.53

Spots on rear window

If you wear polarized sun glasses while driving, you have probably noticed the large spots, usually arranged in patterns, on the rear windows of other cars. What are those spots, and why must you wear the polarized sunglasses to see them? Are the spots colored?

360, pp. 14-16.

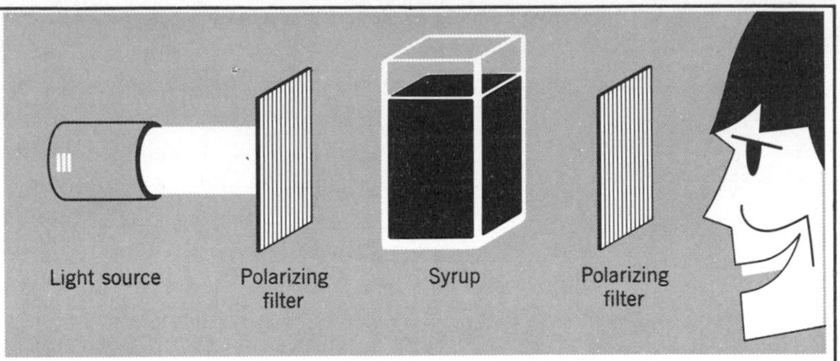

Figure 5.54
Detection of polarization change by syrup.

5.54

Optical activity of Karo syrup

Although you probably have used Karo corn syrup on your pancakes, you most likely are unaware of the syrup's most fascinating property: its optical activity. Try this experiment; between two polarizing filters (they can be from polarized sunglasses), put a glass of Karo syrup. Then place a white light source on one side of the glass and look at the light through the syrup (Figure 5.54). What is responsible for the beautiful colors you see? By turning one of the filters (while leaving the other fixed), find the polarization of the emerging light and thus the polarization change experienced by the light in the syrup. By repeating this procedure for several thicknesses of syrup, you will discover that the polarization change depends on the distance the light travels through the syrup. Why? How much rotation of the polarization is there per centimeter of syrup, and is it clockwise or counterclockwise? Why is the rotation in one direction instead of the other?

155, p. 425; 170, pp. 425-426, 447; 533, p. 198; 1070, pp. 115-118; 1082, pp. 136-144; 1083; 1084.

5.55

Animal navigation by polarized light

Honeybees, ants, and various other creatures use the polarization of the sky* as an aid to navigation. How are they able to detect the polarization angle of the light? And how can they use this ability to navigate?

332, Vol. I, p. 36-7; 1070, p. 98; 1085, Chapter 13; 1086 through 1089; 1557; 1584.

*See Prob. 5.50.

5.56

Magic sun stones

Dichroic crystals are different colors when under light of different polarizations. The crystal may be clear with a faint yellow tinge under light of one polarization, but dark blue when the polarization is changed by 90°.

It is believed that the Vikings used a dichroic crystal (cordierite) to locate the sun when it was not directly visible. At least, according to the tales, they had some kind of magic "sun stone" by which they could find the sun even when it was behind the clouds or below the horizon. Since in the high latitudes the sun can be below the horizon even at noon, such magic stones would have been a very valuable navigational aid.

Why are different colors transmitted through such crystals for different incident polarizations? Can the cyrstals really be used to find the sun even if the sky is cloudy or the sun is below the horizon?

170, pp. 448–449.

5.57

Haidinger's brush

You may not realize it, but you are capable of detecting polarized light with your own eyes. By looking through a polarizing sheet (polarized sunglasses, for example) at a bright light, you will momentarily see a yellow hourglass figure with a blue cloud to each side

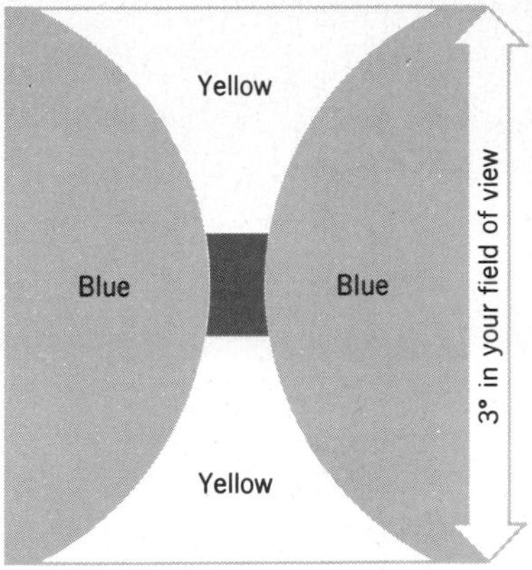

Figure 5.57

(Figure 5.57). Suddenly rotating the filter in its own plane may help you to spot the hourglass easier. This pattern is called Haidinger's brush and is a direct result of the linear polarization caused by the filter. But why? What part of the eye is sensitive to the polarization sense, and why is this particular pattern created? How does the orientation of the hourglass depend on the polarization axis? Why does the pattern fade after a few seconds? I can see the brush fairly well, without a polarizing filter, in the partially polarized light of the sky. Some people see it so clearly that it becomes irritating.

You can also detect circularly polarized light with your eyes: left-circularly polarized light gives a yellow brush tilted to the right at about 45°, whereas the opposite polarization gives a brush titlted to the left at about 45°. Why?

954, pp. 254–257; 1070, pp. 95–97; 1090, pp. 300–304; 1091, Vol. 2, pp. 304–307; 1092 through 1094; 1621.

Scattering

(5.58 through 5.90)

Rayleigh and Mie scattering

diffraction

dispersion

5.58

Sunset colors

All of us too often neglect sunsets, especially physicists who tend to shove the twilight colors under the heading of "Rayleigh scattering" and then forget them. Can you explain the beautiful variety of colors in the twilight sky? (The setting sun may be red, but the sky is certainly not just red.) As the sun sets, the western sky first assumes yellow and orange tints. By the time the sun has turned a fiery red, the afterglow left in the western sky varies upward from the horizon from a yellow-orange to a green-azure. Eventually the area about 25° above the western horizon turns rose-colored (the "purple light" discussed below).

Especially brilliant twilight colors can be seen soon after major volcanic eruptions. What causes such color enhancements?

164, pp. 566–567; 165, pp. 184 ff; 380; 954, Chapter 11; 983, pp. 234–244; 1075; 1102 through 1109; 1526.

5.59

The blue sky

Probably the all-time standard physics question is "Why is the sky blue?" Physicists often toss it aside with a few mutterings about "Rayleigh scattering." Certainly the question deserves better treatment than that. For example, what part of the sky is bluest, and why isn't the entire sky a uniform color? Does the daytime sky color actually follow the Rayleigh prediction? Why isn't the sky blue on nights with a full moon? What is scattering the sunlight to produce the daytime sky color? Would you get a blue sky if the scatterers were much larger or much smaller? Finally, why is the sky on Mars blue only within a few degrees of the horizon, and black overhead?

164, Chapter 7; 165, pp. 192 ff; 170, pp. 559-562; 466, pp. 35 ff; 954, pp. 238-251; 983, Chapter 9; 1075, p. 10; 1079; 1098 through 1102; 1505; 1526.

5.60

Twilight purple light

What causes the purple light (which may be more pink than purple) that first appears in the western sky as the sun sinks beneath the horizon (Figure 5.60)? It is the brightest about 15 to 40 minutes after sunset.

Is the same physics responsible for the "second" purple light that sometimes appears after the common one has vanished and which may last up to two hours after sunset? How could the sun still provide light to the sky after having set an hour or so earlier?

164, p. 567; 165, pp. 184-192; 954, pp. 270-280; 1075; 1102; 1104; 1110.

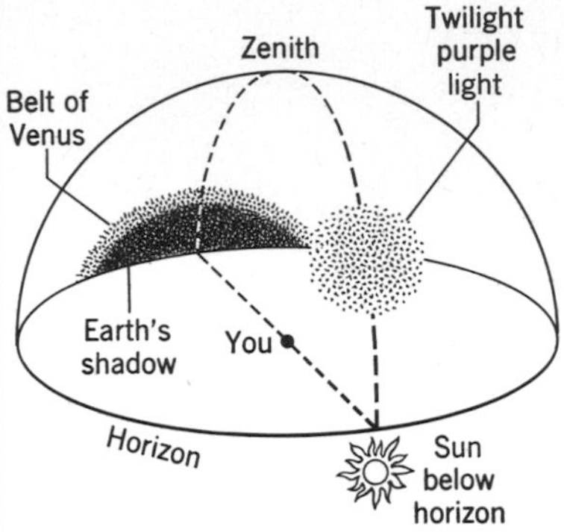

Figure 5.60
Sunset phenomena. (After H. Neuberger, Introduction to Physical Meteorology, Pennsylvania State University.)

5.61

Zenith blue enhancement

Have you ever noticed the zenith (overhead sky) turns a deep blue during sunset (Figure 5.60)? Isn't that strange? Wouldn't you think the zenith would be red, for the same reason the setting sun is red?

466, pp. 207-208; 1075, Chapter 4; 1102.

5.62

Belt of Venus

What causes the twilight's rosy patch ("belt of Venus") that borders the earth's shadow as the shadow rises out of the east (Figure 5.60)?

164, p. 566; 165, pp. 184 ff; 954, pp. 268 ff; 1075.

5.63

Green street lights and red Christmas trees

While flying into a city you may have noticed that many streets are lit by green lights. When you drive through these streets, however, the lights are not green at all, but white. Why is there a color difference in the two situations? Similarly, why is the light from a distant Christmas tree primarily red when in fact the tree is covered with lights of many colors?

1111, pp. 172-173; 1112.

5.64

Brightness of daytime sky

Why is the daytime sky bright? Can you calculate roughly how bright it is?

164, pp. 563-565; 170, pp. 559-562; 466, pp. 35 ff; 954, pp. 245-247; 1075; 1100, p. 33.

vision

5.65
Yellow ski goggles

Although skiers wear yellow-tinted goggles largely to be fashionable, they often claim that the goggles improve their vision on hazy days. Supposedly, a skiier can better distinguish the small snow bumps in his path. Such a claim must have some validity, because the famous polar explorer Vilhjalmur Stefansson also recommended amber glasses for travel across snow and ice fields. Why might yellow glasses help? For example, is there a dominance of yellow in snow-reflected sunlight on hazy days?

354; 1113, pp. 200-202.

5.66
Stars seen through shafts

Ever since Aristotle men have believed that stars can be seen in the daytime if they are viewed through long shafts such as chimneys. A shaft will decrease the total skylight seen, thereby (supposedly) allowing the stars to be distinguished in the small patch of light at the top of the shaft. Your partial dark adaptation (due to the smaller amount of skylight you see) may also aid in the distinction. Do you believe such measures will actually make stars visible in the daytime? Can you verify your belief by calculations and by trying the experiment?

1114; 1115.

reflection

absorption

scattering

5.67
Colors of lakes and oceans

What is the color of a clear, clean mountain lake? Does it matter if the sky is clear or cloudy? How much do the material on the bottom and the depth of the water matter? What is responsible for the different colors of other lakes? What color is the ocean near the shore and far at sea? What colors do you see in ocean waves?

While swimming as deep as possible, hold out a hand horizontally and notice that the top is a different color than the bottom. Why is there a color difference?

360, pp. 17-19; 380; 466, pp. 201-203; 954, pp. 308-335; 992, Chapter 13; 1116 through 1118.

absorption

transmission

scattering

5.68
Color of overcast sky

If you have ever lived in the country, you may have noticed a seasonal change in the color of an overcast sky. Some people claim that an overcast sky is slightly greener in the summer than in the winter. Now I could make the obvious guess about what causes this color change, if it really does happen, but is there any validity to my guess?

1119.

5.69
Seeing the dark part of the moon

When the sun has just set and the new moon appears as a narrow crescent, the "dark" part of the moon can be seen. How is that possible?

466, p. 199; 954, p. 297.

5.70
White clouds

Why are most clouds white? Why aren't they blue like the sky? Why are thunderclouds dark?

332, Vol. I, p. 32-8; 1123; 1124.

5.71
Sunlight scattered by clouds

Why does water scatter so much more sunlight after it has condensed to form clouds than before, when it was just water vapor? Isn't the total number of atoms the same, and shouldn't the scattered light thus be the same?

332, Vol. I, p. 32-8; 1123.

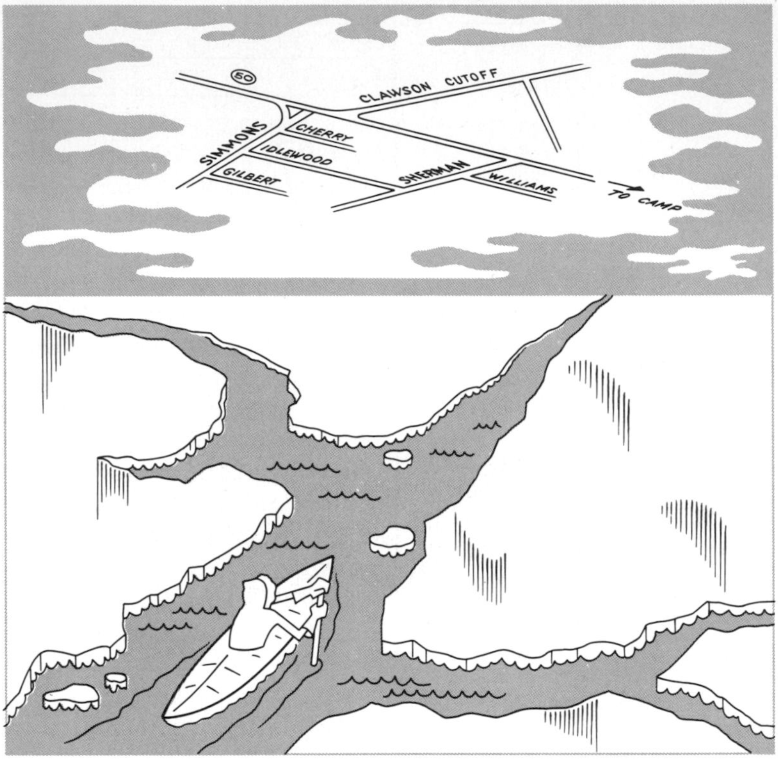

Figure 5.72
A kayaker finding his way through the ice field by the map in the sky.

5.72
Maps in the sky

Over the ice fields in the far north, large maps of the surrounding region sometimes appear at the base of overhanging clouds. These maps, called "ice blink" and "cloud maps," allow the Eskimo to pick a route through the ice field if he is kayaking, or over the ice if he is sledding (Figure 5.72).

> On approaching a pack, field, or other compact aggregation of ice, the phenomenon of the ice-blink is seen whenever the horizon is tolerably free from clouds, and in some cases even under a thick sky. The "ice-blink" consists in a stratum of a lucid whiteness, which appears over ice in that part of the atmosphere adjoining the horizon. . . when the ice-blink occurs under the most favorable circumstances, it affords to the eye a beautiful and perfect map of the ice, twenty or thirty miles beyond the limit of direct vision, but less distant in proportion as the atmosphere is more dense and obscure. The ice-blink not only shows the figure of the ice, but enables the experienced observer to judge whether the ice thus pictured be field or packed ice; if the latter, whether it be compact or open, bay or heavy ice. Field-ice affords the most lucid blink, accompanied with a tinge of yellow; that of packs is more purely white; and of bay-ice, greyish. The land, on account of its snowy covering, likewise occasions a blink, which is more yellow than that produced by the ice of fields (1120).

Can you explain these cloud maps?

1075, p. 8; 1113, p. 220; 1120 through 1122.

Mother-of-pearl clouds

5.73

Not all clouds are white or dark. Mother-of-pearl clouds (nacreous clouds) may have very beautiful, delicate colors. Though they are rare and are usually seen only in the high latitudes and only after sunset, they can occasionally be bright enough to color snow on the ground. What is different about these clouds so that they show such colors? Do the colors arise from a fortuitous particle size? Why are these clouds usually confined to the high latitudes and to an altitude range of about 20 to 30 kilometers?

361, pp. 20-21, 28-29; 362, pp. 74-75; 536, p. 170; 954, pp. 229-230; 1124 through 1129.

interference

5.74

Young's dusty mirror

When you look past a small lamp directly into a dusty mirror, you will find that the reflected image of the lamp is surrounded by distinct colored fringes. A very clean mirror won't make the fringes; you must have a dusty or slightly dirty one. What causes the fringes, and how many fringes of any one color are there? Most of all, why must the mirror be dusty or slightly dirty?

1130; 1131.

illumination

scattering

intensity

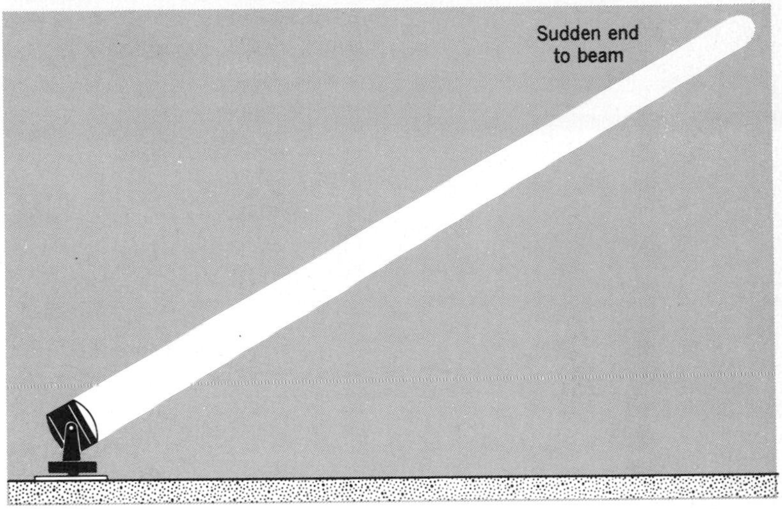

Figure 5.75

5.75

Searchlight beams

Why do searchlight beams (the kind used for airplane detection in World War II but that have now been demoted to signaling supermarket openings) end as abruptly as they do (Figure 5.75)? Wouldn't you expect a gradual fading of the beam?

954, pp. 262-263; 1147.

5.76

Zodiacal light and gegenschein

The next time you find yourself away from city lights on a clear moonless night, search for the zodiacal light and gegenschein. The former is a milky triangle that may be in the west for a few hours after sunset or in the east before sunrise. The triangle is nearly as bright as the Milky Way and is oriented along the plane of the ecliptic.* The gegenschein is a rather faint light seen at the antisolar point in the sky. What is responsible for these lights in the night sky?

954, pp. 290-295; 1143 through 1146.

*The ecliptic plane is the plane in which the earth orbits about the sun.

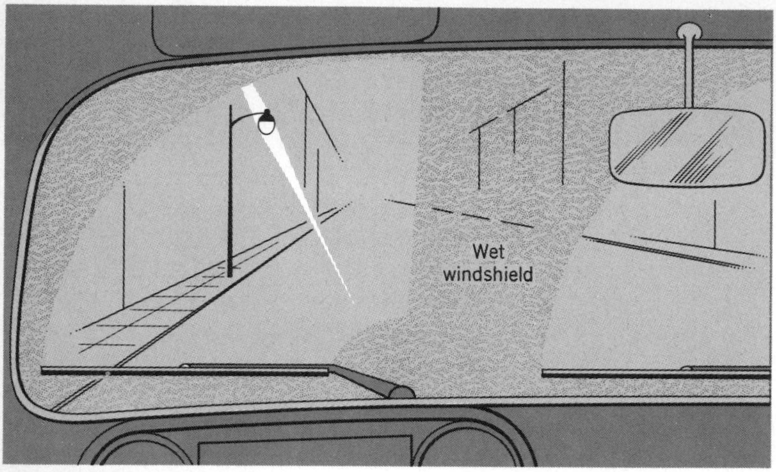

Figure 5.77
Streak of light in windshield from streetlight.

5.77

Windshield light streaks

When driving through rain at night you will find long streaks of light on your front windshield due to the lights outside your car (Figure 5.77). Each streak appears to run through the lights source, and the smaller sources (such as streetlights) give more pronounced streaks. As you move, the light streaks move too. If you step outside or look through any of the car's other windows, however, you won't see them. What causes these streaks? Are they as prevalent when it's not raining?

1148.

5.78

Color of a city haze

If you've lived in a large city, you almost certainly have spent part of your life in a haze. Why are such hazes brown? Is it due to some kind of selective absorption of the light? If so, by what? Or is it due to dispersive scattering of the light? Might it depend on what you're looking at through the haze?

1112; 1163; 1164.

"You win a little and you lose a little. Yesterday the air didn't look as good, but it smelled better."

5.79

Glory

If you stand on a mountain with your back to the sun and peer into a thick mist below you, there may be a series of colored rings around the shadow of your head. This set of colored rings, which may even be full circles, is called the glory (as well as the anticorona or brocken bow). You may momentarily feel divine when you notice that this beautiful and saintly display is around your head but not around a companion's. What causes this seemingly divine selection?

Glories are now most often seen from airplanes. Next time you fly, sit on the side away from the sun, and watch for the glory around the plane's shadow on the clouds or mist below. I have seen three full spectrums at once, and as many as five have been observed and photographed.

What causes the glory? Why does it surround the shadow of your head? What is the color sequence in each ring? How does the glory depend on the size of the particles in the mist?

164, p. 555; 165, pp. 180–184; 360, pp. 68–70; 361, pp. 4–5; 362, pp. 138–139; 380; 536, p. 131; 954, pp. 224–225; 983, Chapter 7; 1016; 1017; 1019; 1028, p. 130; 1149 through 1156; 1499; 1626.

5.80

Corona

Why are the sun and moon sometimes surrounded by bright bands, called coronas? Usually there is a single white band, but occasionally there will be blue, green, and red bands outside the white one. If you're lucky, you may even see two such spectrums. What causes the brightening, and why can you distinguish colors only occasionally? What determines the corona's width Can you predict the color arrangement?

164, pp. 547 ff; 165, pp. 178 ff; 360, pp. 78-79; 536, pp. 130-131; 954, pp. 214-219; 983, Chapter 5.

5.81

Frosty glass corona

Walking past a frosty store window on a cold winter night, you may find the interior lights of the store surrounded by colored rings. At first thought, these colored rings seem to be the same as in the solar and lunar coronas. In the store window, though, the image of the light is surrounded by a black band, not a white band as in the coronas discussed above. Why is there a difference? And again, what is responsible for the colored rings?

954, pp. 219-221; 983, p. 157 ff.

5.82

Bishop's Ring

A different type of corona (and a much larger one, being about 15° in angular radius) is the white and red-brown Bishop's Ring caused by volcanic dust spewed into the atmosphere. (After some volcanic eruptions the twilight sun turns a beautiful gold, the twilight sky colors take on a brilliant richness, and one can also see a second purple light* which lasts for hours after sunset.) What size particles are responsible for the red-brown color if that color is present? Will the Bishop's Ring be colored if there is a large range of particle sizes?

164, p. 555; 165, pp. 178, 191; 536, p. 130; 954, p. 282; 983, pp. 167, 243; 1104 through 1108; 1109, pp. 430-434, 441; 1110.

*See Prob. 5.60.

5.83

Streetlight corona

On your nighttime walk you may also be struck by the colored rings around the streetlights you pass. Is the same physics responsible for this corona as for the solar and lunar coronas and the store-light coronas? There is a simple test to show that there is at least some difference. If you screen off the streetlight, a store's interior lights, and the moon or sun, do the coronas in all three cases remain? If any one disappears, then you should explain why it is different from the others.

954, pp. 221-223; 1091, pp. 224-225; 1157; 1158.

5.84

Blue moons

My grandmother is from Aledo, Texas, where the population is about 100 people, dogs, and chickens. According to her, excitement comes to Aledo only once in a blue moon. But how often does a blue moon come? In fact, why would the moon ever be blue? Can there be blue suns too? Is either the moon or sun ever green?

536, p. 121; 954, pp. 298-299; 983, p. 242; 991; 1014, p. 421-423; 1101; 1159 through 1162.

5.85

Yellow fog lights

Why are car fog lights yellow? Does it really help to have them yellow? Does it matter whether you're driving in the city or in the countryside?

983, p. 244; 1111, p. 40.

5.86

Blue hazes

There is a colorful but mysterious haze that appears over vegetated areas relatively free from man-made contamination. The Blue Ridge Mountains of Tennessee and the Blue Mountains of Australia are both well known for their beautiful blue haze. What causes this type of haze? Smoke? No, because the haze is found in relatively uninhabited areas. Windswept dust? No, because the haze has the deepest blue during very light winds. Finally, the haze cannot be fog, because the blue is most common during warm summers. What, then, causes the haze, and why is it blue?

1112; 1165; 1166.

5.87

Shadows in muddy water

Why can you see your shadow in slightly muddy water but not in clear water? Why can you see shadows of other people only if the water is very muddy?

You might also notice the colors around shadows in slightly muddy water. The edges closest to you are colored differently from those farthest from you. What causes this coloring? Does the color of the edges depend on whether you are facing toward or away from the sun?

954, pp. 332–333; 1565.

5.88

Color of milk in water

After adding a few drops of milk to a glass of water, look through the glass at a white light such as a light bulb. The source will appear to be red or pale orange. Next look at the light reflected from the glass. The light will be blue. Why is there such a remarkable change of color?

360, pp. 60–61.

5.89

Color of cigarette smoke

If you closely examine the smoke rising directly from a cigarette, you'll find that the smoke is slightly blue. If the smoke is inhaled and then blown out, however, the smoke is white. Why is there a change? (It is not due to removal of tar and nicotine.)

155, p. 411; 360, p. 62; 533, p. 147; 536, p. 383; 954, p. 236–237; 983, p. 235.

5.90

Color of campfire smoke

A similar change of color is apparent in campfire smoke. When it is viewed against a dark background (trees, for example), the smoke appears to be blue. Higher up, however, when it is seen against a light sky, the same smoke appears to be yellow. Why does it change its color?

533, p. 147; 954, pp. 235–237, 309.

5.91

Oil slick and soap film colors

Why do oil slicks on the street display colors? How thick are these slicks? Must the street be wet? Can you see them on overcast days or only in direct sunlight? If you can calculate the width of one of the colored rings, compare your number with a measured width. Will the finite size of the sun change the theoretical width of the rings in any way?

Why do you see colors in soap films? How thin are the soap films, and in what thickness range will they show colors? Why that range? Why are some parts of some films black? Finally, why is there such a sharp boundary between the colored and black areas? Shouldn't there be a gradual change?

322; 528 through 531; 533, pp. 139 ff.

5.92

Color effects after swimming

Why do you see colored rings around lights after you've been swimming?

refraction

dispersion

crystal structure

stress

5.93
Liquid crystals

If a deformable container containing a liquid crystal is squeezed, colors appear around the squeezed area. The particular colors you see, however, will depend on your angle of view. How do the angle dependence and color sequence compare with those of an oil slick? If there is a difference, can you explain it?

1081; 1132 through 1137.

5.94
Butterfly colors

Why are the wings of butterflies colored? Are the colors due to pigmentation? In some wings, yes, but in others, such as for the *Morpho* butterfly, the colors do not arise from any pigmentation. A possible clue to their origin may be found by looking at a wing from several different angles: the wing takes on slightly different colors for different viewing angles. Why?

1138 through 1142; 1625.

diffraction

5.95
Dark lines in a fork

You have probably seen the dark line which lies between your finger and thumb when they're almost touching (Figure 5.95). You can see many such dark lines by looking through a fork's prongs as you rotate the fork. What's responsible for these dark lines? Can you predict whether the spacing between the lines will decrease or increase for a given turn of the fork?

170, p. 487.

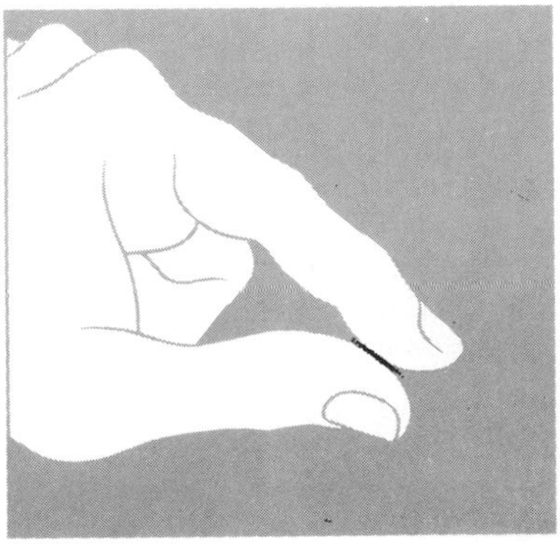

Figure 5.95
Dark line seen between two fingers.

5.96
Eye floaters

What are the tiny, diffuse spots you often find floating in your field of view? Are they illusions? Are they bits of dust on the eye's surface? Or are they objects within the eye? By looking at a bright light source through a pinhole in some opaque material, you'll find a beautiful array of floating concentric circles and long chains (Figure 5.96). If the spots are merely shadows, then why do you see concentric circles and chains? Also, why does a pinhole help you see the structure of the spots?

170, p. 530; 1091, Vol. 1, pp. 204 ff; 1167; 1168;

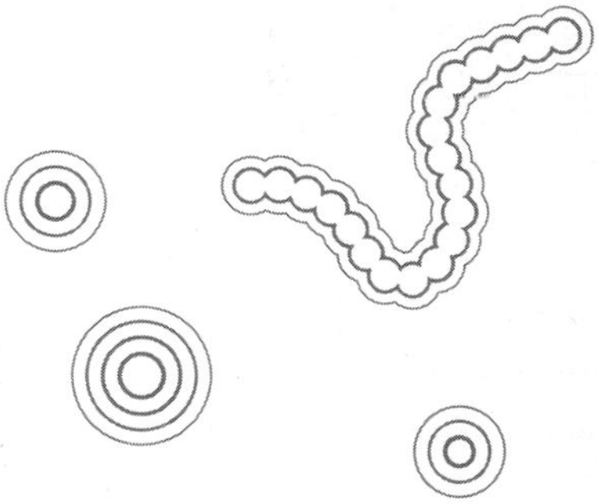

Figure 5.96
Structure of the floaters in your eyes.

5.97
Points on a star

What causes the occasional spiked appearance of car headlights? The cause cannot be entirely physiological since photographs of the headlights also show spikes. Similarly, what causes the spikes found in star photographs? Is it possible to find any number of spikes on a star or a headlight photograph? In particular, can you find a star with an odd number of points?

1169, p. 3.

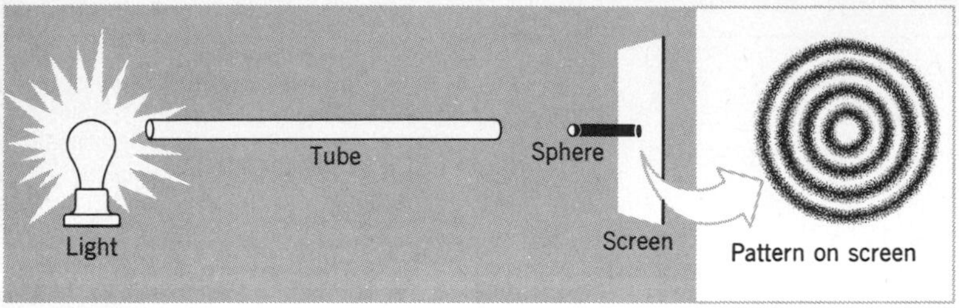

Figure 5.98
Demonstration to show Poisson spot in the shadow of a small sphere.

5.98
Poisson spot

Why is there a bright central spot in the shadow of a small disc or sphere (say, two millimeters in diameter), whereas larger objects give ordinary dark shadows? By using a cardboard tube and a screen as shown in Figure 5.98, you'll find that not only is there a bright central spot in the shadow of the disc or sphere, but the shadow is actually composed of multiple dark and bright rings. What causes the central spot, which is called the Poisson spot,* and the rings? Why aren't they found in your own shadow?

204; 1169, p. 200; 1170, pp. 359–360.

*When Fresnel defended his dissertation before his committee in the 1800s, one of the committee members, Poisson, remarked that if the dissertation were correct, there would be a bright spot in a spherical object's shadow. This result clearly being ridiculous, he concluded that the dissertation must be wrong. But as a matter of fact, the spot had been seen some 50 years earlier and, soon after Poisson's conclusion, Arago rediscovered the central bright spot. In spite of all this, in one of those curious twists in the history of physics, it is the objector's name that is associated with the spot.

refraction

interference

turbulence

5.99
Eclipse shadow bands

For several minutes before and several minutes after a total solar eclipse, dark bands called shadow bands race across the ground. The bands are separated by several centimeters and are about two centimeters wide. What could cause these bands? And why do they appear during an eclipse? Are they produced in our atmosphere, or are they made when the sunlight passes the moon?

1171 through 1181; 1561.

5.100
Sunset shadow bands

Another set of shadow bands has been seen during normal sunsets. Ronald Ives (1182) has reported six observations in 15 years, all of which were from high points looking down on flatlands. These bands were several miles wide and moving at about 40 miles per hour. Are these bands another example of shadow bands? In any case, what causes them?

1182.

5.101
Bands around a lake's reflection

As you fly toward a distant small lake, eventually reaching the angle for optimum reflection of the sun, why are there alternating dark and bright bands around the principal reflection from the lake?

360, p. 12.

refraction

scintillation

turbulent cells

5.102
Star twinkle

My mother taught me to say, "Twinkle, twinkle, little star. . ." Why does a star twinkle? Approximately where is the twinkling produced? Does a star change colors or move around because of the twinkling? Does it twinkle more in the winter or summer? Does a red star twinkle more than a white star? Do you see twinkling when you use a telescope? Do the moon and planets twinkle?

What causes the shimmer of an object when you view it over heated surfaces such as, for example, hot car tops or roadways? How high above the heated surface will your viewing be affected? Is it the air closest to you or farthest from you that dominates the shimmer?

164, pp. 462–466; 165, pp. 166-169; 954, pp. 63–71; 983, pp. 17–19; 1111, pp. 80–81; 1183 through 1188.

photochemistry

5.103
Bleaching by light

How does sunlight fade colored clothing? Does the rate of fading depend on the color? Why does sunlight or fluorescent light cause oil paintings to fade? Why are some foods and beverages, such as beer, shielded from sunlight? Is any particular light frequency most destructive?

466, pp. 214–215.

radiation forces

refraction

5.104
Optical levitation

Earlier in this book we discussed levitation of balls by air currents and water jets* and in both cases there was surprising stability. Light can also levitate and stabilize balls, for light from a relatively powerful laser has lifted and held in suspension transparent glass spheres of about 20 microns in diameter (Figure 5.104). How can light lift such a sphere against gravity? And how is stability against sideward motion provided?

1189 through 1191.

*Probs. 4.20 and 4.22.

Figure 5.104
Glass sphere suspended in an upward directed, expanding laser beam. [After A. Ashkin and J. M. Dziedzic, Appl. Phys. Let., 19 (8), 283 (1971).]

diffraction

5.105
Lights through a screen

Car headlights viewed through a screen look very different than when they are viewed without a screen (Figure 5.105). What, in detail, causes the difference?

533, p. 163.

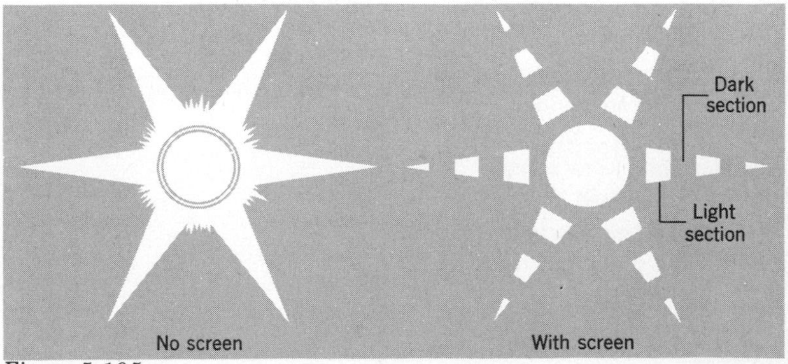

Figure 5.105
The change in the appearance of a light when viewed through a window screen.

blackbody emission

atmospheric transmission

vision

5.106

Star color

Some stars look red. Some look white. Are there blue stars? Or green stars?

5.107

Luminous tornado

There have been many reports, including published accounts, describing mysterious lights associated with tornadoes. Though they are generally dismissed as illusions, there has been at least one published photograph (1192, 1193) that apparently shows luminous columns in two nocturnal tornadoes. Eyewitness accounts of these particular tornadoes gave exciting descriptions of the light emission.

> **The beautiful electric blue light that was around the tornado was something to see, and balls of orange and lightning came from the cone point of the tornado (1193).***

In another tornado occurence an observer saw the following:

> **I was looking. . . up at the clouds when I saw something that looked like a searchlight beam extend out of the cloud and reach to the skyline. It seemed a bit brighter than the cloud background. Edges were very sharp, overall intensity even, sides parallel. Width about a degree of arc. No movement or turbulence evident. The phenomenon was interesting enough so I took out my Polaroid glasses and observed this "ray" through them, twisting the lens to look for polarization. No polarization was noted. This ray was obvious enough so that passersby on the street were staring at it. All this took, say, 60 to 120 seconds (or more). Then abruptly the ray was instantly replaced by a normal tornado funnel. No transition stage was noted. The funnel *did not* descend from the cloud layer. It appeared over all, in situ (1193).***

Although these phenomena are poorly understood, can you suggest causes for them, perhaps making some rough numbers to support your suggestions?*

224; 225; 1192; 1193.

*See Prob. 6.35 also.

triboluminescence

5.108

Sugar glow

Late one night I was stirring some dry granulated sugar in a glass, which is kind of a late-at-night type of thing to do. Suddenly the lights went out. As I continued to stir, I saw brief flashes of light through the side of the glass. How did the mechanical stress and strain of my stirring cause the light emission?

1194, pp. 121, 292, 378–387; 1195.

5.109

Suntans and sunburns

What actually causes suntans and sunburns? Is the same wavelength range of light responsible for both? Why is it more difficult to get sunburned once you have a tan? Can naturally dark skin become sunburned as easily as lighter skin? What do suntain oils, lotions, and creams do to prevent sunburn and promote suntan? The pertinent point is, of course, whether they really do what the advertising claims. If they inhibit whatever causes sunburn, won't they inhibit suntan also?

Why are burning and tanning less likely when the sun is low or when you're behind glass? Why are they more likely at the beach than in a grassy backyard?

344, pp. 19–22; 466, p. 212; 1203; 1512.

photochemistry

5.110
Fireflies

Catching fireflies at my grandmother's house was one of the most enjoyable times of my childhood (Figure 5.110). I have read that the *synchronous* flashing of Asiatic fireflies is even more fascinating.

Imagine a tree thirty-five to forty feet high thickly covered with small ovate leaves, apparently with a firefly on every leaf and all the fireflies flashing in perfect unison at the rate of about three times in two seconds, the tree being in complete darkness between the flashes. . . . Imagine a tenth of a mile of river front with an unbroken line of. . .trees with fireflies on every leaf flashing in unison, the insects on the trees at the ends of the line acting in perfect unison with those between. Then, if one's imagination is sufficiently vivid, he may form some conception of this amazing spectacle (1196).

What mechanism produces the light we see? That light is often referred to as cold light, implying there is no energy lost to heating.

By permission of John Hart. Field Enterprises.

(An incandescent bulb, on the other hand, is a hot light.) Is the firefly 100 percent efficient in converting energy to the form of light? What color is the light? Why that color? Finally, how do the Asiatic fireflies lock themselves into a chorus of synchronous flashing?

1090, Chapter 4; 1194, pp. 538-554; 1196 through 1201; 1458; 1585; 1624.

photochemistry

5.111
Other luminescent organisms

Many other organisms produce their own light, too. The Brazilian railroad worm, for example, has a red light on its head and green lights down its side. Another type of luminescent organism, the dinoflagellates, will "set the sea on fire" when disturbed during the day (by a boat, say) but during the night they will respond with a blue glow. One type of crustacean, when dried, can be made to glow by moistening. Such a light source was used by World War II Japanese soldiers when a stronger light was too dangerous. Spitting on a bit of dried crustacean would give off enough light to read a map.

There have been many other, but less common, examples of

natural luminescence. In one case, cut potatoes glowed sufficiently that one could read by them in an otherwise dark room. There has even been a case in which a corpse glowed in the dark. But the most disturbing, especially if one is relying on darkness to conceal a slight indiscretion, have been the times in which urine glowed in the dark.

In the case of the dinoflagellates, why do they glow red during the day but blue during the night? In all these various examples, what causes the luminescence?

1194, pp. 457-492; 1200 through 1202; 1458.

photochemistry

transmission

5.112
Photosensitive sunglasses

Some sunglasses are clear indoors but darken immediately upon exposure to sunlight. The change is reversed soon after the sunlight is eliminated. What causes this reversible change in the transmission properties of the glass?

984; 1204.

fluorescence

5.113
Black-light posters

How does a black-light poster work? Why does the same physics allow some soap manufacturers to claim that their products get clothes "whiter than white"?

1205, p. 70.

fluorescence

phosphorescence

5.114
Fluorescent light conversion

How is ultraviolet light created and then converted to visible light in a fluorescent lamp? How fast should the conversion be? You don't want it so fast that the lamp's output shows the 60 cycles per second of the line voltage used to excite the tube. But then again, you don't want the lights to stay on long after you've turned off the switch.

466, pp. 233-240; 1205, p. 76.

Vision
(5.115 through 5.141)

coherence

interference

5.115
Speckle patterns

If you look at a smooth, flat-black piece of paper at a 45° angle in direct sunlight, you will see a grainy speckle pattern of various colors dancing on the paper. Similar patterns are more commonly made with laser light, but sunlight is certainly more convenient. In either case, the pattern will move if you move your head, but whether it moves in the same or opposite direction as your head will depend on whether you have normal, nearsighted, or farsighted vision. What causes these speckle patterns, and why are there colors in the sunlit patterns? Finally, can you explain the movement of the pattern and its dependence on your vision?

1206 through 1209; 1560.

stroboscopic effect

5.116
Humming and vision

If you hum while watching television from a distance, horizontal lines will appear on the screen, and you can make them migrate up or down or remain stationary by humming at the appropriate pitch. In a similar demonstration, a black-and-white-sectored disc is rotated on a turntable. If you use a stroboscope to illuminate the disc, you can freeze the rotating sectors or make them slowly migrate one direction or the other by choosing a suitable flashing frequency. However, you can also do this by merely humming at the proper pitch. Why does humming effect your vision in this way?

1210; 1211.

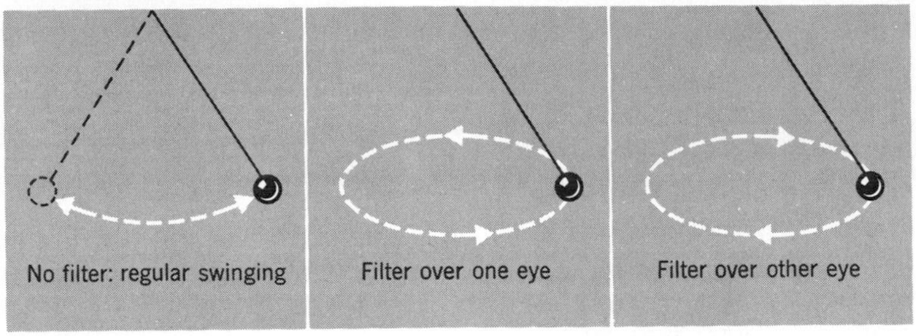

Figure 5.117
A normal pendulum swing changes to a circular motion if a polarioid filter is placed over one eye.

5.117

Sunglasses and motion distortion

With a dark filter over one eye (say, half a pair of sunglasses), watch the swing of a simple pendulum. Even though you know the pendulum's motion is in one plane, the pendulum appears to revolve in an ellipse when the filter is in place (Figure 5.117). To the uninitiated, the surprising result can be quite striking. . . and mysterious. The apparent three-dimensional motion can be enhanced by hanging a string from the pendulum's pivot, for then the string acts as a reference object and the pendulum appears to turn about it.

If you should drive while wearing only half a pair of sunglasses, a car passing on your left will seem to have a considerably different speed than one passing on your right even if they actually have the same speed. In neither case is the apparent speed the correct one. In addition, the apparent distances of objects in the landscape will be wrong and even dependent on which side of the car the objects are.

What causes the apparent three dimensional motion of the pendulum? What exactly does the filter have to do with this motion and the distortion of a car's speed and the distance of objects in the landscape?

1212 through 1222; 1541 through 1543.

5.118

Top patterns before TV screen

If a flat top with a surface design is spun before a TV screen (with a stable picture and in an otherwise dark room), psychedelic patterns appear on the top's surface. Undoubtedly the pattern stems from the top's surface design, but why is the light of a TV needed?

170, p. 36.

5.119

A stargazer's eye jump

Why do you have a better chance of seeing a dim star neighboring a bright star if you jump your eyes to one side of the stars?

332, Vol. 1, p. 35–3; 412, p. 439.

5.120

Retinal blue arcs

Blue arcs of the retina are another physiological problem currently receiving attention. Purkinje first reported seeing them from glowing tinder as he was kindling a fire. For about 30 seconds he saw two blue arcs extending from the tinder. You can see them yourself* under controlled circumstances

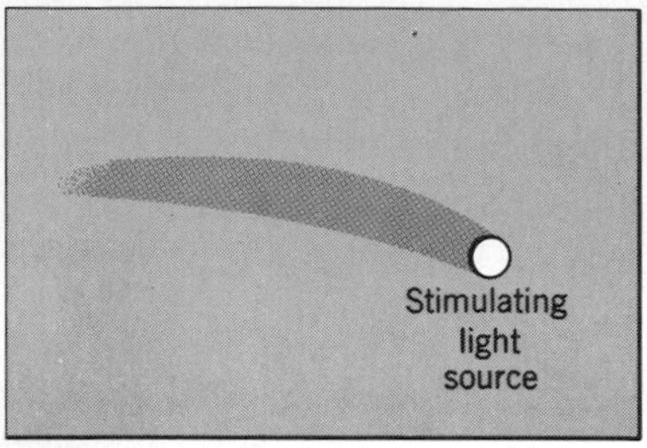

Figure 5.120
Blue arc in your left eye's field of view. [After J. D. Moreland, Vision Research, 8, 99 (1968).]

by using small holes punched into a card that is then placed over a light. After sitting in the dark about a minute (don't wait too long), switch on the light. Depending on the hole's shape, various shaped blue arcs (e.g., Figure 5.120) can be seen for up to a second.

What causes these arcs—scattered light inside the eye? Why, then, are they always blue? Shouldn't they depend on the color of the scattered light? Perhaps they are due to bioluminescence. Or maybe they could be due to a secondary electrical stimulation of nerve fibers or neurons by other active nerve fibers. If the latter is true, the shape of the arcs as a function of stimulus shape should tell us something about the retinal topography. In any case, we still have to explain why the arcs are blue.

1224 through 1227.

*One of Moreland's several papers (1224) describes in further detail how to optimize the observation and how to demonstrate it to a small audience.

5.121

Phosphenes

Prisoners confined to dark cells see brilliant light displays (the "prisoner's cinema") in their perfect darkness. Truck drivers also see such displays after staring at snow-covered roads for long periods. In fact, whenever there is a lack of external stimuli, these displays—called "phosphenes"—appear. They can be made at will, however, by simply pressing your fingertips against closed eyelids, and some hallucinogenic drugs apparently give magnificent phosphene shows. They can also be produced by an electrical shock. In fact, it was high fashion in the eighteenth century to have a phosphene party (even Benjamin Franklin once took part) in which

visual latency

light intensity

5.122

Streetlamp sequence

Sometime when you're driving at dusk, watch the streetlights turn on: they brighten in sequence down the street. Does it really take electricity that much time to travel from lamppost to lamppost? If there are intersections with several streetlamps, you'll find those lamps turning on sooner than the lamps between intersections (Figure 5.122). This certainly can't be due to a lag of electricity. Why, then, are there time lags between the lamps?

1212 through 1222.

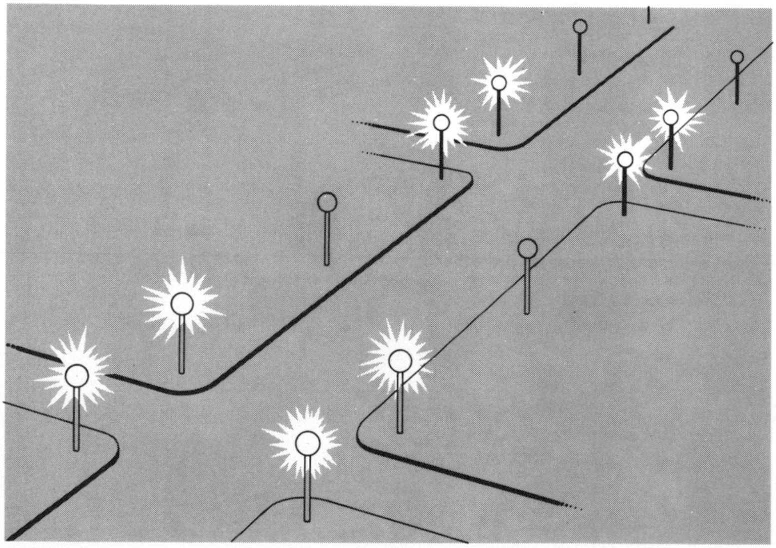

Figure 5.122

a circle of people holding hands would be shocked by a high-voltage electrostatic generator, phosphenes being created each time the circuit was completed or broken.

*In 1819 the Bohemian physiologist Johannes Purkinje published the most detailed account of phosphenes. He applied one electrode to his forehead and the other to his mouth, and by rapidly making and breaking the current with a string of metal beads, he was able to induce stabilized phosphene images (1223).**

Phosphene research is no longer so academic, because recent work has shown that those blind people who experience phosphene displays may someday be given artificial vision by use of those diplays. A miniature TV camera, placed inside an artificial eye, would send its electrical signals to a small computer located inside a pair of eyeglasses. The computer would in turn stimulate the brain by a network of electrodes that had been placed adjacent an occipital lobe. When the TV camera detected an object in its left field of view, for example, the computer would stimulate the electrode that would produce a phosphene image in the left portion of the person's field of view. The person would therefore see the external world.

Why are such visual displays produced under electrical and pressure stimulations or when there is an absence of external stimuli?

1223; 1572; 1573.

*From "Phosphenes" by Gerald Oster. Published in *Scientific American*.

5.123

Spots before your eyes

If you stare at a clear sky you will find your entire field of view covered with moving specks. Those specks are always present, but usually you don't notice them. (Why is that?)

Although the jerking motion appears to be random, if you feel your pulse while watching the specks, you will find the motion correlated with your pulse and also see that the specks always follow certain routes in your field of view. What are the specks, and what causes the jerking along those particular routes?

1091, Vol. 1, pp. 222–223; 1168; 1233, pp. 407–408.

5.124

Purkinje's shadow figures

Close your eyes, place a hand over one, turn to face a bright light, and wave your other hand back and forth across your face so that the shadows of your fingers repeatedly cross over your closed, but exposed, eyelid. In the center of your field of vision you'll see a checkerboard array of dark and bright squares, and down from the center there will be either hexagons or just irregular figures. If you're using the sun as the light source, you'll also see eight-pointed stars and various spiral lines. What causes these several designs?

1091, Vol. 2, pp. 256–257; 1234.

5.125

Early morning shadows in your eyes

If you stare at a sunlit room immediately upon opening your eyes after a night's sleep, why will you briefly see dark images in your field of view? If the images are shadows of objects in your eye, then why don't you see the shadows all the time, and why do they fade so quickly after this early morning glimpse?

1091, Vol. 1, pp. 212 ff; 1168; 1233, pp. 406–407; 1235.

color perception

5.126

Purkinje color effect

In dim light a particular blue may be brighter than a particular red, but in good illumination the relative brightness may be reversed. Why should the relative brightness of reds and blues depend on the illumination level?

332, Vol. 1, p. 35–2.

5.127

Mach bands

How sharp is your shadow's edge when you stand in a strong light such as sunlight? If you look carefully, you will see two shadows, the darker one neatly inside the other. The inside contour of the lighter shadow has a dark band; the outside contour has a bright band. There is nothing unique about your body, because every shadow has such edge patterns (though more than one light source will complicate the patterns, of course). Figure 5.127 shows how the edge pattern can be seen with a piece of cardboard held in front of a fluorescent lamp. Why are the bright and dark bands and your half-shadow present? Can they be photographed?*

954, pp. 129–132; 1228, Chapter 2; 1229 through 1232.

*In early attempts to measure the X-ray wavelength, some physicists used what they thought were X-ray diffraction patterns resulting from the passage of X rays through common diffraction slits. They did find light and dark patterns on their films, and using those patterns they calculated the wavelength. Unfortunately, later work revealed that these patterns you see in your own shadow and were not at all indicative of X-ray diffraction (1228).

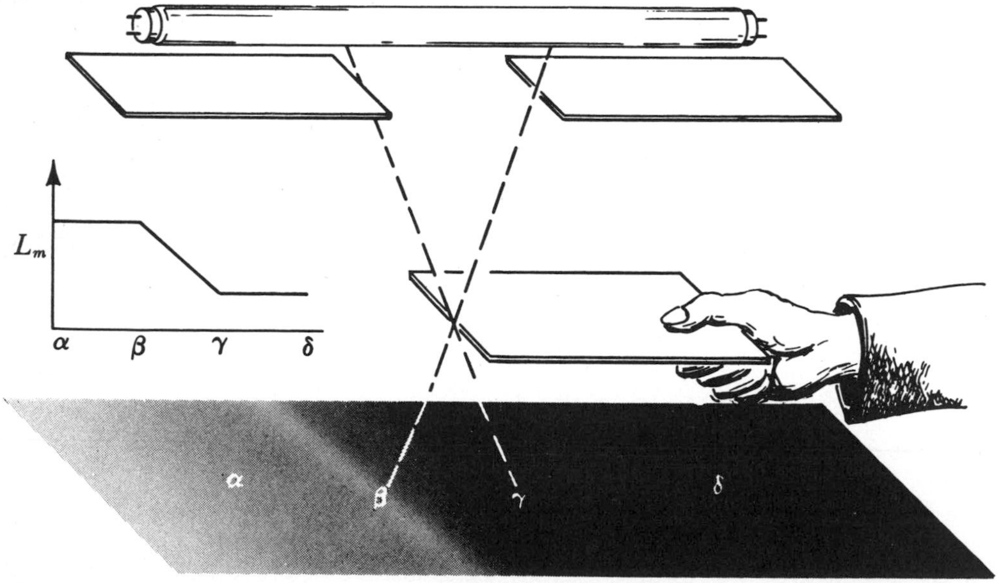

Figure 5.127
Mach bands can be seen in the card's shadow with this arrangement. If the lamp is one foot above a white sheet of paper, then place the card one or two inches above the paper. Small horizontal motions of the card may help you see the bands better. The graph shows the luminosity for various points on the paper. (Figures from Mach Bands: Quantitative Studies on Neural Networks in the Retina by Floyd Ratliff, published by Holden–Day, Inc.)

Land color effect

color perception

5.128

Seeing the colors of your mind

If an object looks blue, blue light must come from the object, right? In fact, each color you see corresponds to light with a certain frequency or a combination of several frequencies. This seems very reasonable, but Edwin Land threw a wrench into the explanation with a few simple experiments that you can do yourself.

What do you have after making two black-and-white slides of a colored scene, using a red filter for one slide and a green filter for the other? Why, two black-and-white pictures of course. How can you get anything else with black-and-white film?

But now, using two projectors, simultaneously project those slides onto a screen (Figure 5.128). Use a red filter with the slide made with the red filter; the normal white projector light is sufficient for projecting the "green" slide. What do you see on the screen? Although each slide is only black and white, and the only colored light you use is red, the superimposed projection gives the full range of color in the original scene.

There is nothing special about the filters used. All you need are two different colors or even one color and one white light. Both slides can even be made in a single color as long as the light frequency used for one slide is at least slightly different from the light frequency used for the other.

What causes this recreation of the color of the original scene even though the color information is seemingly lost in the individual slides? Once again, if an object looks blue, must blue light necessarily be coming from that object?

1236 through 1239; 1566; 1567.

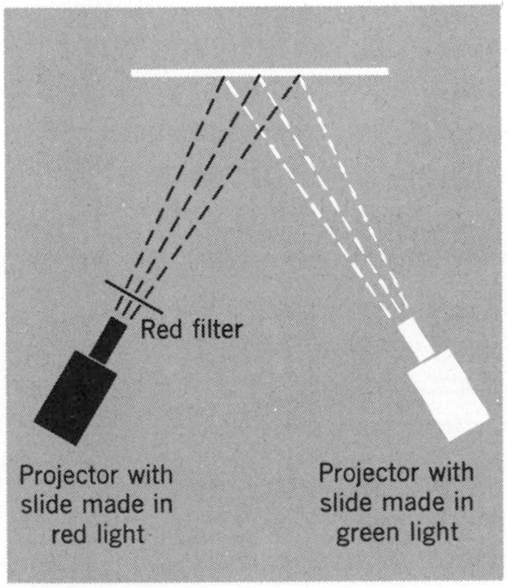

Figure 5.128
Projection arrangement to show Land color effect.

chromatic aberration

5.129

Making colors with a finger

Watching with only one eye, move a finger across your view of a sunlit window that is across the room from you. When the finger first begins to block and distort the image of the window, the side of the image nearest the finger turns yellow-red (Figure 5.129). As your finger reaches the opposite side of the image, that opposite side turns blue. (You can see the same thing using an incandescent bulb, but the blue is fainter.) Why do the colors appear, and why are the opposite sides of the window's image colored differently?

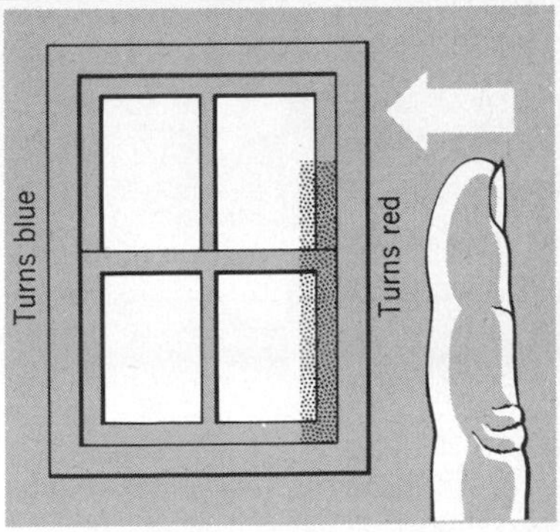

Figure 5.129
[After S. F. Jacobs and A. B. Stewart, Amer, J. Phys., 20, 247 (1952).]

533, pp. 104–105; 1091, Vol. 1, pp. 175–176; 1516.

color perception

5.130

Colors in a black and white disc

Is it possible to see colors in black and white surfaces? Normally, it isn't, but try the following: construct a disc of alternating black and white sectors, and then as the disc is spun at low speed, concentrate on it (but ignore the individual sectors). After a few minutes you'll find that the leading edges of the white sectors will turn red, the trailing edges blue. (Different shades will be seen for different illumination levels.) At a faster speed the whole white sector will be pink-red, and a green-blue will cover part of the black section. With a still-faster speed, the colors cannot be distinguished but little sparks of violet-pink and green-

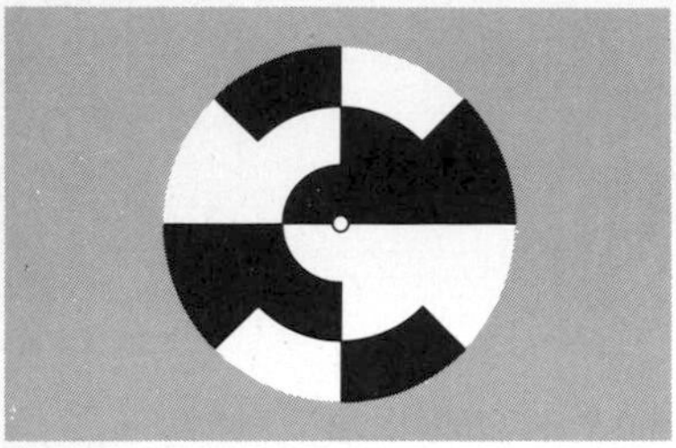

Figure 5.130
Disc that shows colors when spun.

gray light seem to jump about. The disc in Figure 5.130 will give all three effects simultaneously. Why do you see those colors? Why must you watch the disc for several minutes before the colors appear?

332, Vol. 1, p. 36-1; 1091, Vol. 2, pp. 255 ff; 1231; 1240; 1241.

stroboscope

fluorescence

phosphorescence

5.131

Color effect from fluorescent lights

If the disc described above is rotated faster (about 5 to 15 rps), the color effects disappear. But if it is put under a fluorescent light, a new color effect will appear: you will see two concentric rings that are composed of alternating red, blue, and yellow bands. You can also see colored fringes—yellow or orange, depending on the background—if you watch a spinning coin under a fluorescent lamp. Why does the fluorescent lighting cause these color effects? Can they be photographed?

1242 through 1246.

5.132

Floating TV pictures

While watching TV in an otherwise dark room, quickly run your eyes from about a foot to the left of the screen to about a foot to the right. You will see a bright, detailed, ghostlike image of the TV picture floating in space to the right of the screen (Figure 5.132). You may even see three or four images, all right-tilted parallelograms. Why are these ghost images formed, and what's responsible for the tilt? Do you see the same sense of tilt if you move your eyes in the opposite direction? Are there ghost images for a rapid vertical scan of your eyes?

1247.

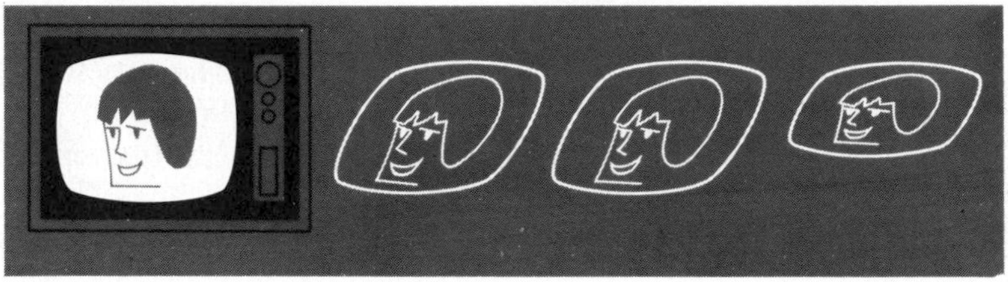

Figure 5.132
Ghost images of TV picture.

5.133

3-D movies, cards, and posters

There are two methods of making commercial three-dimensional movies and comic books. One method involves printing pictures in two colors, red and green, and then using cheap glasses with red cellophane over one eye and green over the other. The other method employs polarizers in the glasses and in front of the two projection cameras, and the cameras project simultaneously onto the screen. How do these methods give a stereoscopic illusion? As you probably know, three-dimensional movies have not gained widespread popularity, which means that there must be some drawbacks. Other than the annoyance of wearing the glasses, what are the problems?

How is the 3-D effect gained in 3-D baseball cards and postcards? Some bright red and blue posters, paperbacks, etc. give an impression of depth if red letters are printed on a blue background: the letters appear to be closer to the viewer than the background. Why? Do such depth illusions with different colors depend on the illumination level? What other methods can produce a stereoscopic illusion?

533, pp. 105-106; 1070, pp. 107-110; 1092; 1213; 1255 through 1260; 1591 through 1607.

5.134

Enlarging the moon

Probably the most striking illusion in the natural landscape is the apparent enlargement of the moon when it is near the horizon (Figure 5.134). Is this illusion brought about by atmospheric conditions, or is it a psychological effect? Can you estimate the apparent enlargement?

165, pp. 154-155; 533, pp. 62-63; 954, pp. 155-166; 1248 through 1253.

By permission of John Hart. Field Enterprises.

Figure 5.135

5.135

Rays of Buddha

Occasionally you will see a sunset in which brilliant rays of light emerge from the setting sun, fanning out across the western sky (Figure 5.135). This display is caused by mountains or clouds blocking part of the sunlight. What color are the rays? What color is the sky against which you see the rays? Not as frequently you will see rays of light converging to the antisolar point in the east. Very rarely you may see those rays emerging from the solar point in the west, arcing across the entire sky and converging to the antisolar point in the east. But wait. How could a cloud or mountain block part of the sunlight to give a fan display? After all, the sun is very far away from us, and the sun's rays should all be parallel.

164, pp. 452, 567; 165, p. 185; 954, pp. 275-277; 1513.

5.136

Moon-to-sun line

Sometime when you find a crescent moon in the daytime sky, mentally draw a line along its symmetry axis (Figure 5.136). Does the line point to the sun? Shouldn't it?

165, pp. 149 ff; 954, pp. 151-166; 1250 through 1254.

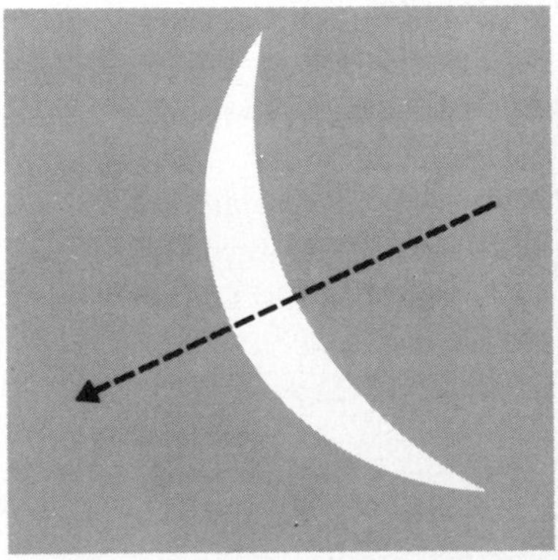

Figure 5.136
Shouldn't the line through the crescent moon point to the sun?

5.137

Bent search beams

When seen from the side, searchlight beams appear to bend over. Does the beam really get scattered or refracted downward by the atmosphere?

165, pp. 149 ff; 954, pp. 151-166; 1250 through 1254.

5.138

Rear lights and a red light

If, while driving at night, you should be about a block behind a car approaching a red light, the rear lights of that car may appear to stop somewhere beyond the intersection. When you reach the red light yourself, however, you find the other car waiting as it should be in front of the red light. What causes this particular illusion?

1261.

light flux

perception

5.139

Snowblindness

What causes snowblindness (whiteout)? After long exposure to the white light of snow and ice fields your eyes feel as though they were full of sand. Intense pain may follow for days. Is snowblindness more likely to occur on a sunny or a cloudy day? In his diary and stories of five years of polar expeditions, Vilhjalmur Stefansson recalls:

> *. . .it might be inferred that snowblindness is most likely to occur on days of clear sky and bright sun. This is not the case. The days most dangerous are those when the clouds are thick enough to hide the sun but not heavy enough to produce what we call heavily overcast or gloomy weather. . .everything looks level. . .You may collide against a snow-covered ice cake as high as your waistline and, far more easily, you may trip over snowdrifts a foot or so in height . . .(1113).**

In such conditions you can't even distinguish the horizon. What role do the clouds play in increasing the probability of snowblindness?

1113, pp. 149, 199-202; 1122; 1262.

*Vilhjalmur Stefansson, *The Friendly Arctic*, copyright © 1921 by the Macmillan Company. Permission granted by McIntosh and Otis, Inc.

5.140

Resolution of earth objects by astronauts

What are the smallest objects orbiting astronauts can distinguish on the earth's surface? In particular, can they see large cities in the day or night or other large objects such as the pyramids? The early Mars fly-bys were disappointing to many people, especially non-scientists, because their pictures showed no signs of intelligent life. What signs of intelligence could you see on the earth if your photos had a resolution of, say, one kilometer, which is a typical value for weather satellite photos? If that resolution is not sufficient, how much *is* required to see signs of life?

360, pp. 182-184; 1263 through 1265; 1498.

reflection

ray optics

resolution

5.141

A Christmas ball's reflection

A shiny Christmas tree ball can give you a picture of nearly the entire room in its reflection. How will it reflect a point source of light in an otherwise dark room? Hold a ball about 10 centimeters from one eye and catch the reflection of the point source. (A pinhole punched in foil that covers a lamp provides a good point source.) The reflected image is an extended line of light,

not a point. But, immediately after switching on the room lights, the line of light quickly contracts to an undistorted image of the point source. First, why is there a distortion of the point image in the dark room? Second, why does the distortion depend on the illumination of the room?

1266.

5.142

Moiré patterns

If two similar patterns with slightly different periodicities are superimposed, a larger pattern, called a Moiré pattern, appears. You can easily see this by folding over a sheer curtain or by looking through a comb held at arm's length in front of a mirror. In the comb example, the comb and its image merge to form a larger comb-tooth pattern. For a more quantitive observation, place one metal sheet covered with circular holes several inches behind another such sheet to get a resultant circular Moiré pattern when the screens are viewed together from a distance. How does the observed Moiré pattern change with your distance from the screens? How does it vary with changes in the separation distance of the screens? Which way and how fast does the Moire pattern move as you walk parallel to the screens? Finally, does the motion of the pattern depend on your distance from the screens?

954, pp. 85–87; 1267 through 1272.

6

The electrician's evil and the ring's magic

Bioelectricity

(6.1 through 6.5)

joule heating

fibrillation

power

6.1

Electrocution

What exactly happens to you if you touch a live wire? What is it that can hurt or kill you? The voltage? The current? Both? Are you burned? Is your heart's rhythm disturbed? How does the danger depend on the frequency of the current? In particular, why is Europe's 50 cycles per second supposedly safer than America's 60 cycles per second? Is direct current more dangerous than alternating current, or does it just depend on circumstances?

You may not be killed outright, but if you continue to hold on to the electrical component, you may eventually die: the longer you wait, the lower your body's resistance becomes, and thus, you get closer to a lethal dose of current. Why does your body's resistance change with time?

1273; 1274.

Exceptionally good references: Singer (1349), Uman (786), Schonland (301), Malan (300), and Corliss (1611).

6.2

Frog legs

In a classic experiment dealing with the nature of nerves and muscles, Galvani (1780s) employed a deceptively simple arrangement. A frog leg was hung from a bronze support that was bolted into an iron railing (Figure 6.2). The hanging leg could also touch part of the railing, but everytime it did, it contracted and was thrown into spasms. When the spasms died out, the leg would droop and touch the railing, and the contraction and spasms would begin again. What caused this reaction? Can you produce some numbers to support your answer?

1275.

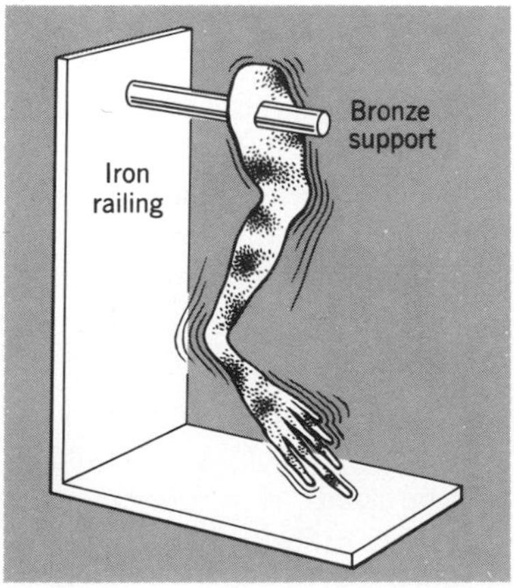

Figure 6.2
Frog leg sent into spasms when it touches the iron railing.

6.3

Getting stuck to electric wire

If you should grab a "live" wire that passes about 25 milliamps through your hand, you probably won't be able to release the wire? Why not? [Do *not* grab such a wire on purpose, for it may lead to your death (see Prob. 6.1)].

1273.

6.4

Electric eel

How can an electric eel shock you? A healthy eel can produce something like one amp at 600 volts. What could possibly be the source of such enormous power? Does the eel continuously discharge in the sea water? Why doesn't it shock itself?

The navigational ability of aquatic animals has long been unexplained. Recent work, however, suggests that some of the animals may be able to detect electric fields created by ocean currents moving through the earth's magnetic field. These fields would supposedly help the animals orient themselves. First of all, can you show how the electric fields would be produced by the moving water? Next, can you explain how an animal could possibly detect such a small field?

1276 through 1282.

Figure 6.5
"When I was a little girl we didn't have microwave ovens, and sometimes it took a whole hour to prepare a meal."

absorption

electric field

electric current

thermoluminescence

6.5
Microwave cooking

An ordinary gas oven will cook a roast from the outside inward, but a microwave oven will cook the interior first. Hence, your microwave-cooked roast may be well done inside and pink outside. Should you be caught in front of a large, active radar dish or hold your hand inside a microwave oven, you too may be well done inside and pink outside. Why do microwaves cook this way? As a matter of fact, how do microwaves cook meat at all?

1492.

6.6
Time to turn on light

When you turn on a light switch, how long does it take for the light to come on? Must you wait for the electrons in the wires to reach the light bulb? Once the current is flowing, how soon does the bulb begin emitting visible light?

1312.

Electrostatics
(6.7 through 6.18)

charge separation

electric field

discharge

6.7
Shocking walk on rug

Being shocked after walking across a rug or sliding across a car seat is a common experience. Granted that you must be building up charge somehow, can you explain more about what's happening? For instance, why must you walk across the rug—why doesn't the charge build up if you merely stand still? Why does the effect depend on the season?

This electrifying experience is normally part of a physics class at some point: glass rods are vigorously rubbed with cat's fur—or something like that. Why are they *rubbed*? Will they charge less rapidly if they are rubbed less vigorously? Does friction actually have anything to do with the charging? And why does the polarity of the rod depend on what's rubbed against it? Finally, why is the charge decreased if the rod is held in the smoke of a match?

300, pp. 168–170; 537; 1288 through 1297.

6.8
Kelvin water dropper

Another common physics demonstration is the Kelvin water dropper (Figure 6.8). Briefly, water drips through two tin cans, the cans being wired together as shown. After a short time, one connected pair of cans becomes positive while the other pair becomes negative. Why? The apparatus is seemingly symmetric. How, then, do the two pairs develop opposite charges? In particular, can you explain how the charging first begins?

155, pp. 261–262.

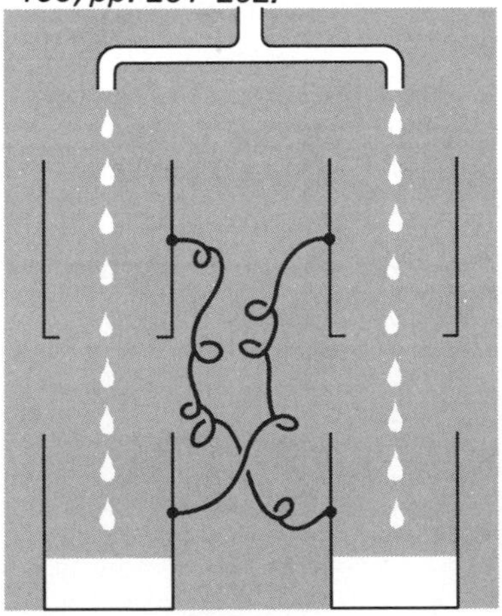

Figure 6.8
Kelvin water dropper.

6.9

Electrical field and water streams

Water streams, while initially well defined, eventually break up into drops. You can stop that breakup very easily by holding a charged object near the stream. If the object is fairly strongly charged, the stream will also be attracted to it. Can you explain these results? Of course, you really should first explain why the water stream normally breaks up.

322, pp. 86–87, 91–95; 1283 through 1287.

6.10

Snow charging wire fences

Electrical shocks are often associated with blowing sand and snow. For example, with snow blowing in the Colorado Rocky area, "wire fences on the plains near the mountains frequently accumulate charges strong enough to knock over men or cattle, and sometimes spit sparks to nearby grounded objects. Plains residents occasionally report sparks that jump as much as a yard from their fences" (354). (One jumping an inch will knock you down and leave you sick for several hours.) Why does the blowing snow charge the fences?

354, pp. 704–705; 1298 through 1301; 1527.

6.11

Scotch tape glow

If you unroll scotch tape in a dark room, you'll see a brief glow along the line where the tape is being ripped from the roll. What causes the light emission? Does it have any particular color? If so, why that color?

1194, p. 252; 1302.

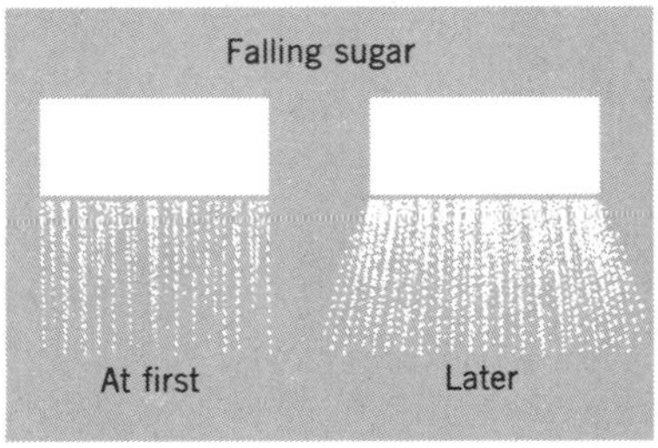

Figure 6.12

6.12

Sifting sugar

One day as I was sifting confectioner's sugar for a cake frosting, a curious thing happened to the sugar. When I started, the sugar would fall straight down, but gradually more and more of the sugar would be thrown to the side (Figure 6.12). Why was it deflected?

6.13

Gas truck chains

Why in the past were chains dragged beneath gasoline trucks? Should you drag a chain from your car?

1303 through 1306.

6.14

Charge in shower

When you take a shower, the splashing water produces negative charges in the room's air and electric fields of up to 800 volts per meter. Similar negative fields are found near natural waterfalls. In addition, when large crude-oil carriers are cleaned with high-velocity water jets, electric fields of up to 300 kilovolts per meter can be created. What is the cause of such fields? In the case of the supertankers that is not merely an academic question, for there have been several large explosions during the cleaning of those ships.

539; 1296; 1307 through 1311.

6.15

Happiness and negative charge

It is thought that if you enter a negatively charged atmosphere, such as the bathroom discussed above, a feeling of well-being will come over you. Being charged negatively makes you happy; being charged positively makes you ill at ease. So, perhaps your feeling good after a shower has as much to do with the negative charge in the bathroom as with feeling clean. Can you explain why negative and positive charge might affect you this way?*

1307; 1408.

*Also see Prob. 3.18 on the Chinook.

6.16

Fall through the floor

Why don't you fall through the floor? Fundamentally, what supports you?

6.17

Sand castles and crumbs

If you want to make a sand castle at the beach, you use wet sand, not dry. Common table salt shows the same tendency to be much more cohesive when wet. Other powders such as cocoa and chalk, however, are cohesive even when dry. What forces are responsible for the cohesiveness of a powder? Why does it matter whether a powder such as sand or salt is wet? Do you think a fine powder should be more or less cohesive than a coarse one?

Crumb formation is essential for maintaining a fertile soil, yet if the soil is misused, a useless dust ball may develop. What is responsible for crumb formation in soil? Why don't other things, such as sand and face powder, form crumbs?

1313, pp. 288–290; 1314; 1315.

6.18

Food wrap

Some clear food wraps can be tightly stretched over a container and folded down the sides, and they will retain the tension and completely secure the container. The food wraps "stick." How do they do this?

Magnetism

(6.19 through 6.24)

6.19

Magnetic-field dollar bill

If you hang a dollar bill from one end and bring a large magnet (with a nonuniform field) toward it, the bill will move toward one of the pole faces. Why?

1316.

6.20

Bubbles moved by magnetic field

A large magnet placed near a carpenter's bubble level will force the bubble to move. How does the magnetic field do that? Does the bubble move toward or away from the magnet?

1316.

induction

6.21

Electromagnetic levitation

You can levitate a metal ring on a coil through which a steady AC current passes (Figure 6.21), but if the current is quickly turned on, the ring will jump into the air very dramatically. Why is there a difference in behavior in these two cases? What supports the ring against gravitation when it is being levitated, and what determines the height at which it floats? How stable is the levitation (does the ring sit against the pole and at a tilt)? In predicting the behavior of various rings, your intuition may fail. So, for fun, first try to guess what will happen in the following circumstances and then actually see what does happen. Will a thin ring float at the same height as a thicker ring if the density and diameter are the same? What happens should both rings be on the coil when the current is slowly increased? Finally, what happens if one of the rings is wider than the other?

1317 through 1321.

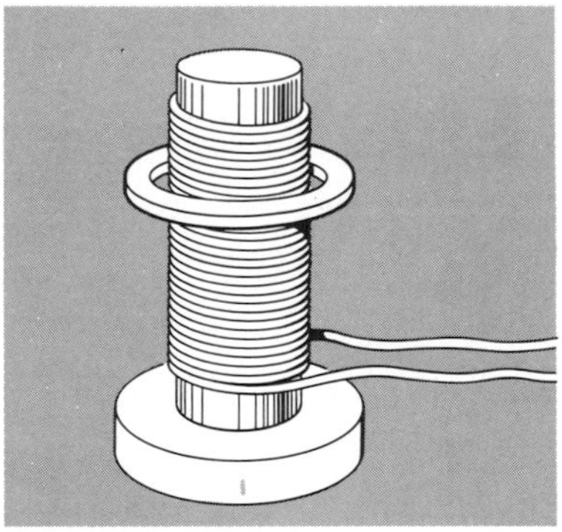

Figure 6.21
Metal ring suspended on coil.

induction

6.22

Turning in the shade of a magnetic field

Can partially shading a magnetic field from a copper disc cause the disc to rotate? Over one of the poles on an alternating magnet, place a copper disc that is free to rotate (Figure 6.22). The disc is repelled but shows no desire to rotate. But now insert another copper sheet between the disc

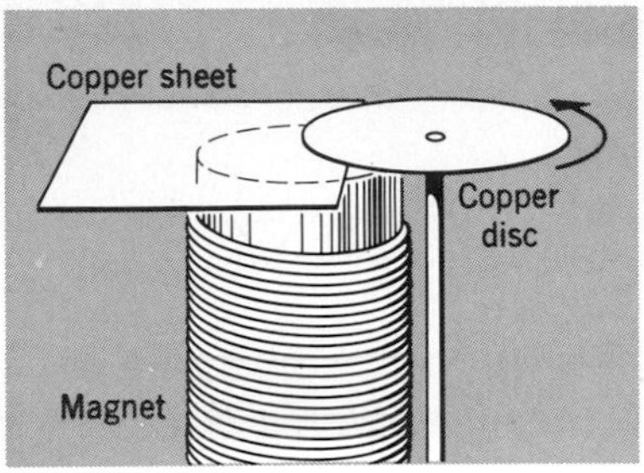

Figure 6.22
Disc rotates when the copper sheet partially shields it.

and the magnet, partially shading the disc from the magnetic field. Immediately the disc begins to turn. Can you explain why?

1321, pp. 82 ff.

induction

6.23

Car speedometer

Will a horseshoe magnet attract aluminum? No, normally it won't. (Why not?) There is a special arrangement, however, in which a magnet will move aluminum. Suspend a horseshoe magnet on a string above an aluminum disc

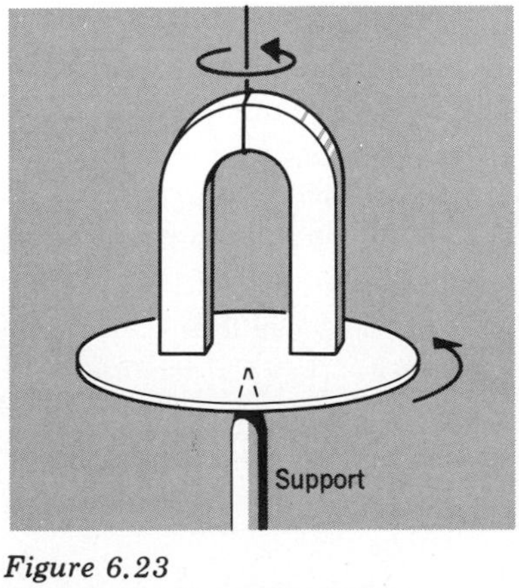

Figure 6.23
Aluminum disc turns underneath turning magnet.

(Figure 6.23). Somehow suspend the disc so that it's free to rotate about its center. If the magnet is set spinning, the disc will spin also. Will the disc turn in the same sense as the magnet? Why is aluminum only affected in this case?

This is basically how your car's speedometer works, except that in your car the rotating magnet is inside an aluminum can to which a pointer is attached and the can is restrained by a spring.

155, p. 344; 592, p. 87.

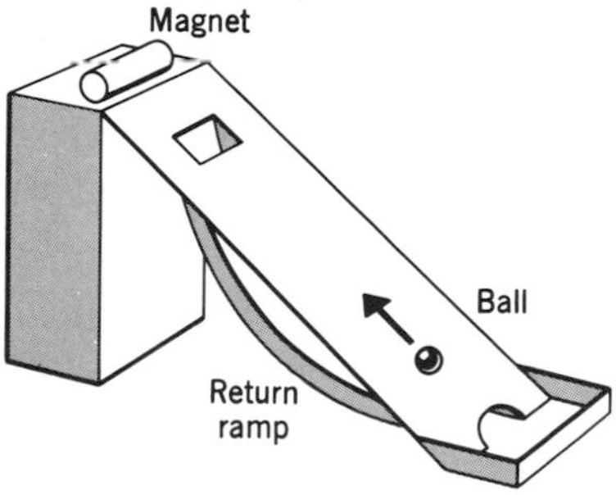

Figure 6.24
Ball undergoes perpetual motion.

6.24

Perpetual magnetic motion

Of the many fascinating perpetual machines proposed through history,* that of the Bishop of Chester (1670s) is one of the simplest (Figure 6.24). The magnet that was fixed on the column was to draw the iron ball up the ramp until the ball reached the hole in the ramp. The ball would then fall and be returned to the ramp's base, and the procedure would begin again. Very straightforward, right? Shouldn't it work?
1325.

*See Refs. 1322 through 1324.

Radio and ionosphere physics

(6.25 through 6.31)

ionospheric physics

plasma frequency

electromagnetic waves

6.25

Radio, TV reception range

There are several things about radio that have always puzzled me. For example, why can AM stations be received at night over a much larger range than during the day? Sometimes you can pick up a station halfway across the United States on a cheap transistor radio. (One consequence of this is that the FCC requires most AM stations to cut their power or even to leave the air at dusk.) When Marconi transmitted the first wireless signals across the Atlantic, many people were amazed. Why didn't those signals go directly into space instead of following the curving surface of the earth as they did?

FM and TV stations, however, hardly even get their reception areas out of the city. Occasionally, such as during a meteor shower, these signals do travel surprising distances; at other times, such as during major solar flares, they are tremendously reduced, worldwide communication being almost destroyed. First, why is there such a difference between the ranges of TV and FM on one hand and AM on the other? Next, why are there occasionally such dramatic changes in the transmission ranges of TV and FM?

170, pp. 138–139; 215, pp. 43 ff; 1326; 1327.

resonance

6.26
Crystal radio

The crystal radio of my boyhood was very simple, being only an antenna wire, a capacitor, a long wire coil, earphones, and finally, a crystal (Figure 6.26). Do you understand how it worked? For example, why did moving the contact on the wire coil change stations? Why was the crystal necessary?

Every now and then there are stories about people who can hear local radio stations on their teeth fillings, on their bedsprings, etc. Could there be any truth to these stories? If so, then what is it in these strange radio sets that is taking the place of the crystal in the crystal set?

158, pp. 577–578; 211, pp. 417-418; 253, p. 409.

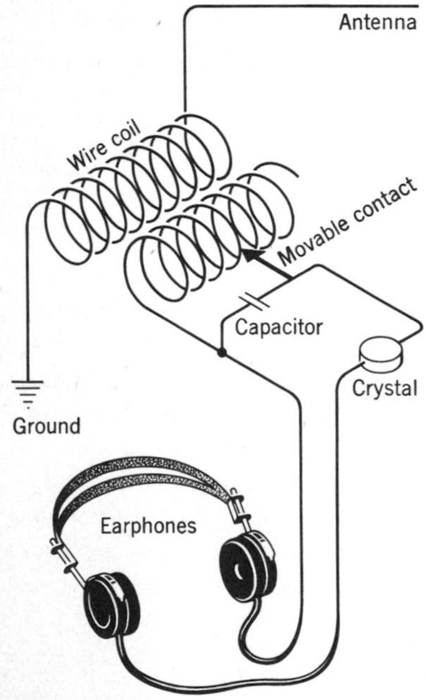

Figure 6.26
Crystal radio.

6.27
Airplane interference with TV

How does a nearby airplane interfere with your TV picture?

6.28
AM car antenna

Why are AM radio antennas mounted outside a car and usually vertically? How much does it matter if they are mounted in the windshield glass?

6.29
Multiple stations on radio

Normally I hear one local station for a given setting on my car radio. Yet when I drive near a radio station's antenna, I can sometimes hear that station plus another for one setting. Why? Sometimes I can even get one station at many settings of my radio dial. Again, why?

charged particles in magnetic field

atomic and molecular excitation

6.30
Auroral displays

"After darkness has fallen, a faint arc of light may sooner or later be seen low on the north horizon, or centered somewhat to the east of north. Gradually it rises in the sky, and grows in brightness. As it mounts in the sky, its ends, on the horizon, advance to the east and west. Its light is a transparent white when faint, and commonly pale yellow–green when bright—rather like the tender color of a young plant that germinates in the dark. The breadth of the arc is perhaps thrice that of a rainbow. The lower edge is generally more definite than the upper. The motion upward toward the zenith may be so slow that the scene is one of repose. As the arc rises, another may appear beyond it, and follow its rise. At times four, five, or even more arcs may thus appear. They rise together, and some of them may cross the zenith and pass onwards into the southern half of the sky.

Figure 6.30
"But we went to see the northern lights last week." (Chicago Tribune Magazine.)

"This may be all that appears on some nights. But on others the aurora enters after a while on a new and distinctly different phase, much more active and varied. The transition from the quiet to the active phase may be speedy, even sudden. The band becomes thinner, rays appear in it, it begins to fold and also to become corrugated in finer pleats. It becomes a rayed band of irregular changing form, like a great curtain of drapery in the sky. Its color may remain yellow-green, but often a purplish-red border appears along the lower edge, perhaps intermittently. Vivid green or violet or blue colors sometimes appear. At times the rays seem to be darting down, like spears shot from above. Sometimes there seems to be an upward motion along the rays, or motion to the east or west along the band. The curtains may sweep rapidly across the sky as if they were the sport of breezes in the high air; or they may vanish and reappear, in the same place or elsewhere. This grand display may continue for many minutes or even hours, incessantly changing in form, location, color and intensity; or intermissions may occur, when the sky has little or no aurora.

"At times the observer may look up into a great auroral fold nearly overhead, when the rays in its different parts will seem to converge, forming what is called a corona or crown. Often such a corona rapidly fluctuates in form, and its rays may flash and flare on all sides, or roll around the center.

"At the end of an outstanding display the aurora may assume fantastic forms, no longer in connected curtains and bands. There may be a widespread collection of small curtains, stretching over a large part of the sky, which brighten and fade, or, as it is said, pulsate. Finally, the sky may be covered by soft billowy clouds, not unlike a mackerel sky with rather large "scales"; but these "scales" and patches appear and disappear, with periods of not many seconds. At last the sky becomes altogether clear, with no more aurora. But later the whole sequence may begin anew, and continue till dawn pales the soft auroral light (1328)."

Explaining the aurora in detail is still a matter of current research, but can you explain in general why the aurora is formed and why some of these colors and wavelike structures appear? Why are auroral displays so much more frequent at high latitudes? Why are there more displays over northern Canada than, for example, over Siberia at the same (geographical) latitude?

219, pp. 242–246; 1328; 1329.

refraction

dispersion

6.31
Whistlers

In World War I the Germans eavesdropped on Allied field telephone messages by detecting the small leakage from the telephone wires into the ground. The initial pickup was by two metallic probes driven into the ground a couple hundred yards apart and at some distance from the telephone wires. Once the signals were fed into a high-gain amplifier, they became audible to the German intelligence personnel. But during such monitorings, the Germans also heard mysterious, relatively strong whistlings whose pitch would steadily fall. These sounds have since been associated with ionospheric phenomena appropriately called "whistlers" and other sounds such as clicks, tweeks, chinks, and a whistling of rapidly rising pitch called the "dawn chorus" have been detected. Can you explain the sources of these sounds?

219, pp. 302-304; 1330; 1331.

Atmospheric discharge

(6.32 through 6.49)

discharge

electric field

electric potential

6.32
*Lightning**

Lightning is so familiar that its beauty runs the risk of being overlooked. So, before we get into some of the strange or paradoxical features of lightning, let's ask some simple questions about its common properties. In a lightning discharge there are at least two strokes: usually there is first a "leader," then a "return." Which do you see, and why don't you see both?† Why do you even see one—what produces the visible light? Does the visible stroke go up or down? Why is it so crooked? How much current is involved in a flash? How bright is a flash? Approximately how wide is the lightning channel you see? One hundred meters? One meter? Several millimeters? How long does the flash last? Several seconds? Several milliseconds? A microsecond or so?

220; 299, pp. 110-123; 300; 301; 332, Vol. II, Chapter 9; 1332; 1333; 1550; 1590.

*Suggestions for photographing lightning flashes are given by Orville (1334). The first lightning photograph ever taken is reproduced in Jennings (1335).

† If you are driving through rain at night, a multiple-flash stroke can give several stroboscopic images of your moving windshield wiper (1336).

6.33
Earth's field

The big question, however, is why there is lightning at all? What is responsible for the electric field that is between the earth's surface and the clouds? Outdoors there is a 200-volt difference between the heights of you nose and feet. Why aren't you shocked by that voltage difference (Figure 6.33)? Can motors be driven by this electric field? In some cases, yes.

299, pp. 97-109; 300, pp. 105-106; 332, Vol. II, Chapter 9; 1296; 1337 through 1339; 1548; 1549; 1568.

Figure 6.33
The earth's electric field.

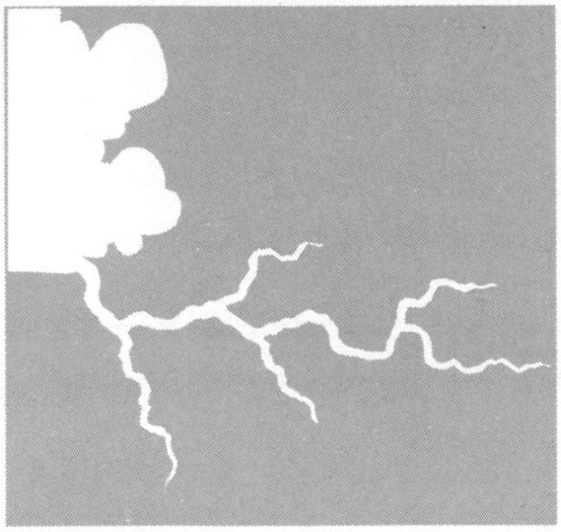

Figure 6.34a
Cloud-to-air stroke.

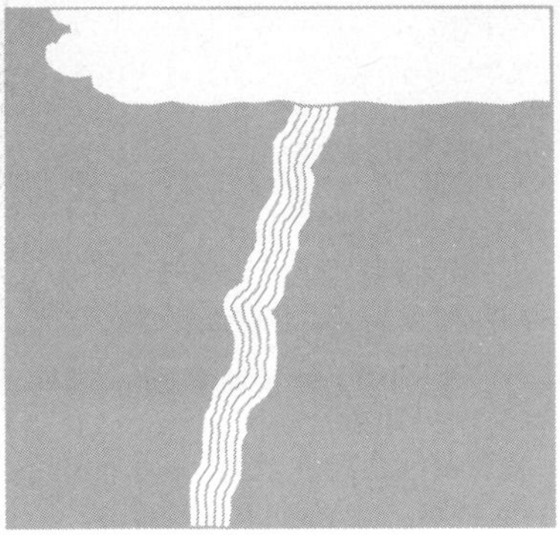

Figure 6.34b
Ribbon lightning.

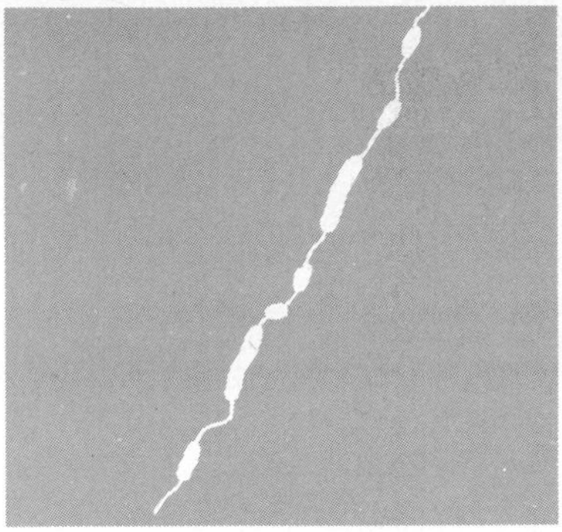

Figure 6.34c
Bead lightning.

6.34
Lightning forms

The cloud-to-ground lightning stroke is not the only type of lightning. The cloud-to-air stroke, for example, terminates in midair (Figure 6.34*a*). If the cloud is too distant to be seen, you may suddenly be awed by such a "bolt from the blue." Under some circumstances you will see several parallel strokes that give the impression of a ribbon hanging from the clouds (Figure 6.34*b*) The most exciting stroke, however, is probably "bead lightning" (Figure 6.34*c*), which appears to be a series of brilliant beads tied to a crooked string. In these several examples what causes the strokes or the special appearance of the strokes? In the cloud-to-air case, where does the current of the discharge go?

299, pp. 128–129; 300, p. 5; 301, p. 45; 1340; 1341; 1611, Section GL; 1623.

6.35
Ball lightning

One of the most controversial subjects in physics is whether or not ball lightning exists. This argument persists in spite of the enormous number of sightings and many published accounts. Perhaps as much as 5 percent of the world's population have seen it (1350, 1351), yet many will argue vigorously that it is an illusion, such as an afterimage resulting from having seen a bright flash of light.

The luminous, silent balls of light reportedly float through the air or slowly dance about for several seconds. They can sometimes pass through window glass without a trace of damage; at other times, the glass is shattered. They are seen in all manner of structures (even in metal airplanes) as well as outdoors. Though they are usually silent, their demise is accompanied with a pop. Finally, they are deadly. G. W. Richmann was apparently a victim while trying to repeat the results of Franklin's kite experiment. A pale blue fireball about the size of a fist left the lightning rod in his lab, floated quietly to Richmann's face, and exploded. With a red spot on his forehead and two holes in one of his shoes, Richmann was left dead on the floor.

In reviewing the many explanations of ball lightning, can you identify those with any real possibility of being correct? Can you also devise other explanations or argue that ball lightning can only be an illusion?

299; pp. 130–133; 1349 through 1370; 1611, Section GLB.

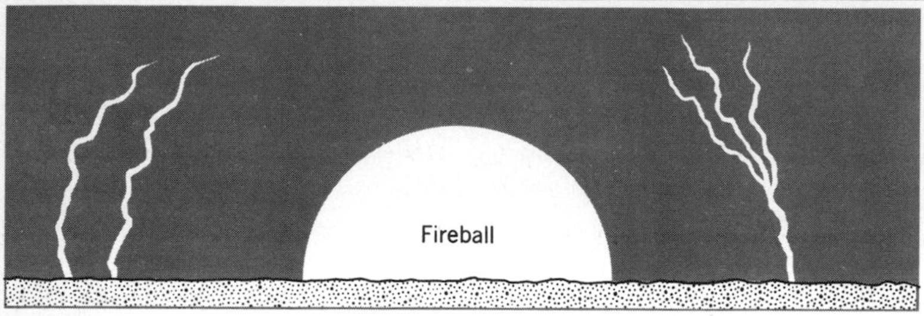

Figure 6.36

6.36

H-bomb lightning

Lightning flashes were also photographed surrounding another catastrophic event, the fireball of the 10-megaton thermonuclear device triggered in 1952 at Eniwetok Atoll. The strokes propagated upward from the surface of the sea, and the branching was also upward (Figure 6.36). As the fireball expanded and reached the points where the lightning channels were previously visible (the visible flashes had disappeared by then), the tortuous channels were again visible against the backdrop of the fireball. The charge production for the lightning must have been set up very rapidly, but precisely what caused it is still not well known. Can you suggest possible explanations? Can you also explain why the channels became visible again against the fireball background?

1347; 1348.

6.37

Volcanic lightning

When the volcano that formed the new islet, Surtsey, rose furiously from the Icelandic sea in 1963, brilliant lightning displays danced in the volcano's dark clouds. What provided the tremendous charging for the lightning? One possible mechansim was sea water striking the molten lava. How would that produce the charge?

1342 through 1346.

6.38

Earthquake lightning

Should an earthquake produce lightning discharges? The Japanese have learned that lightning discharges in a clear sky are signs of impending earthquakes. Indeed, earthquakes there and in other areas are sometimes associated with both normal lightning and ball lightning. Why should there be any connection between the two phenomena?

1371; 1372.

6.39

Franklin's kite

Benjamin Franklin's kite experiment was probably first introduced to you somewhere in elementary school, but do you understand all the little points about what Franklin did, and why he was not killed? The following is Franklin's letter to a friend describing the experiment:

> To the top of the upright stick of the [kite's] cross is to be fixed a very sharp pointed wire, rising a foot or more above the wood. To the end of the twine, next the hand, is to be tied a silk ribbon, and where the silk and twine join, a key may be fastened. This kite is to be raised when a thunder gust appears to be coming on, and the person who holds the string must stand within a door or window or under some cover, so that the silk ribbon may not be wet; and care must be taken that the twine does not touch the frame of the door or window. As soon as any of the thunder clouds come over the kite, the pointed wire will draw the electric fire from them, and the kite, with all the twine, will be electrified, and the loose filaments of the twine will stand out

every way, and be attracted by an approaching finger. And when the rain has wet the kite and twine, so that it can conduct the electric fire freely, you will find it stream out plentifully from the key on the approach of your knuckle. At the key the phial* may be charged, and from electric fire thus obtained, spirits may be kindled, and all the other electric experiments be performed, which are usually done by the help of the rubbed globe or tube, and thereby the sameness of the electric matter with that of lightning completely demonstrated.

Why did he put a pointed wire on the kite's top? Why the silk ribbon between the key and his hand? Why the key? Why was the twine attracted to his finger, and why did loose filaments stand out? What caused the light emission he saw when his knuckle was brought close to the key? Why wasn't Franklin killed? If a lightning stroke had hit the kite or string, would he have survived? In Europe G. W. Richmann *was* killed in trying to repeat the Franklin experiments† so don't you try it, even with Franklin's precautions.

299, pp. 37–44; 301, Chapter 2; 1373; 1374.

*An early form of capacitor.

†See Prob. 6.35.

6.40

Lightning rod

My grandmother's lightning rod has a sharp point, stands several feet taller than the house, and is buried several feet into the ground. Why are those features desirable? What is the rod really supposed to accomplish? There has been considerable debate over this question ever since Benjamin Franklin's invention of the lightning rod. Some claim that the rod helps discharge a cloud as it passes overhead, thereby avoiding the catastrophic breakdown of lightning. Others claim that the rod merely provides a safe route to ground for any flash near the rod.

There have also been many misconceptions and controversies about the performance and installation of lightning rods. For a while after their first introduction, strong arguments were made for a top with a round metal knob or even a glass knob. Convincing arguments were also made that the lower part should be attached to the top soil only, for an explosion could occur if the flash were carried deep into moist ground. Recently a company was fitting its rods with a radioactive source at top. That source was to aid in ionizing the air, thereby further seducing the flash to strike the rod rather than the protected building. Would a radioactive source really be of any aid?

299, pp. 188 ff; 300, Chapter 15; 301, Chapters 2, 6; 1296; 1373 through 1381.

6.41

Lightning and trees

There's an old wives tale about lightning seeking out oak trees. In fact, a strikingly high proportion of trees shattered by lightning are oaks. It is hard to believe, however, that lightning knows the difference between an oak and any other type of tree. Why then is there such preferential shattering? How exactly does the lightning stroke make a tree explode, anyway? Of course, a strike does not always result in an explosion. For example, Orville (1389, 1390) has published a remarkable photograph of a direct hit sustained by a European ash tree. Upon close examination the following day, the tree bore no indication of its experience.

How does lightning start forest fires? Why aren't fires started in all lightning strikes in wooded areas?

299, pp. 177–187; 300, p. 151; 301, p. 60; 1382 through 1390.

6.42

Lightning strikes to aircraft

Lightning strikes to aircraft are frequent, but it is very rare that there is any damage other than perhaps several tiny holes in the fuselage. Cars, buses, and other such vehicles also enjoy immunity from damage. Soon after lift-off Apollo 12 was struck twice by lightning with no apparent ill effects to the spacecraft or its crew. In each of these cases why is there no damage to the vehicle or injury to the occupants? Indeed, the occupants may never even be aware of the strike.*

299, pp. 232-235, 249 ff; 300, pp. 151-152; 301, pp. 51-54; 1296; 1379; 1391, p. 22; 1392 through 1397.

*An alert airplane passenger may foresee a lightning strike by noticing a sudden increase in St. Elmo's fire (see Prob. 6.47) on the wing tips and other pointed objects. The luminous streamers may be 10 or 15 feet long and half a foot wide (301).

6.43

Rain gush after lightning

Perhaps you have noticed sudden gushes of rain or hail moments after lightning strokes in thunderstorms. Is there any connection between the gush and the stroke or the thunder? Or is this just a coincidence?

164, pp. 358-359; 300, pp. 165-166; 301, p. 152; 1398 through 1400; 1619.

6.44

Clothes thrown off

If you're struck by lightning, you may very well have your clothing and shoes thrown off. What causes that?

301, p. 131.

6.45

Ground fields in lightning hit

If you are caught in a thunderstorm you should not stand under a tree, and you should keep your head lower than your surroundings. Why is the tree dangerous? As long as you stand away from the trunk, aren't you safe enough?

Should you ever lie down? That would give your head the minimum possible elevation, but is there any additional danger encountered in lying down? Cows are often killed or hurt by lightning. Not only do they commonly stay outdoors and often seek shelter under trees, but the separation of their hind legs from their front legs increases the danger (Figure 6.45). They are thus similar to a man lying down. Again, why is this dangerous?

299, p. 223; 301, pp. 61-64; 1350, p. 279; 1391, pp. 282-283.

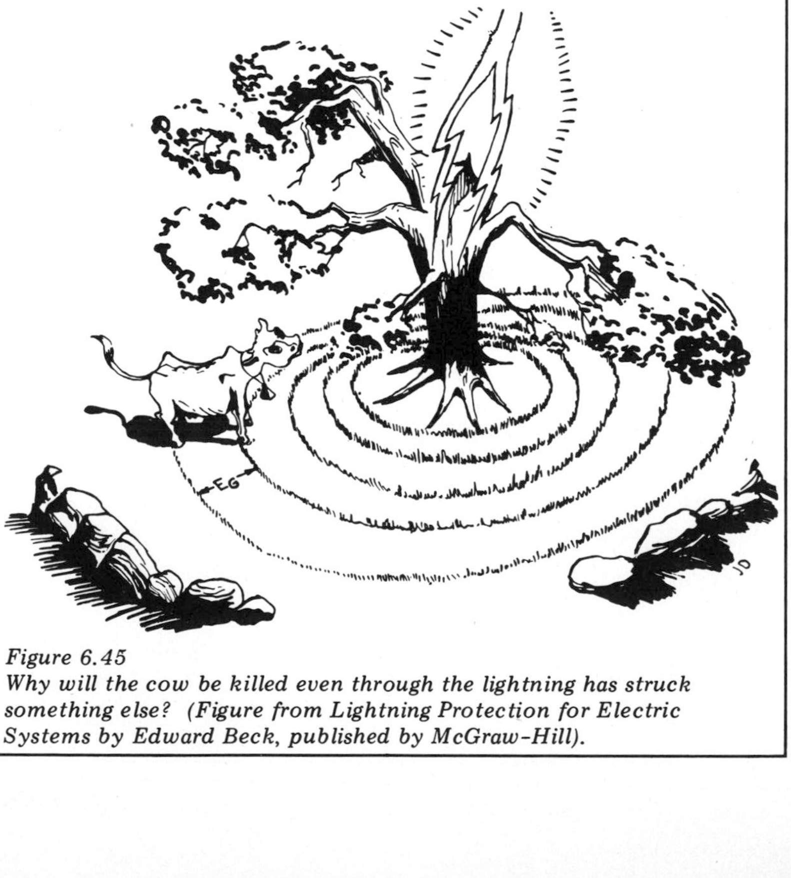

Figure 6.45
Why will the cow be killed even through the lightning has struck something else? (Figure from Lightning Protection for Electric Systems by Edward Beck, published by McGraw-Hill).

6.46

St. Elmo's fire

St. Elmo's fire is a fairly continuous luminous discharge seen from such things as masts of ships, wing tips of airplanes, and even bushes. There is a crackling noise associated with the blue, green, or violet color of the light. Can you explain, first of all, what causes this light, and second, why those particular colors?

> A favorite stunt of mountain guides, when the air is throughly charged, is to imitate Thor by waving an ice-axe over their heads. The metal parts of the ice-axe draw down an impressive display of electrical polytechnics. A geological hammer will sometimes spit long hot sparks in one position, but if the head is turned at right angles to the former position, the sparking stops. . . They usually detect charged air by raising a finger above their heads. When the air is heavily charged, sparks will sizzle from the fingertip, making a noise like frying bacon (354).

Another example, somewhat different in appearance, is the electric sparks, several meters long, which may spring up from the tops of sand dunes during thunderstorms. In this case, the blowing sand must contibute to the sparking, but how?

165, p. 233; 301, pp. 47–50; 354, p. 744; 961; 1402, p. 219; 1403.

6.47

Living through lightning

There are many cases of people living through direct and indirect lightning hits. There are even cases where the lightning has stopped a person's breathing for perhaps 20 minutes, yet the person has fully recovered with no apparent brain damage due to electrical shock or oxygen starvation. It has been suggested (1401) that such a shock momentarily changes the brain's crucial need for oxygen. In any case, shouldn't the victim be severely burned and his heartbeat halted? How much energy (or power) is deposited in such a victim?

299, pp. 226–230; 301, p. 131; 1401.

6.48

Andes glow

Single flashes of light and continuous glows can be seen over the peaks of certain mountain ranges. They have been described as "not only clothing the peaks, but producing great beams, which can be seen miles out to sea" (1404). Generally these mysterious lights are called Andes glow, though this doesn't mean they are restricted to the Andes. What causes this glow? St. Elmo's fire from many points on a peak? St. Elmo's fire is usually only a few centimeters long, so how could it be seen miles away?

165, p. 233; 1404 through 1406.

6.49

Electrical pinwheel

A demonstration sometimes seen in physics classes involves a pinwheel that is made to rotate by a high DC voltage (Figure 6.49). Why this happens was a point of controversy over the last two centuries, but recently the device has been somewhat neglected. Does the pinwheel turn because of something that it throws off or pulls on or for some other reason? Will it work in a vacuum or in a dust-free environment? Why does the color of the discharge depend on the polarity of the pinwheel? Why do the tips need to be sharp? Finally, can you calculate how fast the pinwheel will turn under given conditions?*

155, pp. 434–435; 1407.

*Also see Prob. 6.33.

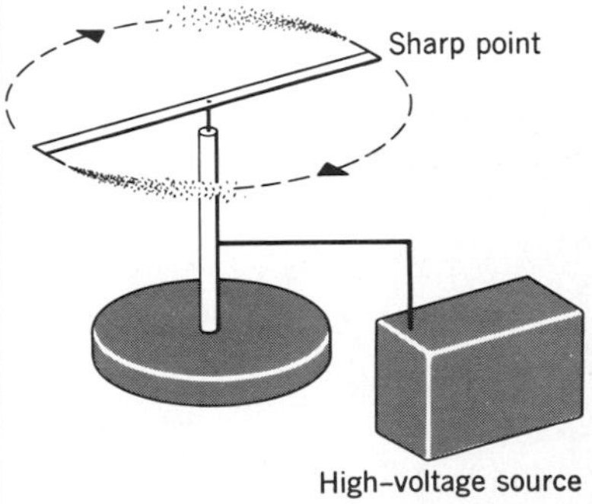

Figure 6.49
Rotating pinwheel driven by electrical discharge.

6.50

Power-line blues

In order to transmit electrical power more efficiently, some electrical companies have erected "extra-high-voltage" (765,000 volt) transmission lines. Such lines may be beneficial on the whole, but they have worried those people living near the lines. Disturbingly, the lines often glow an erie blue and can cause disconnected fluorescent tubes to light mysteriously. More threatening, however, is that numerous people have received shocks when touching metallic objects in the vicinity of the extra-high-voltage lines.

In a recent survey, 18 families living near Ohio Power Co.'s line reported being shocked by touching farm machinery, wire fences or even damp clotheslines. Two women complained of shocks received while on the toilet. Other complaints were bad TV reception and the sizzling sound of the electrical discharge. Said C. B. Ruggles, whose farm is split by the line: "You'd swear we were living near a waterfall" (1558).

How would a powerline such as this cause objects in its vicinity to give shocks? I have heard that some people run electrical motors by connecting them to antennas surreptitiously buried near the power lines. *Is* it possible to get power this way?

1558.

7
The walrus has his last say and leaves us assorted goodies

7.1

UFO propulsion

When "your gravity fails
and negativity don't pull
you through."
---Bob Dylan, "Just Like
Tom Thumb's Blues"*

In light of physical laws, let's reconsider the possibility that the UFOs sighted during the last few decades are intelligently controlled craft. Consider the method of propulsion, for instance. No local destruction has ever been noted at the site of a landing or lift-off. For objects as large as space ships, is this possible with any kind of chemical or nuclear power? How much energy would be involved with those sources? Could the vehicle somehow use the earth's electric or magnetic field? If so, how much acceleration would be possible, and would there be an altitude limitation?

One of the most popular propulsion mechanisms in science fiction has been gravitational shielding. H. G. Wells used it long ago to get men to the moon. Suppose a craft could suddenly shield itself from the earth's gravitational field. Would it lift off? If it did, how fast would it move? In particular, would it move at anywhere near the fast speeds reported for UFOs?

1409.

7.2

Violating the virgin sky

Cyrano de Bergerac uses the most incredible physics ever recorded to keep the villainous de Guiche from Roxanne's house while she is being married. Dropping from a branch into de Guiche's path, Cyrano swears he has just fallen from the moon.

CYRANO: From the moon,
the moon! I fell out of the
moon!

DE GUICHE: The fellow
is mad—

. . .

CYRANO (Rapidly):
You wish to know by
what mysterious
means
I reached the moon?
. . .
I myself
Discovered not one
scheme merely, but six—
Six ways to violate the
virgin sky!

(De Guiche has succeeded in passing him, and moves toward the door of Roxanne's house. Cyrano follows, ready to use violence if necessary.)

DE GUICHE (Looks around.): Six?

CYRANO (With increasing volubility):
As for instance—Having
stripped myself
Bare as a wax candle,
adorn my form
With Crystal vials filled
with morning dew,
And so be drawn aloft,
as the sun rises
Drinking the mist of dawn!

DE GUICHE (Takes a step toward Cyrano.):
Yes—that makes one.

CYRANO (Draws back to lead him away from the door; speaks faster and faster.):
Or, sealing up the air
in a cedar chest,
Rarefy it by means
of mirrors, placed
In an icosadedron.

DE GUICHE (Takes another step.): Two.

Figure 7.2
Self-motivation (Goofy, "Victory Vehicles," © Walt Disney Prod.).

CYRANO (Still retreating):
Again,
I might construct a rocket, in the form
Of a huge locust, driven by impulses
Of villainous saltpetre from the rear,
Upward, by leaps and bounds.

DE GUICHE (Interested in spite of himself, and counting on his fingers.):
Three.

CYRANO (Same business):
Or again,
Smoke having a natural tendency to rise,
Blow in a globe enough to raise me.

DE GUICHE (Same business, more and more astonished.): Four!

CYRANO: Or since Diana, as old fables tell,
Draws forth to fill her crescent horn, the marrow
Of bulls and goats—to anoint myself therewith.

DE GUICHE (Hypnotized):
Five!—

CYRANO (Has by this time led him all the way across the street, close to a bench):
Finally—seated on an iron plate,
To hurl a magnet in the air—the iron
Follows—I catch the magnet—throw again—
And so proceed indefinitely.

DE GUICHE: Six!—
All excellent,—and which did you adopt?

CYRANO (Coolly): Why none of them. . .A seventh.
. . .
The ocean!. . .
What hour its rising tide seeks the full moon,
I laid me on the strand, fresh from the spray,
My head fronting the moonbeams, since the hair
Retains moisture—and so I slowly rose
As upon angel's wings, effortlessly, Upward.*

*From *Cyrano de Bergerac* by Edmond Rostand, translated by Brian Hooker, published by Holt, Rinehart and Winston, Inc.

cosmology

7.3
Olbers' paradox

Some have argued that the universe is infinitely large and contains an infinite number of stars. Olbers' paradox is that "if the universe is infinite in extent and contains an infinite number of stars evenly distributed, the sky should be blazing all over in brilliant light" (1414). Of course, the intensity of the light from distant stars will be less than from nearby stars. But if the stars are evenly distributed, then their number increases with distance from the earth just enough to

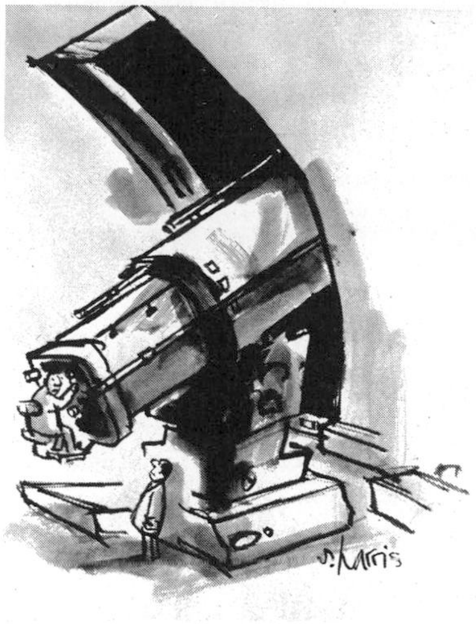

Figure 7.3
"I'm not sure, but it looks like infinity." (Phi Delta Kappan.)

balance the decrease in light intensity from each star. Hence, the total light coming from any given distance should be the same as from any other distance. With an infinite number of stars, the nighttime sky should be bright and evenly lit. Why, instead, is the nighttime sky relatively dark?

1410 through 1416; 1587.

atmospheric physics

gravity waves

7.4
Noctilucent clouds

Shortly after a summer sunset in the high latitudes, ghostly, silvery-blue clouds may appear against the dark sky. They are called noctilucent clouds (luminous night clouds), and their origin is still highly controversial. They may be associated with extraterrestrial dust entering the atmosphere, but this has not yet been proved. Why are they visible only after sunset? Since they are seen when the sky is dark, about how high are they? Why are they usually seen only in the high latitudes and only in the summer? Why do they often appear in a wavy pattern, as though the clouds were the surface of a sea?

362, pp. 150-151; 954, pp. 284-287; 1417 through 1423.

7.5

Water witching

Some people claim they can locate underground water by walking over the area with a forked stick, rod, or something similar (Figure 7.5*a*). When directly over water the instrument reportedly dips to indicate the (unseen) water (Figure 7.5*b*).

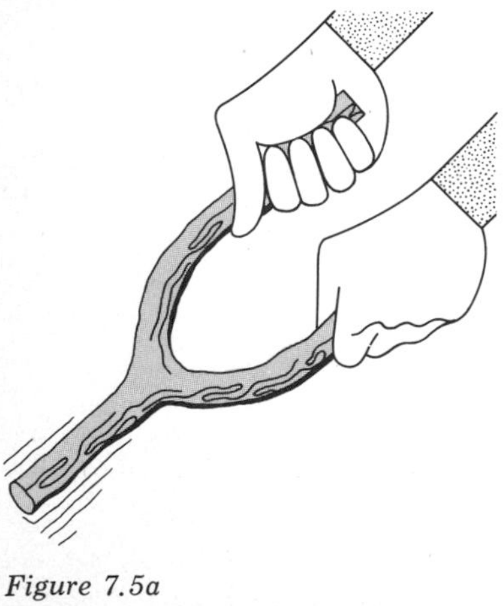

Figure 7.5a
A water witch's forked stick.

Figure 7.5b
(By permission of John Hart. Field Enterprises.)

This procedure—called dowsing, water witching, or divining—is controversial: there are many success stories on the one hand but a complete absence of explanation in physical terms on the other. What could possibly be the force that influences either the instrument itself or the person holding it? Is there some clue, perhaps even a subconscious one, that tips off the water witch to the presence of water?

1520 through 1523.

shock waves

energy transfer

7.6

Snow waves

A footstep in a field of snow may set off a snowquake that propagates away from the site and causes a lowering of the snow level and a swishing sound. If the disturbance encounters a barren area, it will be reflected back through its origin, and the swishing of the second passage can be heard. What causes these snowquakes to propagate, and what determines their speed? Why does their passage lower the snow level and cause a swishing sound? Finally, why will a barren area reflect them?

1426; 1427; 1455.

7.7

Fixed-point theorem

If you stir a cup of coffee and then let it come to rest, at least one point on the coffee's surface will be back in its original place. (The stirring must be smooth, with no splashing.) If you were to rip out this page, crumple it, wad it, and then lay the wadded ball back in the book, at least one point on the page will be directly over its original position. Why is this guaranteed in these two cases every time?

1428 through 1430.

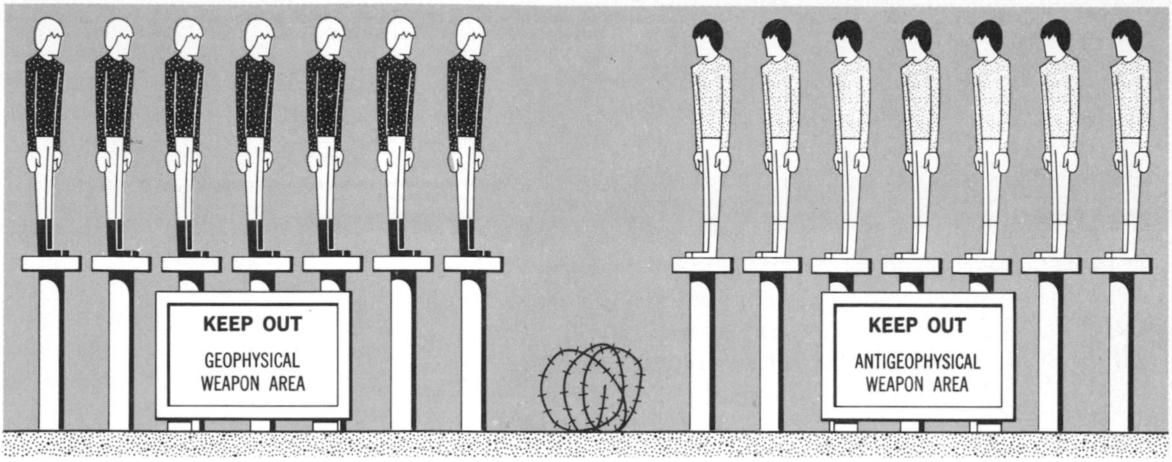

Figure 7.8
Should we worry about a geophysical weapons gap?

7.8

The great leap downward

The Republic of China commands an awesome new weapon—a geophysical weapon. It has been suggested that should all of its 750 million people leap simultaneously from 6 1/2-foot-high platforms, they would set up shock waves in the earth. By jumping again each time the shock waves pass through China, the Chinese could build the waves up to the point that they could destroy parts of the United States, especially California, which is already endangered by earthquakes.

What path would such a shock wave take through the earth? How frequently should the Chinese jump to amplify the wave, and how much energy is added to it by each jump? Is there any way another country's population could defend itself against this geophysical weapon, for example, by some appropriate type of retaliatory jumping (Figure 7.8)? Does it matter *how* the Chinese jump? For example, one writer has argued it is essential the Chinese jump with stiff knees, for bent-knee jumping would impart far less energy to the ground. Is that true?

1424; 1425.

7.9

Beating and heating egg whites

Why does beating egg whites change them from a fluid to a thick foam? For instance, in making meringues the egg whites are beaten until they peak, (when the beater is lifted out, the substance is stiff enough that it is left in a peak). What does the beating do to the egg white to cause it to stiffen? Similarly, what is physically responsible for transforming the egg white—initially a colorless, transparent fluid—into a white solid when, for example, you fry an egg?

316, pp. 123-126, 87-90; 1431.

Scotch tape rheology

stresses

7.10

Pulling off Scotch tape

Scotch tape cannot really get into the surface irregularities of whatever it is being applied to, yet it holds well when you try to peel it off. The adhesion is partly due to

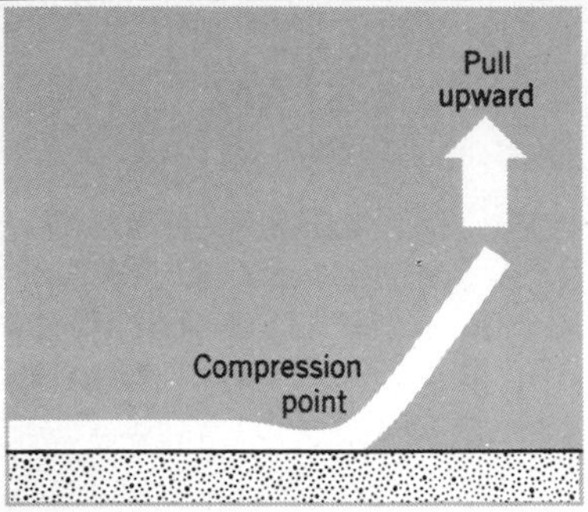

Figure 7.10
Compression point in tape being pulled upward.

a line of compression that runs ahead of the line of separation as you peel the tape (Figure 7.10). The line of compression can be seen if you stick two tape strips together and slowly separate them. What causes the compression?

950; 1432.

shear

stress

7.11
Footprints in the sand

Have you ever strolled along the beach as the water was receding? As you step onto the firm sand, the sand around your foot immediately dries out and turns white. The popular explanation for the whitening is that the water is squeezed out of the sand by your weight. That, however, is not the case, because sand does not behave at all like a sponge. So, what does cause the whitening? Does it last as long as you stand there?

924, p. 373; 937; 938; 1313, pp. 288-294; 1433, pp. 624-626; 1434; 1435.

stress

7.12
Balloon filled with water and sand

Partially fill a rubber balloon with sand and water so there is more than enough water to cover the sand but not enough to fill the entire balloon. Then tie up the top and try squeezing the balloon. Pretty easy at first, isn't it? As you continue to compress the balloon, however, you'll suddenly find a point where the balloon just refuses to bulge even though you squeeze for all you're worth. What causes this sudden and determined resistance to further squeezing?

924, p. 373; 938; 1313, pp. 288-294; 1433, pp. 624-626; 1434; 1435.

7.13
Buying a sack of corn

In the days when shucked corn was sold by volume rather than by weight, vendors would make the corn assume as much volume as possible. Hence, a bag of corn, while appearing full, may have had less corn in it than another bag of the same size sold by a more-honest merchant. Faced with this problem, should the buyer have tried to press a bag so as to make the corn denser? Does the corn's volume decrease if you press on the bag? Actually, pressing is exactly the wrong thing to do. Why?

938; 1433, pp. 624-626; 1434; 1435.

cosmic rays

solar flares

particle reactions

7.14
Radiation levels in an airplane

Do solar flares and galactic radiation present a real danger to people in high-altitude jets? When an airplane takes off and begins its ascent, why does the net radiation level it experiences decrease for the first 1500 feet and then begin to increase with altitude? If there are significant variations in the extraterrestrial radiation, what cause those variations?

1296, pp. 392-393; 1436; 1437.

ionization and excitation

Cerenkov radiation

7.15
Flashes seen by astronauts

Astronauts on the lunar missions saw white, starlike flashes when they were in space. The flashes occurred about once or twice a minute and were seen with eyes both open and closed. Apparently cosmic rays caused the flashes but how? Why did the astronauts see point flashes (sometimes with fuzzy tails) rather than a glow over the whole field of vision? Can a passenger in a high altitude jet see the flashes? (Figure 7.15.)

1438 through 1451.

Figure 7.15
(By permission of John Hart. Field Enterprises.)

X ray, UV and IR interaction with matter

7.16
X rays in the art museum

Ultraviolet light, infrared light, and X rays are often used to find oil paintings over which second paintings have been made. A painter's modifications to a picture can thus be traced, and lost paintings may be found. The technique has also been used to expose forgeries. For example, the famous art forger Hans van Meegeren would paint his imitation over an old but worthless painting so that the old canvas would lend authenticity to the counterfeit. X-ray analysis revealed van Meegeren as a fraud.

If ultraviolet and infrared light and X-rays will interact with the bottom painting, surely they must also interact with the top one. How, then, are the two paintings distinguished?

110, pp. 190–193; 1452 through 1454.

7.17
Nuclear-blast fireball

What exactly causes the fireball, that brilliant ball of light, in a nuclear blast? That is, what produces the light? How long does the fireball last, and what causes its decay? Finally, why is it initially red or reddish-brown and later white?

219, pp. 306–309; 371, pp. 20 ff; 1459.

7.18
Defensive shields in *Dune*

In *Dune* (1460), a classic science fiction novel by Frank Herbert, people wear personal shields that set up some type of "force field" that will only pass slowly moving objects. Hence, the shield would protect you from bullets and knife attacks but still allow you fresh air to breathe. Is such a protective shield physically possible?

Explaining material science to my grandmother
(7.19 through 7.24)

7.19
Friction

Can you explain friction to my grandmother? I don't mean with any really sophisticated ideas, but with some simple model. Is it caused by surface irregularities that jam and mesh together? Or is it due to electrostatic forces? Do molecular forces bring about local adhesion? Or does the harder surface penetrate the softer one, causing them to stick? This subject is so old, so commonplace, and so thoroughly investigated that surely there is a simple explanation.

3; 1462 through 1465.

7.20
The flowing roof

The National Cathedral in Washington, D.C., was built to imitate the cathedrals of medieval England. The roof was made of lead because England, with her abundance of lead, had put lead roofs on her cathedrals. Unfortunately, when the roof on the National Cathedral was only a few years old, it was discovered that "the beautiful, delicately colored, lead roof was slipping inexorably downward, sliding past the nails and battens" (1461). Apparently this was due to two factors: the latitude of

Washington and the high purity of modern lead. How do these factors explain the slipping of the lead?

1461.

7.21

Cracks

Diamond cutting is the art of fracturing a crystal in precisely the right way. Sculpture also requires good control over fracturing. If you have ever cut glass tubing, you have probably used the trick of first putting a small scratch on one side and then snapping the tube. This procedure avoids a jagged edge.

What determines where a crack will go? Why does one start and propagate at all? I can fracture a piece of glass with a stress that is much less than that needed to break the atomic bonding, but the bonding is nevertheless broken. How is the atom-atom ripping accomplished with such relatively small applied forces?

1466 through 1474.

7.22

Chrome corrosion

Your car's chrome finish may corrode with time, although recently that problem has become much less likely. Corrosion would set in at the defects in the outer layer of chromium (Figure 7.22), so in the past car engineers did their best to make a continuous, thick chromium layer to reduce

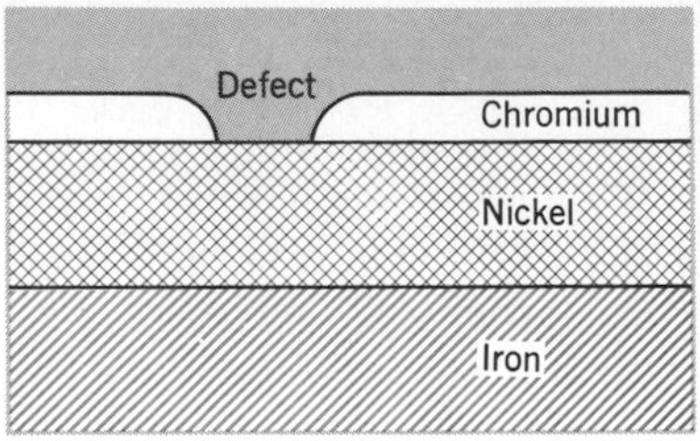

Figure 7.22
Chrome corrosion develops at the defects in the chromium.

the possibility of such defects. However, blemishes were still bound to occur through normal car usage. Then it was noticed that corrosion became much less likely if the chrome finish were *full* of many small defects. So now small defects are put in on purpose. Why does a defect in the chromium layer lead to corrosion, and why do more defects lead to less corrosion?

1475.

7.23

Polishing

Laborious polishing, say of silver utensils, is the curse of many a person. What does the rubbing do? Does it cause fine scale abrasion of the surface, melt the surface, or smear the hills into the valleys on the surface? Actually, "the nature of the polishing process has been an unsettled question ever since Isaac Newton attempted to explain the physics of the process three centuries ago" (1477),* although recent work has shed more light on it. What is meant by a "smooth surface"? Smooth compared to what? What happens to the surface, on the molecular level, if the polishing is either abrasion, melting, or smearing?

1476; 1477.

*From "Polishing," by E. Rabinowicz.

7.24

Sticky fingers

How do adhesives stick? That's a simple question to ask but a very difficult one to answer. You may be tempted to dismiss it by mumbling something about intermolecular forces, but don't, for there are inherent difficulties in such a quick answer.

For example, what holds my coffee cup together? Intermolecular forces? Suppose I crack it in two and then carefully piece it back together. I'll do such a good job that the crack will hardly be visible. Will the two pieces stay together? Aren't the intermolecular forces involved the same?

Glue, paste, or some other adhesive would help here, but exactly how? Does the adhesive *have* to be sticky? Does it *have* to be fluid? Why will some adhesives work in this case whereas others will not? Are there some materials that cannot be made to adhere with any adhesive?

There are cases in which one really should worry about two materials spontaneously adhering without an adhesive. In the early days of manned space exploration there was a real concern that an astronaut's metal-soled boots would spontaneously stick to the metal space capsule. What prompted the concern? We should be thankful such ready adhesion isn't common, for otherwise the world would have long ago ground itself down into a sticky mess.

950; 1432; 1478 through 1480.

Bibliography

1 Spurr, R. T., "Frictional Oscillations," *Nature*, *189*, 50 (1961).

2 Bristow, J. R., "Kinetic Boundary Friction," *Proc. Roy. Soc. Lond.*, *A189*, 88 (1947).

3 Rabinowicz, E., "Stick and Slip," *Sci. Amer.*, *194*, 109 (May 1956).

4 Kirkpatrick, P., "Batting the Ball," *Am. J. Phys.*, *31*, 606 (1963).

5 Jorgensen, T., Jr., "On the Dynamics of the Swing of a Golf Club," *Am. J. Phys.*, *38*, 644 (1970).

6 Graham, L. A., "The Dynamics of a Golf Club" in "The Amateur Scientist," C. L. Stong, ed., *Sci. Amer.*, *210*, 131 (Jan. 1964).

7 Crane, H. R., "Problems for Introductory Physics," *Phys. Teacher*, *7*, 371 (1969).

8 Sutton, R. M., "An Introvert Rocket or Mechanical Jumping Bean," *Phys. Teacher*, *1*, 108 (1963).

9 Routh, E. J., *Dynamics of a System of Rigid Bodies*, Part 1, Dover, New York (1960).

10 Offenbacher, E. L., "Physics and the Vertical Jump," *Am. J. Phys.*, *38*, 829 (1970).

11 Kirkpatrick, P., "Notes on Jumping," *Am. J. Phys.*, *25*, 614 (1957).

12 Kirkpatrick, P., "Bad Physics in Athletic Measurements," *Am. J. Phys.*, *12*, 7 (1944).

13 Heiskanen, W. A., "The Earth's Gravity," *Sci. Amer.*, *193*, 164 (Sept. 1955).

14 Osgood, W. F., *Mechanics*, Macmillan, New York (1937).

15 Miller, J. S., "Observations on a Pile Driver," *Am. J. Phys.*, *22*, 409 (1954).

16 Gray, A., and J. G. Gray, *A Treatise on Dynamics*, Macmillan, London (1920).

17 Reid, W. P., "Weight of an Hourglass," *Am. J. Phys.*, *35*, 351 (1967).

18 Mellen, W. R., "Superball Rebound Projectiles," *Am. J. Phys.*, *36*, 845 (1968).

19 Class of W. G. Harter, "Velocity Amplification in Collision Experiments Involving Superballs," *Am. J. Phys.*, *39*, 656 (1971).

20 Kerwin, J. D., "Velocity, Momentum, and Energy Transmission in Chain Collisions," *Am. J. Phys.*, *40*, 1152 (1972).

21 Evans, G., "The Prodigious Jump of the Click Beetle," *New Scientist*, *55*, 490 (1972).

22 Fox, G. T., "On the Physics of Drag Racing," *Am. J. Phys.*, *41*, 311 (1973).

23 Metzger, E., "An Unusual Case of Simple Harmonic Motion," *Am. J. Phys.*, *40*, 1167 (1972).

24 Sommerfeld, A., *Mechanics*, Academic Press, New York (1964).

25 Bayes, J. H., and W. T. Scott, "Billiard-Ball Collision Experiment," *Am. J. Phys.*, *31*, 197 (1963).

26 Routh, E. J., *Dynamics of a System of Rigid Bodies, Advanced Part*, Macmillan, London (1930).

27 MacMillan, W. D., *Dynamics of Rigid Bodies*, McGraw-Hill, New York (1936).

28 Williamson, C., "Starting an Automobile on a Slippery Road," *Am. J. Phys.*, *11*, 160 (1943).

29 Fountain, C. R., "The Physics of Automobile Driving," *Am. J. Phys.*, *10*, 322 (1942).

30 Bothamley, F. H., "Vehicle Braking" in *Vehicle Equipment*, J. G. Giles, ed., Iliffe Books, London (1969).

31 Garwin, R. L., "Kinematics of an Ultraelastic Rough Ball," *Am. J. Phys.*, *37*, 88 (1969).

32 Strobel, G. L., "Matrices and Superballs," *Am. J. Phys.*, *36*, 834 (1968).

33 Byers, C., *Cowboy Roping and Rope Tricks*, Dover, New York (1966).

34 Caughey, T. K., "Hula-Hoop: An Example of Heteroparametric Excitation," *Am. J. Phys.*, *28*, 104 (1960).

35 Jones, D. E. H., "The Stability of the Bicycle," *Phys. Today*, *23*, 34 (Apr. 1970).

36 Crabtree, H., *An Elementary Treatment of the Spinning Tops and Gyroscopic Motion*, Chelsea, New York (1967).

37 Gray, A., *Treatise on Gyrostatics and Rotational Motion*, Dover, New York (1959).

38 Scarborough, J. B., *The Gyroscope: Theory and Applications*, Wiley-Interscience, New York (1958).

39 Gordon, J. M., "Pedalling Made Perfect," *Engineering*, *211*, 526 (1971).

40 Synge, J. L., and B. A. Griffith, *Principles of Mechanics*, 2nd ed., McGraw-Hill, New York (1949).

41 Wilson, S. S., "Bicycle Technology," *Sci. Amer.*, *228*, 81 (Mar. 1973).

42 Haag, J., *Oscillatory Motions*, Wadsworth, Calif. (1962).

43 Jones, A. T., "Physics and Bicycles," *Am. J. Phys.*, *10*, 332 (1942).

44 Pohl, R. W., *Physical Principles of Mechanics and Acoustics*, Blackie & Son, London (1932).

45 Sindelar, J., "The Physics of the Kayaker's Eskimo Roll," *Am.

Whitewater J., *16*, (4) (winter 1971).

46 Jones, A. T., "The Skidding Automobile," *Am. J. Phys.*, *5*, 187 (1937).

47 Smith, R. C., "Static vs. Spin Balancing of Automobile Wheels," *Am. J. Phys.*, *40*, 199 (1972).

48 Wright, E. H., and K. K. Kirston, in "The Amateur Scientist," C. L. Stong, ed., *Sci. Amer.*, *219*, 112 (Aug. 1968); part is contained in C. L. Stong, *The Amateur Scientist*, Simon & Schuster, New York (1960), pp. 561-563.

49 Van Riper, W., in "Amateur Scientist," C. L. Stong, ed., *Sci. Amer.*, *198*, 136 (Apr. 1958).

50 Lyall, G., ed., *The War in the Air: The Royal Air Force in World War II*, Ballantine Books, New York York (1970), p. 287.

51 Sommerfeld, A., *Mechanics of Deformable Bodies*, Academic Press, New York (1950).

52 *The Way Things Work*, Simon & Schuster, New York (1967).

53 Stong, C. L., ed., "The Amateur Scientist," *Sci. Amer.*, *204*, 177 (Apr. 1961).

54 "Photographs of a Tumbling Cat," *Nature*, *51*, 80 (1894).

55 Shonle, J. I., and D. L. Nordick, "The Physics of Ski Turns," *Phys. Teacher*, *10*, 491 (1972).

56 Kofsky, I. L., "Yo-Yo Technics in Teaching Kinematics," *Am. J. Phys.*, *19*, 126 (1951).

57 Johnson, P. B., "Leaning Tower of Lire," *Am. J. Phys.*, *23*, 240 (1955).

58 Sutton, R. M., "A Problem of Balancing," *Am. J. Phys.*, *23*, 547 (1955).

59 Eisner, L., "Leaning Tower of the Physical Reviews," *Am. J. Phys.*, *27*, 121 (1959).

60 Bundy, F. P., "Stresses in Freely Falling Chimneys and Columns," *J. Appl. Phys.*, *11*, 112 (1940).

61 Jones, A. T., "The Falling Chimney," *Am. J. Phys.*, *14*, 275 (1946).

62 Reynolds, J. B., "Falling Chimney," *Science*, *87*, 186 (1938).

63 Sutton, R. M., "Concerning Falling Chimneys," *Science*, *84*, 246 (1936).

64 Synge, J. L., and B. A. Griffith, *Principles of Mechanics*, 3rd ed., McGraw-Hill, New York (1959).

65 Hess, F., "The Aerodynamics of Boomerangs," *Sci. Amer.*, *219*, 124 (Nov. 1968).

66 Walker, G. T., "On Boomerangs," *Phil. Trans. Roy. Soc. Lond.*, *A190*, 23 (1897).

67 "Boomerang," *Encyclopaedia Britannica*, William Benton, Chicago (1970), Vol. 3, pp. 945-946.

68 Marion, J., *Classical Dynamics of Particles and Systems*, Academic Press, New York (1965), pp. 348-349.

69 Burns, G. P., "Deflection of Projectiles Due to Rotation of the Earth," *Am. J. Phys.*, *39*, 1329 (1971).

70 Muller, E. R., "A Note on: 'Deflection of Projectiles Due to Rotation of the Earth,'" *Am. J. Phys.*, *40*, 1045 (1972).

71 Kirkpatrick, P., "Effects of Form and Rotation of the Earth upon Ranges of Projectiles," *Am. J. Phys.*, *11*, 303 (1943).

72 McDonald, J. E., "The Coriolis Force," *Sci. Amer.*, *186*, 72 (May 1952).

73 Einstein, A., "The Cause of the Formation of Meanders in the Courses of Rivers and the So-Called Beer's Law" in *Essays in Science*, Philosphical Library, New York (1955), pp. 85-91.

74 Bauman, R. P., "Visualization of the Coriolis Force," *Am. J. Phys.*, *38*, 390 (1970).

75 Burns, J. A., "More on Pumping a Swing," *Am. J. Phys.*, *38*, 920 (1970).

76 Gore, B. F., "Starting a Swing from Rest," *Am. J. Phys.*, *39*, 347 (1971).

77 Gore, B. F., "The Child's Swing," *Am. J. Phys.*, *38*, 378 (1970).

78 Tea, P. L., Jr., and H. Falk, "Pumping on a Swing," *Am. J. Phys.*, *36*, 1165 (1968).

79 Siegman, A. E., "Comments on Pumping on a Swing," *Am. J. Phys.*, *37*, 843 (1969).

80 McMullan, J. T., "On Initiating Motion in a Swing," *Am. J. Phys.*, *40*, 764 (1972).

81 Poynting, J. H., and J. J. Thomson, *A Textbook of Physics*, Griffin, London (1920).

82 Jones, A. T., *Sound*, D. Van Nostrand, New York (1937).

83 Corben, H. C., and P. Stehle, *Classical Mechanics*, 2nd ed., Wiley, New York (1960), pp. 67-69.

84 Ness, D. J., "Small Oscillations of a Stabilized, Inverted Pendulum," *Am. J. Phys.*, *35*, 964 (1967).

85 Blitzer, L., "Inverted Pendulum," *Am. J. Phys.*, *33*, 1076 (1965).

86 Phelps, F. M., III, and J. H. Hunter, Jr., "An Analytical Solution of the Inverted Pendulum," *Am. J. Phys.*, *33*, 285 (1965).

87 Kalmus, H. P., "The Inverted Pendulum," *Am. J. Phys.*, *38*, 874 (1970).

88 Tomaschek, R., *A Textbook of*

Physics, Vol. II, Heat and Sound, Blackie & Son, London (1933).

89 Mather, K. B., "Why Do Roads Corrugate?" *Sci. Amer., 208*, 128 (Jan. 1963).

90 Condon, E. U., and P. E. Condon, "Effect of Oscillations of the Case on the Rate of a Watch," *Am. J. Phys., 16*, 14 (1948).

91 Rinehart, J. S., "Waterfall-Generated Earth Vibrations," *Science, 164*, 1513 (1969).

92 Klopsteg, P. E., "Physics of Bows and Arrows," *Am. J. Phys., 11*, 175 (1943).

93 Leonard, R. W., "An Interesting Demonstration of the Combination of Two Linear Harmonic Vibrations to Produce a Single Elliptical Vibration," *Am. J. Phys., 5*, 175 (1937).

94 Miller, J. S., "The Notched Stick," *Am. J. Phys., 23*, 176 (1955).

95 Laird, E. R., "A Notched Stick," *Am. J. Phys., 23*, 472 (1955).

96 Scott, G. D., "Control of the Rotor on the Notched Stick," *Am. J. Phys., 24*, 464 (1956).

97 Jacobs, J. A., "Note on the Behavior of a Certain Symmetrical Top," *Am. J. Phys., 20*, 517 (1952).

98 Synge, J. L., "On a Case of Instability Produced by Rotation," *Philosophical Mag.* (Series 7) *43*, 724 (1952).

99 Braams, C. M., "On the Influence of Friction on the Motion of a Top," *Physica, 18*, 503 (1952).

100 Hugenholtz, N. M., "On Tops Rising by Friction," *Physica, 18*, 515 (1952).

101 Fokker, A. D., "The Rising Top, Experimental Evidence and Theory," *Physica, 8*, 591 (1941).

102 Pliskin, W. A., "The Tippe Top (Topsy-Turvy Top)," *Am. J. Phys., 22*, 28 (1954).

103 Freeman, I. M., "The Tippe Top Again," *Am. J. Phys., 24*, 178 (1956).

104 Del Camp, A. R., "Tippe Top (Topsy-Turnee Top) Continued," *Am. J. Phys., 23*, 544 (1955).

105 Braams, C. M., "The Tippe Top," *Am. J. Phys., 22*, 568 (1954).

106 Johnson, F., "The Tippy Top," *Am. J. Phys., 28*, 406 (1960).

107 Hart, J. B., "Angular Momentum and Tippe Top," *Am. J. Phys., 27*, 189 (1959).

108 Stewartson, K., "On the Stability of a Spinning Top Containing Liquid," *J. Fluid Mech., 5*, 577 (1959).

109 Perry, J., *Spinning Tops and Gyroscopic Motions*, Dover, New York (1957).

110 Richardson, E. G., *Physical Science in Art and Industry*, 2nd ed., English Univ. Press, London (1946).

111 Goldreich, P., "Tides and the Earth-Moon System," *Sci. Amer., 226*, 42 (Apr. 1972).

112 Mills, B. D., Jr., "Satellite Paradox," *Am. J. Phys., 27*, 115 (1959).

113 Bacon, R. H., "On the Retardation of a Satellite," *Am. J. Phys., 27*, 69 (1959).

114 Rhee, J. W., "Simple Derivation of Satellite Paradox," *Am. J. Phys., 34*, 615 (1966).

115 Rubin, S. in "Amateur Scientist," C. L. Stong, ed., *Sci. Amer., 198*, 134 (Apr. 1958).

116 Blitzer, L., "Satellite Orbit Paradox: A General View," *Am. J. Phys., 39*, 882 (1971).

117 Sutton, R. M., "Cider from the Newtonian Apple," *Am. J. Phys., 13*, 203 (1945).

118 Sawyer, W. W., *Mathematician's Delight*, Pelican-Penguin Books, Baltimore (1954), p. 51.

119 Routh, E. J., *Analytical Statics*, Vol. II, Cambridge Univ. Press, Cambridge (1922), pp. 17-19.

120 Mason, B. S., *Roping*, Ronald Press, New York (1940).

121 Chapman, S., "Catching a Baseball," *Am. J. Phys., 36*, 868 (1968).

122 Chapman, S., "Should One Stop or Turn in Order to Avoid an Automobile Collision?" *Am. J. Phys., 10*, 22 (1942).

123 Seifert, H. S., "The Stop-Light Dilemma," *Am. J. Phys., 30*, 216 (1962).

124 Wood, A., *The Physics of Music*, University Paperbacks, Methuen, London (1962).

125 Richardson, E. G., ed., *Technical Aspects of Sound*, Vol. 1, Elsevier, New York (1953).

126 Richardson, E. G., "Mechanical Music Instruments" in Ref. 125, Chapter 18.

127 Wood, A. A., *A Textbook of Sound*, Macmillan, New York (1941).

128 Josephs, J. J., *The Physics of Musical Sound*, D. Van Nostrand, New Jersey (1967).

129 Bragg, W. L., and G. Porter, eds., *The Royal Institution Library of Science: Physical Sciences*, Elsevier, New York (1970).

130 Andrade, E. N. da C., "Sound, Sand, and Smoke: New Light on Old Problems" in Ref. 129, Vol. 9, pp. 297-311.

131 "The Amateur Scientist," *Sci. Amer., 194*, 120 (Jan. 1956).

132 Waller, M. D., "Interpreting

Chladni Figures," *Am. J. Phys.*, *25*, 157 (1957).

133 Jensen, H. C., "Production of Chladni Figures," *Am. J. Phys.*, *25*, 203 (1957).

134 Jensen, H. C., "Production of Chladni Figures on Vibrating Plates Using Continuous Excitation," *Am. J. Phys.*, *23*, 503 (1955).

135 Pierce, W. M., "Chladni Plate Figures," *Am. J. Phys.*, *19*, 436 (1951).

136 Miller, J. S., "Some Observations on Chladni Figures," *Am. J. Phys.*, *18*, 534 (1950).

137 Magrab, E. B., "Vibration of Plates" in Ref. 140, pp. 81–95.

138 Ramakrishna, B. S., "Normal Modes of Vibration of Circular Membranes" in Ref. 140, pp. 115-127.

139 Olmsted, D., *An Introduction to Natural Philosophy*, 4th ed., Charles Collins and The Baker and Taylor Co., New York (1891).

140 Albers, V. M., ed., *Suggested Experiments for Laboratory Courses in Acoustics and Vibrations,* Pennsylvania State Univ. Press, Univ. Park, Penn. (1972).

141 Tyndall, J., *The Science of Sound*, Philosophical Library, New York (1964).

142 Stephens, R. W. B., and A. E. Bate, *Wave Motion and Sound*, Edward Arnold & Co., London (1950).

143 Spandöck, F., "Sound Recording" in Ref. 125, pp. 384–385.

144 Bagnold, R. A., *The Physics of Blown Sand and Desert Dunes*, Methuen, London (1941).

145 Richardson, E. G., *Sound*, 2nd ed., Edward Arnold & Co., London (1935).

146 Bagnold, R. A., "The Shearing and Dilatation of Dry Sand and the 'Singing' Mechanism," *Proc. Roy Soc. Lond.*, *A295*, 219 (1966).

147 Ridgway, K., and R. Rupp, "Whistling Sand of Porth Oer, Caernarvonshire," *Nature*, *226*, 158 (1970).

148 Takahara, H., "Frequency Analysis of Singing Sand," *J. Acoust. Soc. Am.*, *39*, 402 (1966).

149 Ridgway, K., and J. B. Scotton, "Whistling Beaches and Seabed Sand Transport," *Nature*, *238*, 212 (1972).

150 Richardson, E. G., "Sound and the Weather," *Weather*, *2*, (Part 1), 169, (Part 2), 205 (1947).

151 Taylor, C. A., *The Physics of Musical Sounds*, American Elsevier, New York (1965).

152 Backus, J., *The Acoustical Foundations of Music,* W. W. Norton & Co., New York (1969).

153 Schellng, J. C., "The Bowed String and the Player," *J. Acoust. Soc. Am.*, *53*, 26 (1973).

154 Freeman, I. M., "Acoustic Behavior of a Rubber String," *Am. J. Phys.*, *26*, 369 (1958).

155 Sutton, R. M., *Demonstration Experiments in Physics*, McGraw-Hill, New York (1938).

156 Reynolds, O., *Papers on Mechanical and Physical Subjects*, Vol. II, Cambridge Univ. Press, Cambridge (1901).

157 Reynolds, O., "Experiments Showing the Boiling of Water in an Open Tube at Ordinary Temperatures" in Ref. 156, pp. 578-587.

158 Henderson, W. D., *The New Physics in Everyday Life*, Lyons and Carnahan, New York (1935).

159 Bragg, W., *The World of Sound,* Dover, New York (1968).

160 Osborn, F. A., *Physics of the Home*, McGraw-Hill, New York (1929).

161 Apfel, R. E., "Acoustic Cavitation" in Ref. 140, pp. 202-207.

162 Minnaert, M., "On Musical Air-Bubbles and the Sounds of Running Water," *Philosophical Mag., J. Science* (Series 7), *16*, 235 (1933).

163 Strasberg, M., "Gas Bubbles as Sources of Sound in Liquids," *J. Acoust. Soc. Am.*, *28*, 20 (1956).

164 Humphreys, W. J., *Physics of the Air*, Dover, New York (1964).

165 Neuberger, H., *Introduction to Physical Meteorology*, Pennsylvania State Univ., College of Mineral Industires, Penn. (1957).

166 Plumb, R. C., "Squeak, Skid and Glide—The Unusual Properties of Snow and Ice" in "Chemical Principles Exemplified," *J. Chem. Ed.*, *49*, 179 (1972).

167 Watson, R. B., "On the Propagation of Sound over Snow," *J. Acoust. Soc. Am.*, *20*, 846 (1948).

168 "Why Knuckles Crack," *Time*, *98*, 45 (Aug. 16, 1971).

169 Urick, R. J., "The Noise of Melting Icebergs," *J. Acoust. Soc. Am.*, *50*, 337 (1971).

170 Crawford, F. S., Jr., *Waves* (Berkeley Physics Course, Vol. 3), McGraw-Hill, New York (1968).

171 Chedd, G., *Sound*, Doubleday Science Series, Garden City, New York (1970).

172 Young, R. W., "Dependence of Tuning of Wind Instruments on Temperature," *J. Acoust. Soc. Am.*, *17*, 187 (1946).

173 Farrell, W. E., D. P. McKenzie, and R. L. Parker," On the Note Emitted from a Mug while Mixing Instant Coffee," *Proc.*

Cambridge Phil. Soc., *65*, 365 (1969).

174 West, F. M., "Change of Pitch in Certain Sounds with Distance," *Nature*, *65*, 129 (1901).

175 Van Gulik, D., "Change of Pitch in Certain Sounds with Distance," *Nature*, *65*, 174 (1901).

176 Heyl, P. R., "Change of Pitch of Sound with Distance," *Nature*, *65*, 273 (1902).

177 Erskine-Murray, J., "A New Acoustical Phenomena," *Nature*, *107*, 490 (1921).

178 Hartridge, H., H. S. Rowell, and W. B. Morton, letters about "A New Acoustical Phenomena," *Nature*, *107*, 586 (1921).

179 Shakespear, G. A., "A New Acoustic Phenomena," *Nature*, *107*, 623 (1921).

180 West, F. M., "A New Acoustic Phenomena," *Nature*, *107*, 652 (1921).

181 Crawford, F. S., "Culvert Whistlers," *Am. J. Phys.*, *39*, 610 (1971).

182 Rinard, P. M., "Rayleigh, Echoes, Chirps, and Culverts," *Am. J. Phys.*, *40*, 923 (1972).

183 Lawrence, A., *Architectural Acoustics*, Elsevier, New York (1970).

184 Kinsler, L., and A. R. Frey, *Fundamentals of Acoustics*, 2nd ed., Wiley, New York (1962).

185 Stewart, G. W., and R. B. Lindsay, *Acoustics*, D. Van Nostrand Co., New York (1930).

186 Winstanley, J. W., *Textbook on Sound*, Longmans, Green and Co., New York (1957).

187 Randall, R. H., *An Introduction to Acoustics*, Addison-Wesley, Mass. (1951).

188 Shankland, R. S., "Rooms for Speech and Music," *Phys. Teacher*, *6*, 443 (1968).

189 Shankland, R. S., and H. K. Shankland, "Acoustics of St. Peter's and Patriarchal Basilicas in Rome," *J. Acoust. Soc. Am.*, *50*, 389 (1971).

190 Shankland, R. S., "Quality of Reverberation," *J. Acoust. Soc. Am.*, *43*, 426 (1968).

191 Shankland, R. S., "The Development of Architerctural Acoustics," *Am. Scientist, 60*, 201 (1972).

192 Morse, P. M., and K. U. Ingard, *Theoretical Acoustics*, McGraw-Hill, New York (1968), Sec. 9.5.

193 Purkis, H. J., "Room Acoustics and the Design of Auditoria" in Ref. 125, pp. 122–156.

194 Kingsbury, H., "Measurement of Reverberation Time" in Ref. 140, pp. 183–186.

195 Meyer, W., and H. Kuttruff, "Progress in Architectural Acoustics" in Ref. 196, Chapter 5.

196 Richardson, E. G., and E. Meyer, eds., *Technical Aspects of Sound*, Vol. III, Elsevier, New York (1962).

197 Sabine, W. C., *Collected Papers on Acoustics*, Dover, New York (1964).

198 Rayleigh, Lord, *The Theory of Sound, Dover*, New York (1945).

199 Sato, Y., "Normal Mode Interpretation of the Sound Propagation in Whispering Galleries," *Nature*, *189*, 475 (1961).

200 Rulf, B., "Rayleigh Waves on Curved Surfaces," *J. Acoust. Soc. Am.*, *45*, 493 (1969).

201 Keller, J. B., and F. C. Karal, Jr., "Surface Wave Excitation and Propagation," *J. Appl. Phys.*, *31*, 1039 (1960).

202 Dorsey, H. G., "Acoustics of Arches," *J. Acoust. Soc. Am.*, *20*, 597 (1948).

203 Jones, A. T., "The Echoes at Echo Bridge," *J. Acoust. Soc. Am.*, *20*, 706 (1948).

204 Rayleigh, Lord, "Shadows" in Ref. 129, Vol. 6, pp. 54–61.

205 Raman, C. V., and G. A. Sutherland, "On the Whispering-Gallery Phenomenon," *Proc. Roy. Soc. Lond.*, *100*, 424 (1922).

206 Crawford, F. S., "Douglas Fir Echo Chamber," *Am. J. Phys.*, *38*, 1477 (1970).

207 Hall, W., and O. M. Mathews, *Sound*, Edward Arnold & Co., London (1951).

208 Crawford, F. S., Jr., "Chirped Handclaps," *Am. J. Phys.*, *38*, 378 (1970).

209 Colby, M. Y., *Sound Waves and Acoustics*, Henry Holt, New York (1938).

210 Brown, R. C., *Sound*, Longmans, Green and Co., New York (1954).

211 Spinney, L. B., *A Textbook of Physics*, 3rd ed., Macmillan, New York (1925).

212 Reynolds, O., "On the Refraction of Sound by the Atmosphere" in Ref. 594, pp. 157–169.

213 Stewart, G. W., *Introductory Acoustics*, D. Van Nostrand, New York (1937).

214 Ratcliffe, J. A., *Physics of the Upper Atmosphere,* Academic Press, New York (1960).

215 Ratcliffe, J. A., *Sun, Earth and Radio*: *An Introduction to the Ionosphere and Magnetosphere*, World University Library of McGraw-Hill, New York (1970).

216 Botley, C. M., "Abnormal Audibility," *Weather*, *21*, 232 (1966).

217 Richardson, E. G., "Propagation of Sound in the Atmosphere," in *Technical Aspects of Sound*, Vol. II, E. G. Richardson, ed., Elsevier, New York (1957), pp. 1-28.

218 Cox, E. F., "Atomic Bomb Blasts," *Sci. Amer., 188*, 94 (Apr. 1953).

219 Fleagle, R. G., and J. A. Businger, *An Introduction to Atmospheric Physics*, Academic Press, New York (1963).

220 Uman, M. A., *Lightning*, McGraw-Hill, New York (1969).

221 Fleagle, R. G., "The Audibility of Thunder," *J. Acoust. Soc. Am., 21*, 411 (1949).

222 Stokes, G. G., "On the Effect of Wind on the Intensity of Sound," in *Mathematical and Physical Papers*, Vol. IV, Johnson Reprint Corp., New York (1966), pp. 110–111.

223 Tufty, B., *1001 Questions Answered about Storms and Other Natural Air Disasters*, Dodd, Mead & Co., New York (1970).

224 Vonnegut, B., and C. B. Moore, "Giant Electrical Storms" in *Recent Advances in Atmospheric Electricity*, Pergamon Press, New York (1959), pp. 399–411.

225 Vonnegut, B., "Electrical Theory of Tornadoes," *J. Geophys. Res., 65*, 203 (1960).

226 Justice, A. A., "Seeing the Inside of a Tornado," *Mon. Weather Rev., 58*, 205 (1930).

227 Businger, J. A., "Rabelais' Frozen Words and Other Unidentified Sounds of the Air," *Weather, 23*, 497 (1968).

228 Hunter, J. L., *Acoustics*, Prentice-Hall, New Jersey (1962).

229 Urick, R. J., *Principles of Underwater Sound for Engineers*, McGraw-Hill, New York (1967), pp. 100–109, 113–115.

230 Frank, P. G., P. G. Bergmann, and A. Yaspan, "Ray Acoustics" in Ref. 233, pp. 65-68.

231 Spitzer, L., Jr., "Deep-Water Transmission" in Ref. 233, pp. 89-90.

232 Herring, C., "Transmission of Explosive Sound in the Sea" in Ref. 233, pp. 200-206.

233 Bergmann, P. G., and A. Yaspan, eds., *Physics of Sound in the Sea*, Part 1, Gordon and Breach Science Publ., New York (1968).

234 Capstick, J. W., *Sound*, 2nd ed., Cambridge Univ. Press, Cambridge (1922).

235 Watson, F. R., *Sound*, Wiley, New York (1935).

236 Rayleigh, Lord (John William Strutt), "On the Production and Distribution of Sound," *Philosophical Mag., 6*, 289 (1903); included in *Scientific Papers*, Vol. V, Cambridge Univ. Press, Cambridge (1912), pp. 126–141.

237 Gulick, W. L., *Hearing: Physiology and Psychophysics*, Oxford Univ. Press, New York (1971).

238 Van Bergeijk, W. A., J. R. Pierce, and W. E. David, Jr., *Waves and the Ear*, Doubleday Anchor, New York (1960).

239 Members of the Transmission Research Dept. of Bell Telephone Labs, Inc. (Murray Hill, New Jersey), "Speech" in Ref. 125, Chapter 10.

240 Pockman, L. T., "The Resonance Frequency of a Trash Can," *Am. J. Phys., 15*, 359 (1947).

241 Troke, R. W., "Tube-Cavity Resonance," *J. Acoust. Soc. Am., 44*, 684 (1968).

242 Benade, A. H., "The Physics of Brasses," *Sci. Amer., 229*, 24 (July 1973).

243 Seifert, H. S., "A Miniature Kundt Tube," *Am. J. Phys., 7*, 421 (1939).

244 Hammond, H. E., "A Variation of Kundt's Method for the Speed of Sound," *Am. J. Phys., 7*, 423 (1939).

245 Brinker, B. L., "Preparing Rods for Stroking in the Kundt's Tube Experiment," *Am. J. Phys., 18*, 579 (1950).

246 Parsons, K. A., "Exciting a Kundt's Tube with a Siren," *Am. J. Phys., 21*, 392 (1953).

247 Hammond, H. E., "Exciting a Kundt's Tube with a Siren," *Am. J. Phys., 21*, 475 (1953).

248 Carman, R. A., "Kundt Tube Dust Striations," *Am. J. Phys., 23*, 505 (1955).

249 Hastings, R. B., and Y.-Y. Shih, "Experiments with an Electrically Operated Kundt Tube," *Am. J. Phys., 30*, 512 (1962).

250 Hastings, R. B., "Thermistor Explorations in a Kundt Tube," *Am. J. Phys., 37*, 709 (1969).

251 Callaway, D. B., F. G. Tyzzer, and H. C. Hardy, "Resonant Vibrations in a Water-Filled Piping System," *J. Acoust. Soc. Am., 23*, 550 (1951).

252 Davies, H. G., and J. E. F. Williams, "Aerodynamic Sound Generation in a Pipe," *J. Fluid Mech., 32*, 765 (1968).

253 Whitman, W. G., *Household Physics*, 3rd ed., Wiley, New York (1939).

254 Fry, D. B., and P. Denes, "The Role of Acoustics in Phonetic Studies" in Ref. 196, pp. 1-22.

255 Taylor, G. I., *The Scientific Papers of Sir Geoffrey Ingram Taylor*, G. K. Batchelor, ed., Cambridge Univ. Press, Cambridge,

Vol. III (1963), Vol. IV (1971).

256 Jeans, J., *Science of Music*, Cambridge Univ. Press, Cambridge (1953).

257 Richardson, E. G., "Flow Noise" in Ref. 196, Chapter 3.

258 Chanaud, R. C., "Aerodynamic Whistles," *Sci. Amer., 222*, 40 (Jan. 1970).

259 Taylor, G. I., "The Singing of Wires in a Wind," *Nature, 113*, 536 (1924); also in Ref. 255, Vol. III, p. 69.

260 Phillips, O. M., "The Intensity of Aeolian Tones," *J. Fluid Mech., 1*, 607 (1956).

261 Suzuki, S., "Aeolian Tones in a Forest and Flowing Cloudlets over a Hill," *Weather, 13*, 20 (1958).

262 Vonnegut, B., "A Vortex Whistle," *J. Acoust. Soc. Am., 26*, 18 (1954).

263 Powell, A., "On the Edgetone," *J. Acoust. Soc. Am., 33*, 395 (1961).

264 Gross, M. J., "Underwater Edge Tones," *Acustica, 9*, 164 (1959).

265 Curle, N., "The Mechanics of Edge-Tones," *Proc. Roy. Soc. Lond., A216*, 412 (1953).

266 Powell, A., "On Edgetones and Associated Phenomena," *Acustica, 3*, 233 (1953).

267 Powell, A., "Aspects of Edgetone Experiment and Theory," *J. Acoust. Soc. Am., 37*, 535 (1965).

268 Chanaud, R. C., and A. Powell, "Some Experiments Concerning the Hole and Ring Tone," *J. Acoust. Soc. Am., 37*, 902 (1965).

269 Beavers, G. S., and A. Wilson, "Vortex Growth in Jets," *J. Fluid Mech., 44*, 97 (1970).

270 Littler, T. S., *The Physics of the Ear*, Pergamon, New York (1965).

271 Culver, C. A., *Musical Acoustics*, Blakiston Co., Philadelphia (1941).

272 Helmholtz, H. L. F., *Sensations of Tone*, Dover, New York (1954).

273 Olson, D., "Musical Combination Tones and Oscillations of the Ear Mechanism," *Am. J. Phys., 37*, 730 (1969).

274 Jones A. T., "The Discovery of Difference Tones," *Am. J. Phys., 3*, 49 (1935).

275 Williamson, C., "Demonstration of Subjective Harmonics," *Am. J. Phys., 21*, 316 (1953).

276 Bragg, W., "Combination Tones in Sound and Light" in Ref. 129, Vol. 10, pp. 404–413.

277 Smoorenburg, G. F., "Combination Tones and Their Origin," *J. Acoust. Soc. Am., 52*, 615 (1972).

278 Smoorenburg, G. F., "Audibility Region of Combination Tones," *J. Acoust. Soc. Am., 52*, 603 (1972).

279 Wilson, J. P., and J. R. Johnstone," Basilar Membrane Correlates of the Combination Tone $2f_1$-f_2," *Nature, 241*, 206 (1973).

280 Novick, A., "Echolocation in Bats: Some Aspects of Pulse Design," *Am. Scientist, 59*, 198 (1971).

281 Simmons, J. A., "Echolocation in Bats: Signal Processing of Echoes for Target Range," *Science, 171*, 925 (1971).

282 Aidley, D. J., "Echo Intensity in Range Estimation by Bats," *Nature, 224*, 1330 (1969).

283 Griffin, D. R., *Echoes of Bats and Men*, Anchor Books, Doubleday, New York (1959).

284 Griffin, D. R., "More about Bat 'Radar'," *Sci. Amer., 199*, 40 (July 1958).

285 MacLean, W. R., "On the Acoustics of Cocktail Parties," *J. Acoust. Soc. Am., 31*, 79 (1959).

286 Mitchell, O. M. M., C. A. Ross, and G. H. Yates, "Signal Processing for a Cocktail Party Effect," *J. Acoust. Soc. Am., 50*, 656 (1971).

287 Wilson, H. A., "Sonic Boom," *Sci. Amer., 206*, 36 (Jan. 1962).

288 Carlson, H. W., and F. E. McLean, "The Sonic Boom," *Internat. Sci. Tech.* (Jul. 1966) p. 70.

289 Cox, E. F., H. J. Plagge, and J. W. Reed, "Meteorology Directs Where Blast Will Strike," *Bul. Am. Met. Soc., 35*, 95 (1954).

290 Pack, D. H., "Simplification of Method," *Bul. Am. Met. Soc., 39*, 364 (1958).

291 Nicholls, J. M., "Meteorological Effects on the Sonic Bang," *Weather, 25*, 265 (1970).

292 Carlson, H. W., and D. J. Maglieri, "Review of Sonic-Boom Generation Theory and Prediction Methods," *J. Acoust. Soc. Am., 51*, 675 (1972).

293 Reed, J. W., "Atmospheric Focusing of Sonic Booms," *J. Appl. Met., 1*, 265 (1962).

294 Nicholls, J. M., "A Note on the Calculation of 'Cut-Off' Mach Number," *Met. Mag., 100*, 33 (1971).

295 Hayes, W. D., "Sonic Boom" in *Annual Review of Fluid Mechanics*, Vol. 3, M. Van Dyke, W. G. Vincenti, and J. V. Wehausen, eds. Annual Reviews, Calif. (1971), pp. 269–290.

296 Nicholls, J. M., and B. F. James, "The Location of the Ground Focus Line Produced by a Transonically Accelerating Aircraft," *J. Sound Vibration, 20*, 145 (1972).

297 Ribner, H. S., "Supersonic Turns without Superbooms," *J. Acoust. Soc. Am., 52*, 1037 (1972).

298 Pierce, A. D., and D. J. Maglieri, "Effects of Atmospheric Irregularities on Sonic-Boom Propagation," *J. Acoust. Soc. Am.*, *51*, 702 (1972).

299 Viemeister, P. E., *The Lightning Book,* MIT Press, Cambridge, Mass. (1972).

300 Malan, D. J., *Physics of Lightning*, English Universities Press, London (1963).

301 Schonland, B., *The Flight of Thunderbolts*, Clarendon Press, Oxford (1964).

302 Remillard, W. J., "The History of Thunder Research," *Weather*, *16*, 245 (1961).

303 Colgate, S. A., and C. McKee, "Electrostatic Sound in Clouds and Lightning," *J. Geophys. Res.*, *74*, 5379 (1969).

304 Holmes, C. R., M. Brook, P. Krehbiel, and R. McCrory, "On the Power Spectrum and Mechanism of Thunder," *J. Geophys. Res.*, *76*, 2106 (1971).

305 Jones, D. L., G. G. Goyer, and M. N. Plooster, "Shock Wave from a Lightning Discharge," *J. Geophys. Res.*, *73*, 3121 (1968).

306 Jones, A. T., "Secondary Shock Waves and an Unusual Photograph," *Am. J. Phys.*, *15*, 57 (1947).

307 Saunders, F. A., "Visible Sound Waves," *Science*, *52*, 442 (1920).

308 "A Solar Halo Phenomenon," *Nature*, *154* (1944): G. H. Archenhold, p. 433; V. Vand, p. 517; R. Holdsworth, p. 517.

309 Maxim, H. P., "The Whiplash Crack and Bullet Sound Waves," *Sci. Amer.*, *113*, 231 (1915).

310 Benade, A. H., "The Physics of Wood Winds," *Sci. Amer.*, *203*, 144 (Oct. 1960).

311 "Roar in Our Ears," *Phys. Teacher*, *4*, 46 (Jan. 1966).

312 Bekesy, G. von, "The Ear," *Sci. Amer.*, *197*, 66 (Aug. 1957).

313 Craig, R. A., *The Upper Atmosphere: Meteorology and Physics*, Academic Press, New York (1965).

314 Loose, T. C., "Champagne Recompression" in "Chemical Principles Exemplified," R. C. Plumb, ed., *J. Chem. Ed. 48*, 154 (1971).

315 Hateley, R. J., "A Footnote to the Champagne Recompression Exemplum of Henry's Law" in "Chemical Principles Exemplified," R. C. Plumb, ed., *J. Chem. Ed.*, *48*, 837 (1971).

316 Nason, E. H., *Introduction to Experimental Cookery*, McGraw-Hill, New York (1939).

317 Avery, M., *Household Physics*, Macmillan, New York (1946).

318 Keene, E. S., *Mechanics of the Household*, McGraw-Hill, New York (1918).

319 Lawrence, E. N., "Advances in Spelaeo-Meteorology," *Weather*, *27*, 252 (1972).

320 Folsom, F., *Exploring American Caves*, Collier Books, New York (1962), pp. 202, 275-276.

321 Chapin, E. K., "The Strange World of Surface Film," *Phys. Teacher*, *4*, 271 (1966).

322 Boys, C. V., *Soap Bubbles*, Doubleday Anchor Books, New York (1959).

323 Schenck, H., Jr., "Physics and Physiology in Diving Decompression," *Am. J. Phys.*, *21*, 277 (1953).

324 Dodd, L. E., "Physics of Deep-Sea Diving," *Am. J. Phys.*, *8*, 181 (1940).

325 Schenck, H., Jr., "Emergency Ascent of an Undersea Diver from Great Depths," *Am. J. Phys.*, *23*, 58 (1955).

326 Taylor, H. J., "Underwater Swimming and Diving," *Nature*, *180*, 883 (1957).

327 Cooke, E. D., and C. Baranowski, "Scuba Diving and the Gas Laws" in "Chemical Principles Exemplified," R. C. Plumb, ed., *J. Chem. Ed.*, *50*, 425 (1973).

328 Cooperman, E. M. et al., "Mechanism of Death in Shallow-Water Scuba Diving," *Canadian Medical Assoc. J.*, *99*, 1128 (1968).

329 Weld, L. D., *A Textbook of Heat,* Macmillan, New York (1948).

330 Allen, H. S., and R. S. Maxwell, *A Textbook of Heat,* Part I, Macmillan, London (1939).

331 Nelkon, M., *Heat*, Blackie & Son, London (1949).

332 Feynman, R. P., R. B. Leighton, M. Sands, *The Feynman Lectures on Physics*, Addison-Wesley, Mass. (1964).

333 Brown, J. B., "Thermodynamics of a Rubber Band," *Am. J. Phys.* *31*, 397 (1963).

334 Hayward, R. in "Amateur Scientist," *Sci. Amer.*, *194*, 154 (May 1956).

335 Elliott, D. R., and S. A. Lippmann, "The Thermodynamics of Rubber at Small Extensions," *J. Appl. Phys.*, *16*, 50 (1945).

336 Carroll, H. B., M. Eisner, and R. M. Henson, "Rubber Band Experiment in Thermodynamics," *Am. J. Phys.*, *31*, 808 (1963).

337 Paldy, L. G., "Rubber Bands and Cryogenics," *Am. J. Phys.*, *32*, 388 (1964).

338 Knight, C. A., *The Freezing of Supercooled Liquids*, D. Van Nostrand, New Jersey (1967).

339 Reese, H. M., "Freezing in

Water Pipes," *Am. J. Phys.*, *19*, 425 (1951).

340 Welander, P., "Note on the Self-Sustained Oscillations of a Simple Thermal System," *Tellus*, *9*, 419 (1957).

341 Plumb, R. C., "Tire Inflation Thermodynamics" in "Chemical Principles Exemplified," *J. Chem. Ed.*, *48*, 837 (1971).

342 Batt, R., "Pop! Goes the Champagne Bottle Cork" in "Chemical Principles Exemplified," R. C. Plumb, ed., *J. Chem. Ed.*, *48*, 75 (1971).

343 Sommerfeld, A., *Thermodynamics and Statistical Physics,* Academic Press, New York (1964).

344 Landsberg, H. E., *Weather and Health*, Anchor Books, Doubleday, New York (1969).

345 Thomas, T. M., "Some Observations on the Chinook 'Arch' in Western Alberta and North-Western Montana," *Weather*, *18*, 166 (1963).

346 Brinkmann, W. A. R., "What Is a Foehn?" *Weather*, *26*, 230 (1971).

347 Riehl, H., "An Usual Chinook Case," *Weather*, *26*, 241 (1971).

348 Holmes, R. M., and K. D. Hage, "Airborne Observations of Three Chinook-Type Situations in Southern Alberta," *J. Appl. Met.*, *10*, 1138 (1971).

349 "Curing an Ill Wind," *Time*, *97*, 73 (June 14, 1971).

350 McClain, E. P., "Synoptic Investigation of a Typical Chinook Situation in Montana," *Bul. Am. Met. Soc.*, *33*, 87 (1952).

351 Cook, A. W., and A. G. Topil, "Some Examples of Chinooks East of the Mountains in Colorado," *Bul. Am. Met. Soc.*, *33*, 42 (1952).

352 Glenn, C. L., "The Chinook," *Weatherwise*, *14*, 175 (1961).

353 Oki, M., G. M. Schwab, P. E. Stevenson, and R. C. Plumb, "Chinook Winds—The Foehn Phenomenon" in "Chemical Principles Exemplified," *J. Chem. Ed.*, *48*, 154 (1971).

354 Ives, R. L., "Weather Phenomena of the Colorado Rockies," *J. Franklin Institute*, *226*, 691 (1938).

355 Pedgley, D. E., "Weather in the Mountains," *Weather*, *22*, 266 (1967).

356 Virgo, S. E., "Hazards of the Foehn Wind in Switzerland," *Weather*, *21*, 306 (1966).

357 Thambyahpillay, G., "The Kachchan—a Foehn Wind in Ceylon," *Weather*, *13*, 107 (1958).

358 Dordick, I., "The Influence of Variations in Atmospheric Pressure upon Human Beings," *Weather*, *13*, 359 (1958).

359 Plumb, R. C., "The Convertible Effect" in "Chemical Principles Exemplified," *J. Chem. Ed.*, *49*, 285 (1972).

360 Wood, E., *Science for the Airplane Passenger*, Houghton Mifflin Co., Boston (1968).

361 Scorer, R., and H. Wexler, *Cloud Studies in Colour*, Pergamon Press, New York (1967).

362 Scorer, R., *Clouds of the World,* Stackpole Books, Harrisburg, Pennsylvania (1972).

363 Tricker, R. A. R., *The Science of the Clouds*, American Elsevier, New York (1970).

364 Scorer, R. S., *Natural Aerodynamics*, Pergamon Press, New York (1958).

365 Sartor, J. D., "Clouds and Precipitation," *Phys. Today*, *25*, 32 (Oct. 1972).

366 Scorer, R. S., "Lee Waves in the Atmosphere," *Sci. Amer.*, *204*, 124 (Mar. 1961).

367 Ludlam, F. H., "Hill-Wave Cirrus," *Weather*, *7*, 300 (1952).

368 Scorer, R. S., "Forecasting the Occurrence of Lee Waves," *Weather*, *6*, 99 (1951).

369 Monteith, J. L., "Lee Waves in Wester Ross," *Weather*, *13*, 227 (1958).

370 Wallington, C. E., "An Introduction to Lee Waves in the Atmosphere," *Weather*, *15*, 269 (1960).

371 Glasstone, S., *The Effects of Nuclear Weapons*, U. S. Atomic Energy Commission (June 1957).

372 Sutton, O. G., "The Atom Bomb Trial as an Experiment in Convection," *Weather*, *2*, 105 (1947).

373 Richards, J. M., "The Effect of Wind Shear on a Puff," *Quart. J. Roy. Met. Soc.*, *96*, 702 (1970).

374 "Fallstreak Holes," *Weather*, *19*, 90 (1964).

375 Ludlam, F. H., "Fall-Streak Holes," *Weather*, *11*, 89 (1956).

376 Brunt, D., "Patterns in Ice and Cloud," *Weather*, *1*, 184 (1946).

377 Kinney, J. R., "Hole-in-Cloud," *Bul. Am. Met. Soc.*, *49*, 990 (1968).

378 Johnson, H. M., and R. L. Holle, "Observations and Comments on Two Simultaneous Cloud Holes over Miami," *Bul. Am. Met. Soc.*, *50*, 157 (1969).

379 Photographs and comments on hole-in-cloud: *Weatherwise*, *21*, cover photo (Aug. 1968); p. 194 (Oct. 1968); p. 238 (Dec. 1968); *22*, p. 19 (Feb. 1969).

380 Larmore, L., and F. F. Hall, Jr., "Optics for the Airborne Observer," *J. Soc. Photo-Optical Instrumentation Engineers*, *9*, 87 (Feb.-Mar. 1971).

381 Scorer, R. S., and L. J. Davenport, "Contrails and Aircraft Downwash," *J. Fluid Mech.*, *43*, 451 (1970).

382 Subrahmanyam, V. P., and G. Nicholson, "Contrail Shadows," *Weather*, *22*, 244 (1967).

383 Wallington, C. E., "Distrail in a Wave Cloud," *Weather*, *22*, 454 (1967).

384 Dunning, H. H., and N. E. La Seur, "An Evaluation of Some Condensation Trail Observations," *Bul. Am. Met. Soc.*, *36*, 73 (1955).

385 Appleman, H., "The Formation of Exhaust Condensation Trails by Jet Aircraft," *Bul. Am. Met. Soc.*, *34*, 14 (1953).

386 Scorer, R. S., "Condensation Trails," *Weather*, *10*, 281 (1955).

387 Brewer, A. W., "Condensation Trails," *Weather*, *1*, 34 (1946).

388 Byers, H. R., *General Meteorology*, McGraw-Hill, New York (1959).

389 McDonald, J. E., "Homogeneous Nucleation of Vapor Condensation. I. Thermodynamic Aspects," *Am. J. Phys.*, *30*, 870 (1962); "II. Kinetic Aspects," *31*, 31 (1963).

390 Monahan, E. C., "Sea Spray and Whitecaps," *Oceanus*, *14*, 21 (Oct. 1968).

391 Mann, C. R., and G. R. Twiss, *Physics*, Scott, Foresman and Co., Chicago (1910).

392 Brown, S., *Count Rumford: Physicist Extraordinary*, Anchor Books, Doubleday, New York (1962).

393 Wilson, M., "Count Rumford," *Sci. Amer.*, *203*, 158 (Oct. 1960).

394 Achenbach, P. R., "Physics of Chimneys," *Phys. Today*, *2*, 18 (Dec. 1949).

395 Turner, J. S., "A Comparison between Buoyant Vortex Rings and Vortex Pairs," *J. Fluid Mech.*, *7*, 419 (1960).

396 Lilly, D. K., "Comments on Case Studies of a Convective Plume and a Dust Devil,'"; J. C. Kaimal and J. A. Businger, "Reply," *J. Appl. Met.*, *10*, 590 (1971).

397 Birely, E. W., and E. W. Hewson, "Some Restrictive Meteorological Conditions to be Considered in the Design of Stacks," *J. Appl. Met.*, *1*, 383 (1962).

398 Csanady, G. T., "Bent-Over Vapor Plumes," *J. Appl. Met.*, *10*, 36 (1971).

399 Tricker, R. A. R., *Bores, Breakers, Waves and Wakes*, American Elsevier, New York (1965).

400 Corrsin, S., "Turbulent Flow," *Am. Scientist*, *49*, 300 (1961).

401 Whipple, F. J. W., "Modern Views on Atmospheric Electricity," *Quart. J. Roy. Met. Soc.*, *64*, 199 (1938).

402 Turnbull, D., "The Undercooling of Liquids," *Sci. Amer.*, *212*, 38 (Jan. 1965).

403 Chalmers, B., "How Water Freezes," *Sci. Amer.*, *200*, 144 (Feb. 1959).

404 Mason, B. J., "The Growth of Snow Crystals," *Sci. Amer.*, *204*, 120 (Jan. 1961).

405 Kell, G. S., "The Freezing of Hot and Cold Water," *Am. J. Phys.*, *37*, 564 (1969).

406 Deeson, E., "Cooler—Lower Down," *Phys. Ed.*, *6*, 42 (1971).

407 Firth, I., "Cooler?" *Phys. Ed.*, *6*, 32 (1971).

408 Mpemba, E. B., and D. G. Osborne, "Cool?" *Phys. Ed.*, *4*, 172 (1969).

409 Ahtee, M., "Investigation into the Freezing of Liquids," *Phys. Ed.*, *4*, 379 (1969).

410 Wray, E. M., "Cool Origins," *Phys. Ed.*, *6*, 385 (1971).

411 Letters, *New Scientist*, *42*, 655-656 (1969); *43*, 89, 158, 662 (1969); *44*, 205 (1969); *45*, 225 (1970).

412 Tilley, D. E., and W. Thumm, *College Physics*, Cummings, California (1971).

413 Milikan, R. A., H. G. Gale, and and W. R. Pyle, *Practical Physics*, Ginn, Boston (1920).

414 Pounder, E. R., *The Physics of Ice*, Pergamon Press, New York (1965).

415 Plasschaert, J. H. M., "Weather and Avalanches," *Weather*, *24*, 99 (1969).

416 LaChapelle, E. R., "The Control of Snow Avalanches," *Sci. Amer.*, *214*, 92 (Feb. 1966).

417 Atwater, M. M., "Snow Avalanches," *Sci. Amer.*, *190*, 26 (Jan. 1954).

418 Stearns, H. O., *Fundamentals of Physics and Applications*, Macmillan, New York (1956).

419 Reynolds, O., "On the Slipperiness of Ice" in Ref. 156, pp. 734-738.

420 Pupezin, J., G. Jancso, and W. A. Van Hook, "The Vapor Pressure of Water: A Good Reference System?" *J. Chem. Ed.*, *48*, 114 (1971); see footnote 4.

421 Roberts, J. K., and A. R. Miller, *Heat and Thermodynamics*, 5th ed., Wiley-Interscience, New York (1960).

422 Plumb, R. C., "Sliding Friction and Skiing" in "Chemical Principles Exemplified," *J. Chem. Ed.*, *49*, 830 (1972).

423 Adamson, A. W., *Physical Chemistry of Surfaces*, Wiley-Interscience, New York (1960), pp. 335-337.

424 Bowden, F. P., and D. Tabor, *The Friction and Lubrication of Solids*, Oxford Clarendon Press, Oxford (1950), pp. 66-71.

425 Kingery, W. D., and W. H. Goodnow, "Brine Migration in Salt Ice" in *Ice and Snow*, W. D. Kingery, ed., MIT Press, Cambridge, Mass. (1963), pp. 237-247.

426 Glenn, H. T., *Glenn's Auto Repair Manual*, Chilton Books, Philadelphia (1967), p. 214.

427 Thumm, W., and D. E. Tilley, *Physics in Medicine*, Cummings, California (1972).

428 Millington, R. A., "Physiological Responses to Cold," *Weather*, *19*, 334 (1964).

429 Steadman, R. G., "Indices of Windchill of Clothed Persons," *J. Appl. Met.*, *10*, 674 (1971).

430 Plumb, R. C., "Faster Dinner via Molecular Potential Energy" in "Chemical Principles Exemplified," *J. Chem. Ed.*, *49*, 706 (1972).

431 Eastman, G. Y., "The Heat Pipe," *Sci. Amer.*, *218*, 38 (May 1968).

432 Haggin, J., "Heat Pipes," *Chemistry*, *46*, 25 (Jan. 1973).

433 Plumb, R. C., "Physical Chemistry of the Dunking Duck," *J. Chem. Ed.*, *50*, 213 (1973).

434 Gaines, J. L., "Dunking Bird," *Am. J. Phys.*, *27*, 189 (1959).

435 Miller, J. S., "Physics of the Dunking Duck," *Am. J. Phys.*, *26*, 42 (1958).

436 Kolb, K. B., " 'Reciprocating' Engine," *Phys. Teacher*, *4*, 121 (1966).

437 Frank, D. L., "The Drinking Bird and the Scientific Method," *J. Chem. Ed.*, *50*, 211 (1973).

438 Hurst, G. W., "Frost: Aberystwyth 1967 Symposium," *Weather*, *22*, 445 (1967).

439 Bainbridge, J. W., "Stocking Northumbrain Icehouses: An Exercise in Relating Climate to History," *Weather*, *28*, 68 (1973).

440 Weisskopf, V. F., "Modern Physics from an Elementary Point of View," CERN Summer Lectures, Geneva, 1969, pp. 8-10.

441 Barton, A. W., *A Textbook on Heat*, Longmans, Green, and Co., New York (1935), pp. 131-133.

442 Leidenfrost, J. G., "On the Fixation of Water in Diverse Fire," *Internat. J. Heat Mass Transfer*, *9*, 1153 (1966).

443 Gottfried, B. S., C. J. Lee, and K. J. Bell, "The Leidenfrost Phenomenon: Film Boiling of Liquid Droplets on a Flat Plate," *Internat. J. Heat Mass Transfer*, *9*, 1167 (1966).

444 Hickman, K. D., "Floating Drops and Boules," *Nature*, *201*, 985 (1964).

445 Hall, R. S., S. J. Board, A. J. Clare, R. B. Duffey, T. S. Playle, and D. H. Poole, "Inverse Leidenfrost Phenomenon," *Nature*, *224*, 266 (1969).

446 Holter, N. J., and W. R. Glasscock, "Vibrations of Evaporating Liquid Drops," *J. Acoust. Soc. Am.*, *24*, 682 (1952).

447 Gaddis, V. H., *Mysterious Fires and Lights*, David McKay Co., Inc., New York (1967), pp. 133-155.

448 Photos, *National Geographic Mag.*, *114*, 543-545 (1958); *129*, 482-483 (1966).

449 Gibson, W., *The Master Magicains*, Doubleday, New York (1966), p. 204.

450 Rinehart, J. S., "Old Faithful Geyser," *Phys. Teacher*, *7*, 221 (1969).

451 Fournier, R. O., "Old Faithful: A Physical Model," *Science*, *163*, 304 (1969).

452 Muffler, L. J. P., D. E. White, and A. H. Truesdell, "Hydrothermal Explosion Craters in Yellowstone National Park," *Geol. Soc. Am. Bul.*, *82*, 723 (1971).

453 Prandtl, L., *Essentials of Fluid Dynamics*, Blackie & Sons, London (1967).

454 Mackay, R. S., "Boat Driven by Thermal Oscillations," *Am. J. Phys.*, *26*, 583 (1958).

455 Finnie, I., and R. L. Curl, "Physics in a Toy Boat," *Am. J. Phys.*, *31*, 289 (1963).

456 Miller, J. S., "Physics in a Toy Boat," *Am. J. Phys.*, *26*, 199 (1958).

457 Baker, J. G., "Self-Induced Vibrations," *Trans. Am. Soc. Mechanical Engineers*, *55*, APM-55-2 (1933), pp. 5-13.

458 Satterly, J., "Casual Observations on Milk, Pickled Beet-Root, and Dried-Up Puddles," *Am. J. Phys.*, *24*, 529 (1956).

459 Kelley, J. B., "Heat, Cold and Clothing," *Sci. Amer.*, *194*, 109 (Feb. 1956).

460 Fanger, P. O., *Thermal Comfort,* Danish Technical Press, Copenhagen (1970).

461 Cowles, R. B., "Black Pigmentation: Adaptation for Concealment or Heat Conservation," *Science*, *158*, 1340 (1967).

462 Ruchlis, H., *Bathtub Physics*, Harcourt, Brace & World, New York (1967).

463 Irving, L., "Human Adaptation to Cold," *Nature*, *185*, 572 (1960).

464 Irving, L., "Adaptations to Cold," *Sci. Amer.*, *214*, 94 (Jan. 1966).

465 Mazess, R. B., and R. Larsen, "Responses of Andean Highlanders to Night Cold," *Internat. J. Biometeorology*, *16*, 181 (1972).

466 Henderson, S. T., *Daylight and Its Spectrum*, American Elsevier, New York (1970).

467 Frisken, W. R., "Extended Industrial Revolution and Climate Change," *EOS, Trans. Am. Geophys. Union*, *52*, 500 (1971).

468 Lee, R., "The 'Greenhouse' Effect," *J. Appl. Met.*, *12*, 556 (1973).

469 Wood, R. W., "Note on the Theory of the Greenhouse," *Philosophical Mag.* (Series 6), *17*, 319 (1909).

470 Hoyle, F., *The Black Cloud*, Perennial Library, Harper & Row, New York (1957).

471 Bedford, E. A., *General Science*, Allyn and Bacon, New York (1921).

472 Stong, C. L., ed., "The Amateur Scientist," *Sci. Amer.*, *216*, 128 (Jan. 1967).

473 Peterson, E., and A. W. H. Damman, "Convection Plumes from *Ulmus americana L.*," *Science*, *148*, 392 (1965).

474 Ward, D. B., and J. Beckner, "Convection Plumes from Trees," *Science*, *149*, 764 (1965).

475 Hackman, R., "Convection Plumes from Trees," *Science*, *149*, 764 (1965).

476 Drapeau, R. E., "Convection-Plume-Like Phenomenon," *Science*, *150*, 509 (1965).

477 Rigby, M., "Convection Plumes and Insects," *Science*, *150*, 783 (1965).

478 Corbet, P. S., and J. A. Downes, "Convection Plumes from Trees," *Science*, *150*, 1629 (1965).

479 Steyskal, C., "Convection Plumes from Trees," *Science*, *150*, 1629 (1965).

480 Wiersma, J. H., "Convection Plumes and Insects," *Science*, *152*, 387 (1966).

481 Mason, D. T., "Density-Current Plumes," *Science*, *152*, 354 (1966).

482 Plumb, R. C., "Knowing Some Thermodynamics Can Save a Life" in "Chemical Principles Exemplified," *J. Chem. Ed.*, *49*, 112 (1972).

483 Chandler, T. J., "City Growth and Urban Climates," *Weather*, *19*, 170 (1964).

484 Lowry, W. P., The Climate of Cities," *Sci. Amer.*, *217*, 15 (Aug. 1967).

485 Hutcheon, R. J., R. H. Johnson, W. P. Lowry, C. H. Black, and D. Hadley, "Observations of the Urban Heat Island in a Small City," *Bul. Am. Met. Soc.*, *48*, 7 (1967).

486 Preston-Whyte, R. A., "A Spatial Model of an Urban Heat Island," *J. Appl. Met.*, *9*, 571 (1970).

487 Olfe, D. B., and R. L. Lee, "Linearized Calculations of Urban Heat Island Convection Effects," *J. Atmos. Sci.*, *28*, 1374 (1971).

488 Woollum, C. A., "Notes from a Study of the Microclimatology of the Washington, D. C. Area for the Winter and Spring Seasons," *Weatherwise*, *17*, 263 (1964).

489 Mitchell, J. M., "The Temperature of Cities," *Weatherwise*, *14*, 224 (1961).

490 Sundborg, A., "Local Climatological Studies of the Temperature Conditions in an Urban Area," *Tellus*, *2*, 222 (1950).

491 Duckworth, F. S., and J. S. Sandberg, "The Effect of Cities upon Horizontal and Vertical Temperature Gradients," *Bul. Am. Met. Soc.*, *35*, 198 (1954).

492 Bornstein, R. D., "Observations of the Urban Heat Island Effect in New York City," *J. Appl. Met.*, *7*, 575 (1968).

493 Myrup, L. O., "A Numerical Model of the Urban Heat Island," *J. Appl. Met.*, *8*, 908 (1969).

494 Emden, R., "Why Do We Have Winter Heating," *Nature*, *141*, 908 (1938).

495 Bilkadi, Z., and W. B. Bridgman, "When You Heat Your House Does the Thermal Energy Content Increase?" in "Chemical Principles Exemplified," R. C. Plumb, ed., *J. Chem. Ed.*, *49*, 493 (1972).

496 Plumb, R. C., "Footnote to the House Heating Exemplum" in "Chemical Principles Exemplified," *J. Chem. Ed.*, *50*, 365 (1973).

497 Plumb, R. C., "Are We Teaching the Most Useful Ideas about Transport?" in "Chemical Principles Exemplified," *J. Chem. Ed.*, *49*, 112 (1972).

498 Walton, A. G., "Nucleation of Crystals from Solution," *Science*, *148*, 601 (1965).

499 Bragg, W., "Ice" in Ref. 129, Vol. 10, pp. 377 ff.

500 Knight, C., and N. Knight, "Snow Crystals," *Sci. Amer.*, *227*, 100 (Jan. 1973).

501 Tolansky, S., "Symmetry of Snow Crystals," *Nature*, *181*, 256 (1958).

502 "The Six-Sided Snowflake," *Chemistry*, *44*, 19 (Sep. 1971).

503 Mason, B. J., "On the Shapes of Snow Crystals" in *The Six-Cornered Snowflake* by J. Kepler,

Oxford Clarendon Press, London (1966), pp. 47-56.
504 Bentley, W. A., and W. J. Humphreys, *Snow Crystals*, Dover, New York (1962).
505 Nakaya, U., *Snow Crystals*, Harvard Univ. Press, Cambridge (1954).
506 LaChapelle, E. R., *Field Guide to Snow Crystals*, Univ. Washington Press, Seattle (1969).
507 Chapin, E. K., "Two Contrasting Theories of Capillary Action," *Am. J. Phys.*, *27*, 617 (1959).
508 Schwartz, A. M., "Capillarity: Theory and Practice," *Industrial & Engineering Chem.*, *61*, 10 (Jan. 1968).
509 Dempsey, D. F., "Measurement of the Capillary Curve," *Am. J. Phys.*, *26*, 89 (1958).
510 Thomson, W., "Capillary Attraction" in Ref. 129, Vol. 3, pp. 325-349.
511 Walton, A. J., "Surface Tension and Capillary Rise," *Phys. Ed.*, *7*, 491 (1972).
512 Hayward, A. T., "Negative Pressure in Liquids: Can It Be Harnessed to Serve Man?" *Am. Scientist*, *59*, 434 (1971).
513 Zimmermann, M., "How Sap Moves in Trees," *Sci. Amer.*, *208*, 132 (Mar. 1963).
514 Plumb, R. C., "Entropy Makes Water Run Uphill—in Trees," in "Chemical Principles Exemplified," *J. Chem. Ed.*, *48*, 837 (1971).
515 Hayward, A. T. J., "Mechanical Pump with a Suction Lift of 17 Meters," *Nature*, *225*, 376 (1970).
516 Plumb, R. C., and W. B. Bridgman, "Ascent of Sap in Trees," *Science*, *176*, 1129 (1972).
517 Plumb, R. C., and W. B. Bridgman, "Columns of Liquids Bearing a Constrained Chemical Activity Gradient," *J. Phys. Chem.*, *76*, 1637 (1972).
518 Scholander, P. F., "Tensile Water," *Am. Scientist*, *60*, 584 (1972).
519 "On the Ascent of Sap," letters by P. F. Scholander, R. C. Plumb, W. B. Bridgman, H. T. Hammel, H. H. Richter, J. Levitt, T. S. Storvick, *Science*, *179*, 1248 (1973).
520 Krebs, R. D., and R. D. Walker, *Highway Materials*, McGraw-Hill, New York (1971), Chapter 6.
521 Lacy, R. E., and C. D. Ovey, *Weather*, *21*, 456 (1966).
522 Bowley, W. W., and M. D. Burghardt, "Thermodynamics and Stones," *EOS, Trans. Am. Geophys. Union*, *52*, 4 (1971).
523 Corte, A. E., "Particle Sorting by Repeated Freezing and Thawing," *Science*, *142*, 499 (1963).
524 Inglis, D. R., "Particle Sorting and Stone Migration by Freezing and Thawing," *Science*, *148*, 1616 (1965).
525 Kaplar, C. W., "Stone Migration by Freezing of Soil," *Science*, *149*, 1520 (1965).
526 Jackson, K. A., and D. R. Uhlman, "Particle Sorting and Stone Migration Due to Frost Heave," *Science*, *152*, 545 (1966).
527 Gray, D. H., "Prevention of Moisture Rise in Capillary Systems by Electrical Short Circuiting," *Nature*, *223*, 371 (1969).
528 Lawrence, A. S. C., *Soap Films*, G. Bell and Sons, London (1929).
529 Bragg, W., "Liquid Films" in Ref. 129, Vol. 10, pp. 446-457.
530 Dewar, J., "Studies on Liquid Films" in Ref. 129, Vol. 8, pp. 136-178.
531 Satterly, J., "C. V. Boy's Rainbow Cup and Experiments with Thin Films," *Am. J. Phys.*, *19*, 448 (1951).
532 Grosse, A. V., in "The Amateur Scientist," C. L. Stong, ed., *Sci. Amer.*, *229*, 110 (July 1973).
533 Bragg, W., *The Universe of Light*, Dover, New York (1959).
534 Skogen, N., "Inverted Soap Bubbles—A Surface Phenomenon," *Am. J. Phys.*, *24*, 239 (1956).
535 Lamprecht, I., and B. Schaarschmidt, "The Flickering of a Dying Flame," *Nature*, *240*, 445 (1972).
536 Green, H. L., W. R. Lane, and H. Hartley, *Particulate Clouds: Dusts, Smokes and Mists*, 2nd ed., Van Nostrand, New Jersey (1964).
537 Henry, P. S. H., "'Static' in Industry," *Phys. Ed.*, *3*, 3 (1968).
538 Price, D. J., and H. H. Brown, *Dust Explosions*, National Fire Protection Association, Boston.
539 Eden, H. F., "Electrostatic Nuisances and Hazards" in Ref. 540, pp. 425-440.
540 Moore, A. D., *Electrostatics and Its Applications*, Wiley, New York (1973).
541 *The Way Things Work, Volume Two*, Simon and Schuster, New York (1971).
542 Plumb, R. C., "The Critical Mass-Configuration in Chemical Reactions" in "Chemical Principles Exemplified," *J. Chem. Ed.*, *48*, 525 (1971).
543 Kindle, E. M., "Some Factors Affecting the Development of Mud-Cracks," *J. Geology*, *25*, 135 (1917).
544 Longwell, C. R., "Three Common Types of Desert Mud-Cracks," *Am. J. Sci.*, *215*, 136 (1928).
545 Lang, W. B., "Gigantic Drying

Cracks in Animas Valley, New Mexico," *Science*, *98*, 583 (1943).

546 Hewes, L. I., "A Theory of Surface Cracks in Mud and Lava and Resulting Geometrical Relations," *Am. J. Sci.*, *246,* 138 (1948).

547 Willden, R., and D. R. Mabey, "Giant Desiccation Fissures on the Black Rock and Smoke Creek Deserts, Nevada," *Science*, *133*, 1359 (1961).

548 Tomkins, J. Q., "Polygonal Sandstone Features in Boundary Butte Anticline Area, San Juan County, Utah," *Geol. Soc. Am. Bul.*, *76*, 1075 (1965).

549 Neal, J. T., "Polygonal Sandstone Features in Boundary Butte Anticline Area, San Juan County, Utah," *Geol. Soc. Am. Bul.*, *77*, 1327 (1966).

550 Neal, J. T., and W. S. Motts, "Recent Geomorphic Changes in Playas of Western United States," *J. Geology*, *75*, 511 (1967).

551 Neal, J. T., A. M. Langer, and P. F. Kerr, "Giant Desiccation Polygons of Great Basin Playas," *Geol. Soc. Am. Bul.*, *79*, 69 (1968).

552 Kerfoot, D. E., "Thermal Contraction Cracks in an Arctic Tundra Environment," *Arctic*, *25*, 142 (1972).

553 Leffingwell, E. de K., "Ground-Ice Wedges. The Dominant Form of Ground-Ice on the North Coast of Alaska," *J. Geology* *23*, 635 (1915).

554 Chuter, I. H., "Snow Polygons," *Weather, 14,* 139 (1959).

555 Pewe, T. L., "Sand-Wedge Polygons (Tesselations) in the McMurdo Sound Region, Antarctica—A Progress Report," *Am. J. Sci.*, *257*, 545 (1959).

556 Washburn, A. L., "Classification of Patterned Ground and Review of Suggested Origins," *Bul. Geol. Soc. Am.*, *67*, 823 (1956).

557 Conrad, V., "Polygon Nets and Their Physical Development," *Am. J. Sci.*, *244*, 277 (1946).

558 Tallis, J. H., and K. A. Kershaw, "Stability of Stone Polygons in North Wales," *Nature*, *183*, 485 (1959).

559 Morowitz, H. J., *Energy Flow in Biology*, Academic Press, New York (1968).

560 Schroedinger, E., *What Is Life*? Cambridge Univ. Press, New York (1945).

561 Fox, R. F., "Entropy Reduction in Open Systems," *J. Theoretical Biology*, *31*, 43 (1971).

562 Ciures, A., and D. Margineano, "Thermodynamics in Biology: An Intruder?" *J. Theoretical Biology*, *28*, 147 (1970).

563 Mackay, R. S., "To Determine the Greatest Depth in Water at which One Can Breathe through a Tube," *Am. J. Phys.*, *16*, 186 (1948).

564 "Manual on Lock Valves" compiled by Committee on Lock Valves, Waterways Division, Am. Soc. Civil Engineers, Headquarters of Society, New York (1930), p. 67.

565 Deacon, G. E. R., "Physics of the Ocean," *Brit. J. Appl. Phys.*, *12*, 329 (1961).

566 Gardner, M., and D. B. Eisendrath, "Mathematical Games," *Sci. Amer.*, *215*; questions, pp. 96-99 (Aug. 1966); answers, pp. 266-272 (Sep. 1966).

567 Koehl, G. M., "Archimedes' Principle and the Hydrostatic Paradox—Simple Demonstrations," *Am. J. Phys.*, *17*, 579 (1949).

568 Dodd, L., "The Hydrostatic Paradox: Phases I and II," *Am. J. Phys.*, *23*, 113 (1955).

569 Reid, W. P., "Floating of a Long Square Bar," *Am. J. Phys.*, *31*, 565 (1963).

570 Harnwell, G. P., "Submarine Physics," *Am. J. Phys.*, *16*, 127 (1948).

571 Denton, E., "The Buoyancy of Marine Animals," *Sci. Amer.*, *203*, 118 (July 1960).

572 Weltin, H., "A Paradox," *Am. J. Phys.*, *29*, 711 (1961).

573 Fermi, E., *Collected Papers*, Vol. II, Univ. Chicago Press, Chicago (1965).

574 Fermi, E., excerpt from a lecture on Taylor instability, given during the fall of 1951 at Los Alamos Scientific Lab, in Ref. 573, pp. 813-815.

575 Fermi, E., "Taylor Instability of an Incompressible Liquid" in Ref. 573, pp. 816-820.

576 Fermi, E., and J. von Neuman, "Taylor Instability at the Boundary of Two Incompressible Liquids" in Ref. 573, pp. 821-824.

577 Davies, R. M., and G. Taylor, "The Mechanics of Large Bubbles Rising through Extended Liquids and through Liquids in Tubes," *Proc. Roy. Soc. Lond.*, *A200*, 375 (1950); see Part II.

578 Taylor, G., "The Instability of Liquid Surfaces When Accelerated in a Direction Perpendicular to Their Planes. I," *Proc. Roy. Soc. Lond.*, *A201,* 192 (1950).

579 Lewis, D. J., "The Instability of Liquid Surfaces When Accelerated in a Direction Perpendicular to Their Planes. II," *Proc. Roy. Soc. Lond.*, *A202*, 81 (1950).

580 Hidy, G. M., *The Waves: The Nature of Sea Motion*, Van Nostrand Reinhold, New York (1971).

581 Stommel, H., A. B. Arons, D. Blanchard, "An Oceanographical Curiosity: The Perpetual Salt Fountain," *Deep Sea Research*, *3*, 152 (1956).

582 Stern, M. E., "The 'Salt-Fountain' and Themohaline Convection," *Tellus*, *12*, 172 (1960).

583 Stern, M. E., "Optical Measurement of Salt Fingers," *Tellus*, *22*, 76 (1970).

584 Shirtcliffe, T. G. L., and J. S. Turner, "Observations of the Cell Structure of Salt Fingers," *J. Fluid Mech.*, *41*, 707 (1970).

585 Stern, M. E., "Collective Instability of Salt Fingers," *J. Fluid Mech.*, *35*, 209 (1969).

586 Stern, M. E., and J. S. Turner, "Salt Fingers and Convecting Layers," *Deep-Sea Research*, *16*, 497 (1969).

587 Stern, M. E., "Salt Finger Convection and the Energetics of the General Circulation," *Deep-Sea Research*, *16*, 263 (1969).

588 Stern, M. E., "Lateral Mixing of Water Masses," *Deep-Sea Research*, *14*, 747 (1967).

589 Nield, D. A., "The Thermohaline Rayleigh-Jeffreys Problem," *J. Fluid Mech.*, *29*, 545 (1967).

590 Gregg, M. C., "The Microstructure of the Ocean," *Sci. Amer.*, *228*, 64 (Feb. 1973).

591 Martin, S., "A Hydrodynamic Curiosity: the Salt Oscillator," *Geophys. Fluid Dynamics*, *1*, 143 (1970).

592 Swezey, K. M., *Science Magic*, McGraw-Hill, New York (1952).

593 Smith, N. F., "Bernoulli and Newton in Fluid Mechanics," *Phys. Teacher*, *10*, 451 (1972).

594 Reynolds, O., *Papers on Mechanical and Physical Subjects,* Vol. 1, Cambridge Univ. Press (1900).

595 Reynolds, O., "On the Suspension of a Ball by a Jet of Water," in Ref. 594, pp. 1-6.

596 Satterly, J., "Running Water," *Am. J. Phys.*, *24*, 463 (1956).

597 Miller, J. S., "On Demonstrating Bernoulli's Principle," *Am. J. Phys.*, *22*, 147 (1954).

598 Daws, L. F., and R. E. Lacy, "Cowls in the Old Deer Park," *Weather*, *24*, 513 (1969).

599 Kelley, J. B., "From Archimedes to Supersonics," *Phys. Today*, *3*, 20 (Apr. 1950).

600 Price, B. T., "Airflow Problems Related to Surface Transport Systems," *Phil. Trans. Roy. Soc. Lond.*, *A269*, 327 (1971).

601 Vernikov, G. I., and M. I. Gurevich, "Aerodynamic Pressure on a Wall Due to Movement of a High Speed Train," *Fluid Dynamics*, *2*, 88 (Jul.-Aug. 1967).

602 Gurevich, M. I., "Aerodynamic Effect of Train on a Small Body," *Fluid Dynamics*, *3*, 63 (May-June 1968).

603 Webster, D. L., "What Shall We Say about Airplanes," *Am. J. Phys.*, *15*, 228 (1947).

604 Storer, J. H., "Bird Aerodynamics," *Sci. Amer.*, *186*, 24 (Apr. 1952).

605 Wild, J. M., "Airplane Flight," *Phys. Teacher*, *4*, 295 (1966).

606 McMasters, J. H., C. J. Cole, and D. A. Skinner, "Man-Powered Flight," *Am. Institute Aeronautics Astronautics Student J.*, *9*, 5 (Apr. 1971).

607 "They Wanted Wings," *Time*, *98*, 25 (Aug. 23, 1971).

608 Sherwin, K., "Man Powered Flight as a Sport," *Nature*, *238*, 195 (1972).

609 Shenstone, B. S., "Unconventional Flight" in *The Future of Aeronautics*, J. E. Allen and J. Bruce, eds., St. Martin's Press, New York (1970), Chapter 8.

610 Wilkie, D. R., "The Work Output of Animals: Flight by Birds and by Man-power," *Nature*, *183*, 1515 (1959).

611 Herreshoff, H. C., and J. N. Newman, "The Study of Sailing Yachts," *Sci. Amer.*, *215*, 60 (Aug. 1966).

612 Davidson, K. S. M., "The Mechanics of Sailing Ships and Yachts" in *Surveys in Mechanics*, G. K. Batchelor and R. M. Davies, eds., Cambridge Univ. Press, Cambridge (1956), pp. 431-475.

613 Ashton, R. in "The Amateur Scientist," *Sci. Amer.*, *195*, 128 (Aug. 1956).

614 Holford, Lord, "Problems for the Architect and Town Planner Caused by Air in Motion," *Phil. Trans. Roy. Soc. Lond.*, *A269*, 335 (1971).

615 Einstein, A., "The Flettner Ship," *Essays in Science*, Philosophical Library, New York (1955), pp. 92-97.

616 Sutton, R. M., "Baseballs *Do* Curve *and* Drop," *Am. J. Phys.*, *10*, 201 (1942).

617 Rayleigh, Lord, *Scientific Papers*, Vol. II, Dover, New York (1964), pp. 344-346.

618 Verwiebe, F. L., "Does a Baseball Curve?" *Am. J. Phys.*, *10*, 119 (1942).

619 Thomson, J. J., "The Dynamics of a Golf Ball" in Ref. 129, Vol. 7, pp. 104-119.

620 Briggs, L. J., "Effect of Spin and Speed on the Lateral Deflection (Curve) of a Baseball; and the Magnus Effect for Smooth Spheres," *Am. J. Phys.*, *27*, 589 (1959).

621 Daish, C. B., *The Physics of Ball Games*, English Univ. Press, London (1972); Part 1 also pub-

lished as *Learn Science through Ball Games*, Sterling, New York.
622 Lyttleton, R. A., "On the Swerve of a Cricket Ball," *Weather*, *12*, 140 (1957).
623 Putnam, P. C., *Power from the Wind?*, D. Van Nostrand, New York (1948), pp. 98-101.
624 Gottifredi, J. C., and G. J. Jameson, "The Growth of Short Waves on Liquid Surfaces under the Action of a Wind," *Proc. Roy. Soc. Lond.*, *A319*, 373 (1970).
625 Wilson, W. S., M. L. Banner, R. J. Flower, J. A. Michael, and D. G. Wilson, "Wind-Induced Growth of Mechanically Generated Water Waves," *J. Fluid Mech.*, *58*, 435 (1973).
626 Dixon, P. L., *The Complete Book of Surfing*, Ballantine, New York (1969).
627 Draper, L., "'Freak' Ocean Waves," *Oceanus*, *10*, 12 (June 1964); also in *Weather*, *21*, 2 (1966).
628 Bascom, W., "Ocean Waves," *Sci. Amer.*, *201*, 74 (Aug. 1959).
629 King, C. A. M., *Beaches and Coasts*, 2nd ed., St. Martin's Press, New York (1972).
630 Monahan, E. C., "Fresh Water Whitecaps," *J. Atmos. Sci.*, *26*, 1026 (1969).
631 Donelan, M., M. S. Longuet-Higgins, and J. S. Turner, "Periodicity in Whitecaps," *Nature*, *239*, 449 (1972).
632 Acosta, A. J., "Hydrofoils and Hydrofoil Craft" in *Annual Review of Fluid Mechanics*, Vol. 5, Annual Reviews, Inc., California (1973).
633 Tucker, V. A., "Waves and Water Beetles," *Phys. Teacher*, *9*, 10 (1971).
634 Tucker, V. A., "Wave-Making by Whirligig Beetles (Gyrinidae)," *Science*, *166*, 897 (1969).
635 Stoker, J. J., *Water Waves*, Wiley-Interscience, New York (1957).
636 Keith, H. D., "Simplified Theory of Ship Waves," *Am. J. Phys.*, *25*, 466 (1957).
637 Ursell, F., "On Kelvin's Ship-wave Pattern," *J. Fluid Mech.*, *8*, 418 (1960).
638 Havelock, T. H., "The Propagation of Groups of Waves in Dispersive Media, with Application to Waves on Water Produced by a Travelling Disturbance," *Proc. Roy. Soc. Lond.*, *A81*, 398 (1908).
639 Hunter, C., "On the Calculation of Wave Patterns," *J. Fluid Mech.*, *53*, 637 (1972).
640 Hogben, N., "Nonlinear Distortion of the Kelvin Ship-Wave Pattern," *J. Fluid Mech.*, *55*, 513 (1972).
641 Garrett, J. R., "On Cross Waves," *J. Fluid Mech.*, *41*, 837 (1970).
642 Faraday, M., "On the Forms and States Assumed by Fluid in Contact with Vibrating Elastic Surfaces," *Phil. Trans. Roy. Soc.*, *31*, 319 (1831).
643 Faraday, M., Entry 118 for 1 July 1831 and Entry 140 for 5 July 1831, *Faraday's Diary*, Vol. 1, 1820—June 1832, G. Bell and Sons, Ltd., London (1932).
644 Mahony, J. J., "Cross-Waves. Part 1. Theory," *J. Fluid Mech.*, *55*, 229 (1972).
645 Barnard, B. J. S., and W. G. Pritchard, "Cross-Waves, Part 2. Experiments," *J. Fluid Mech.*, *55*, 245 (1972).
646 Faraday, M., *Experimental Researches in Chemistry and Physics*, Richard Taylor and William Francis, London (1859), pp. 352-355.
647 Bowen, A. J., "Rip Currents 1. Theoretical Investigations," *J. Geophys. Res.*, *74*, 5467 (1969).
648 Bowen, A. J., and D. L. Inman, "Rip Currents 2. Laboratory and Field Observations," *J. Geophys. Res.*, *74*, 5479 (1969).
649 Bowen, A. J., D. L. Inman, and V. P. Simmons, "Wave 'Set-Down' and Set-Up," *J. Geophys. Res.*, *73*, 2569 (1968).
650 Sonu, C. J., "Comment on Paper by A. J. Bowen and D. L. Inman, 'Edge Wave and Crescentic Bars'," *J. Geophys. Res.*, *77*, 6629 (1972); reply, *77*, 6632 (1972).
651 Huntley, D. A., and A. J. Bowen, "Field Observations of Edge Waves," *Nature*, *243*, 160 (1973).
652 Russell, R. C. H., "Waves" in *Waves and Tides*, Greenwood Press, Westport, Conn. (1970).
653 Edge, R. D., "The Surf Skimmer," *Am. J. Phys.*, *36*, 630 (1968).
654 Fejer, A. A., and R. H. Backus, "Porpoises and the Bow-Riding of Ships Under Way," *Nature*, *188*, 700 (1960).
655 Scholander, P. F., "Wave-Riding of Dolphins: How Do They Do It?" *Science*, *129*, 1085 (1959).
656 Hayes, W. D., "Wave-Riding Dolphins," *Science*, *130*, 1657 (1959).
657 Scholander, P. F., "Wave-Riding Dolphins," *Science*, *130*, 1658 (1959).
658 Hayes, W. D., "Wave Riding of Dolphins," *Nature*, *172*, 1060 (1953).
659 Perry, B., A. J. Acosta, and T. Kiceniuk, "Simulated Wave-Riding Dolphins," *Nature*, *192*, 148 (1961).

660 Cousteau, J., *Silent World*, Harper, New York (1950), pp. 228–229.

661 Darwin, G. H., *The Tides*, W. H. Freeman, San Francisco (1962).

662 Defant, A., *Ebb and Flow*, Univ. Michigan Press, Ann Arbor (1958).

663 Clancy, E. P., *The Tides*, Anchor Books, Doubleday, New York (1969).

664 Macmillan, D. H., "Tides" in *Waves and Tides*, Greenwood Press, Westport, Conn. (1970).

665 Elmore, W. C., and M. A. Heald, *Physics of Waves*, McGraw-Hill, New York (1969).

666 Runcorn, K., ed., *International Dictionary of Geophysics*, Pergamon Press, New York (1967).

667 Rossiter, J. R., "Tide Generating Forces" in Ref. 666, pp. 1537-1539.

668 Rossiter, J. R., "Tides" in Ref. 666, pp. 1539–1543.

669 Rossiter, J. R., "Tides in Oceans" in Ref. 666, pp. 1547-1549.

670 Rossiter, J. R., "Tides in Seas and Gulfs" in Ref. 666, pp. 1549-1551.

671 Zahl, P. A., "The Giant Tides of Fundy, "*National Geographic Mag.*, *112*, 153 (1957).

672 Jeffreys, H., "Tidal Friction" in Ref. 666, pp. 1535–1536.

673 Newton, R. R., "Secular Accelerations of the Earth and Moon," *Science*, *166*, 825 (1969).

674 Rossiter, J. R., "Tides in Shallow Water" in Ref. 666, pp. 1551–1553.

675 Fenton, J. D., "Cnoidal Waves and Bores in Uniform Channels of Arbitrary Cross-Section," *J. Fluid Mech.*, *58*, 417 (1973).

676 Photos, *National Geographic Mag.*, *142*, 490–491 (1972); *112*, 166 (1957).

677 Olsson, R. G., and E. T. Turkdogan, "Radial Spread of a Liquid Stream on a Horizontal Plate," *Nature*, *211*, 813 (1966).

678 Watson, E. J., "The Radial Spread of a Liquid Jet over a Horizontal Plane," *J. Fluid Mech.*, *20*, 481 (1964).

679 Tani, I., "Water Jump in the Boundary Layer," *J. Phys. Soc. Japan*, *4*, 212 (1949).

680 Glauert, M. B., "The Wall Jet," *J. Fluid Mech.*, *1*, 625 (1956).

681 Taylor, G., "Oblique Impact of a Jet on a Plane Surface," *Phil. Trans. Roy. Soc. Lond.*, *A260*, 96 (1966); also in Ref. 255, Vol. IV, p. 510.

682 Komar, P. D., "Nearshore Cell Circulation and the Formation of Giant Cusps," *Geol. Soc. Am. Bul.*, *82*, 2643 (1971).

683 Russell, R. J., and W. G. McIntire, "Beach Cusps," *Geol. Soc. Am. Bul.*, *76*, 307 (1965).

684 Worrall, G. A., "Present-Day and Subfossil Beach Cusps on the West African Coast," *J. Geology*, *77*, 484 (1969).

685 Dolan, R., "Coastal Landforms: Crescentic and Rhythmic," *Geol. Soc. Am. Bul.*, *82*, 177 (1971).

686 Cloud, P. E., Jr., "Beach Cusps: Response to Plateau's Rule," *Science*, *154*, 890 (1966).

687 Stong, C. L., ed., "The Amateur Scientist," *Sci. Amer.*, *219*, 116 (Dec. 1968).

688 Bascom, W., "Beaches," *Sci. Amer.*, *203*, 80 (Aug. 1960).

689 Schwartz, M. L., "Theoretical Approach to the Origin of Beach Cusps," *Geol. Soc. Am. Bul.*, *83*, 1115 (1972).

690 Bowen, A. J., and D. L. Inman, "Edge Waves and Crescentic Bars," *J. Geophys. Res.*, *76*, 8662 (1971).

691 Gorycki, M. A., "Sheetflood Structure: Mechanism of Beach Cusp Formation and Related Phenomena," *J. Geology*, *81*, 109 (1973).

692 Stewart, R. W., "The Atmosphere and the Ocean," *Sci. Amer.*, *221*, 76 (Sep. 1969).

693 Stommel, H., "An Elementary Explanation of Why Ocean Currents are Strongest in the West," *Bul. Am. Met. Soc.*, *32*, 21 (1951).

694 Stommel, H., "The Westward Intensification of Wind-Driven Ocean Currents," *EOS*, *Trans. Am. Geophys. Union*, *29*, 202 (1948).

695 Baker, D. J., Jr., "Models of Oceanic Circulation," *Sci. Amer.*, *222*, 114 (Jan. 1970).

696 Munk, W., "The Circulation of the Oceans," *Sci. Amer.*, *193*, 96 (Sep. 1955).

697 Morisawa, M., *Streams: Their Dynamics and Morphology*, McGraw-Hill, New York (1968).

698 Garrels, R. M., *A Textbook of Geology*, Harper & Bros., New York (1951).

699 Shelton, J. S., *Geology Illustrated*, W. H. Freeman & Co., Calif. (1966).

700 Goldstein, S., *Modern Developments in Fluid Dynamics*, Oxford, Clarendon Press (1938).

701 Leopold, L. B., and W. B. Langbein, "River Meanders," *Sci. Amer.*, *214*, 60 (June 1966).

702 Callender, R. A., "Instability and River Channels," *J. Fluid Mech.*, *36*, 465 (1969).

703 Schumm, S. A., and H. R.

Khan, "Experimental Study of Channel Patterns," *Nature*, *233*, 407 (1971).

704 Francis, J. R. D., and A. F. Asfari, "Visualization of Spiral Motion in Curved Open Channels of Large Width," *Nature*, *225*, 725 (1970).

705 Tinkler, K. J., "Active Valley Meanders in South-Central Texas and Their Wider Implications," *Geol. Soc. Am. Bul.*, *82*, 1783 (1971).

706 Tinkler, K. J., "Pools, Riffles, and Meanders," *Geol. Soc. Am. Bul.*, *81*, 547 (1970); see also: E. A. Keller *82*, 279 (1971) and Tinkler *82*, 281 (1971).

707 Horlock, J. H., "Erosion in Meanders," *Nature*, *176*, 1034 (1955).

708 Schwartz, M., "How to Construct a Stream Table to Simulate Geological Processes" in "The Amateur Scientist," C. L. Stong, ed., *Sci. Amer.*, *208*, 168 (Apr. 1963).

709 Dury, G. H., ed., *Rivers and River Terraces*, Praeger, New York (1970).

710 Leopold, L. B., and M. G. Wolman, "River Channel Patterns" in Ref. 709.

711 Langbein, W. B., and L. B. Leopold, "River Meanders and the Theory of Minimum Variance" in Ref. 709.

712 Davis, W. M., "River Terraces in New England" in Ref. 709.

713 Shepherd, R. G., "Incised River Meanders: Evolution in Simulated Bedrock," *Science*, *178*, 409 (1972).

714 Wilson, I. G., "Equilibrium Cross-section of Meandering and Braided Rivers," *Nature*, *241*, 393 (1973).

715 Gorycki, M. A., "Hydraulic Drag: A Meander-Initiating Mechanism," *Geol. Soc. Am. Bul.*, *84*, 175 (1973).

716 Baker, D. J., Jr., "Demonstrations of Fluid Flow in a Rotating System II: The 'Spin-Up' Problem," *Am. J. Phys.*, *36*, 980 (1968).

717 Baker, D. J., Jr., "Demonstrations of Fluid Flow in a Rotating System," *Am. J. Phys.*, *34*, 647 (1966).

718 Pritchard, W., "The Motion Generated by a Body Moving along the Axis of a Uniformly Rotating Fluid," *J. Fluid Mech.*, *39*, 443 (1969).

719 Maxworthy, T., "An Experimental Determination of the Slow Motion of a Sphere in a Rotating, Viscous Fluid," *J. Fluid Mech.*, *23*, 373 (1965).

720 Taylor, G. I., "Experiments with Rotating Fluids," *Proc. Roy. Soc. Lond.*, *A100*, 114 (1921); also in Ref. 255, Vol. IV, pp. 17-23.

721 Paget, R., and E. N. Da C. Andrade, "Whirlpools and Vortices," *Proc. Roy. Institution Great Britain*, *29*, 320 (1936).

722 Shapiro, A., "Bath-Tub Vortex," *Nature*, *196*, 1080 (1962).

723 Andrade, E. N. da C., "Whirlpools, Vortices and Bath Tubs," *New Scientist*, *17*, 302 (1963).

724 Shapiro, A. H., film, *Vorticity*, Educational Services, Inc., Watertown, Mass. (1961).

725 Shapiro, A. H. film loop No. FM-15, *The Bathtub Vortex,* Educational Servies, Inc., Watertown, Mass. (1963).

726 Binnie, A. M., "Some Experiments on the Bath-Tub Vortex," *J. Mechanical Engineering Science*, *6*, 256 (1964).

727 Trefethen, L. M., R. W. Bilger, P. T. Fink, R. E. Luxton, and R. I. Tanner, "The Bath-Tub Vortex in The Southern Hemisphere," *Nature*, *207*, 1084 (1965).

728 Sibulkin, M., "A Note on the Bathtub Vortex," *J. Fluid Mech.*, *14*, 21 (1962).

729 Kelly, D. L., B. W. Martin, and E. S. Taylor, "A Further Note on the Bathtub Vortex," *J. Fluid Mech.*, *19*, 539 (1964).

730 Stong, C. L., ed., "The Amateur Scientist," *Sci. Amer.*, *209*, 133 (Oct. 1963).

731 Heighes, J. M., "Origin of Three-Dimensional Vortices," *Weather, 23,* 523 (1968).

732 Morton, B. R., "The Strength of Vortex and Swirling Core Flows," *J. Fluid Mech.*, *38*, 315 (1969).

733 Granger, R. A., "Speed of a Surge in a Bathtub Vortex," *J. Fluid Mech.*, *34*, 651 (1968).

734 Goodman, J. M., "Paraboloids and Vortices in Hydrodynamics," *Am. J. Phys.*, *37*, 864 (1969).

735 Stong, C. L., ed., "How to Make and Investigate Vortexes in Water and Flame" in "The Amateur Scientist," *Sci. Amer.*, *209*, 133 (Oct. 1963).

736 Whiten, A. J., "Concerning Plug-Holes," *Weather*, *18*, 73 (1963).

737 Kuehnast, E. L., and D. A. Haines, "Unusual Features Observed within a Series of Tornado Pictures," *Mon. Weather Rev.*, *99*, 545 (1971).

738 Fujita, T., D. L. Bradbury, and C. F. Van Thullenar, "Palm Sunday Tornadoes of April 11, 1965," *Mon. Weather Rev. 98*, 29 (1970).

739 Dessens, J., Jr., "Influence of Ground Roughness on Tornadoes: A Laboratory Simulation," *J. Appl. Met.*, *11*, 72 (1972).

740 Bathurst, G. B., "The Earliest Recorded Tornado?", *Weather*, *19*, 202 (1964).

741 Golden, J. H., "Waterspouts and Tornadoes over South Florida," *Mon. Weather Rev.*, *99*, 146 (1971).

742 Golden, J. H., "Waterspouts at Lower Matecumbe Key, Florida, September 2, 1967," *Weather*, *23*, 103 (1968).

743 Gordon, A. H., "Waterspouts," *Weather*, *6*, 364 (1951).

744 Rossmann, F. O., "Differences in the Physical Behaviour of Tornadoes and Waterspouts," *Weather*, *13*, 259 (1958).

745 Rossmann, F. O., "Some Further Comments on Waterspouts," *Weather*, *14*, 104 (1959).

746 Roberts, W. O., "We're Doing Something about the Weather!", *National Geographic Mag.*, *141*, 518 (1972).

747 Vonnegut, B., C. B. Moore, and C. K. Harris, "Stabilization of a High-Voltage Discharge by a Vortex," *J. Met.*, *17*, 468 (1960).

748 Colgate, S. A., "Tornadoes: Mechanism and Control," *Science*, *157*, 1431 (1967).

749 Ryan, R. T., and B. Vonnegut, "Miniature Whirlwinds Produced in the Laboratory by High-Voltage Electrical Discharges," *Science*, *168*, 1349 (1970).

750 Brook, M., "Electric Currents Accompanying Tornado Activity," *Science*, *157*, 1434 (1967).

751 Turner, J. S., "Laboratory Models of Evaporation and Condensation," *Weather*, *20*, 124 (1965).

752 Turner, J. S., "Tornado," *Oceanus*, *10*, 14 (Sep. 1963).

753 Turner, J. S., and D. K. Lilly, "The Carbonated-Water Tornado Vortex," *J. Atmos. Sci.*, *20*, 468 (1963).

754 Turner, J. S., "The Constraints Imposed on Tornado-like Vortices by the Top and Bottom Boundary Conditions," *J. Fluid Mech.*, *25*, 377 (1966).

755 Morton, B. R., "Model Experiments for Vortex Columns in the Atmosphere," *Nature*, *197*, 840 (1963).

756 Hallett, J., and T. Hoffer, "Dust Devil Systems," *Weather*, *26*, 247 (1971).

757 Kaimel, J. C., and J. A. Businger, "Case Studies of a Convective Plume and a Dust Devil," *J. Appl. Met.*, *9*, 612 (1970).

758 Lilly, D. K., "Comments on 'Case Studies of a Convective Plume and a Dust Devil'," *J. Appl. Met.*, *10*, 590 (1971).

759 Kaimal, J. C., and J. A. Businger, "Reply," *J. Appl. Met.*, *10*, 591 (1971).

760 Sinclair, P. C., "General Characteristics of Dust Devils," *J. Appl. Met.*, *8*, 32 (1969).

761 Ryan, J. A., and J. J. Carroll, "Dust Devil Wind Velocities: Mature State," *J. Geophys. Res.*, *75*, 531 (1970).

762 Carroll, J. J., and J. A. Ryan, "Atmospheric Vorticity and Dust Devil Rotation," *J. Geophys. Res.*, *75*, 5179 (1970).

763 Cooley, J. R., "Damaging, Mischievous, and Interesting Whirlwinds and Waterspouts," *Mon. Weather Rev.*, *100*, 317 (1972).

764 Cooley, J. R., "Dust Devil," *National Oceanic and Atmospheric Administration* (NOAA), *2*, 19 (Apr. 1972).

765 "The Torrey Canyon Smoke Plume," *Weather*, *22*, 368 (1967).

766 Goldie, E. C. W., "The Torrey Canyon Smoke Plume," *Weather*, *22*, 508 (1967).

767 Heighes, J. M., "Vortices Produced by the Torrey Canyon Smoke Plume," *Weather*, *22*, 508 (1967).

768 Thorarinsson, S., and B. Vonnegut, "Whirlwinds Produced by the Eruption of Surtsey Volcano," *Bul. Am. Met. Soc.*, *45*, 440 (1964).

769 Atallah, S., "Some Observations on the Great Fire of London, 1666," *Nature*, *211*, 105 (1966).

770 Lawrence, E. N., "Meteorology and the Great Fire of London, 1666," *Nature*, *213*, 168 (1967).

771 Graham, H. E., "Fire Whirlwinds," *Bul. Am. Met. Soc.*, *36*, 99 (1955).

772 Dessens, J., "Man-Made Tornadoes," *Nature*, *193*, 13 (1962).

773 Lyons, W. A., and S. R. Pease, "'Steam Devils' over Lake Michigan during a January Arctic Outbreak," *Mon. Weather Rev.*, *100*, 235 (1972).

774 Heighes, J. M., "Comment on 'Picture of the Month—' Steam Devils' over Lake Michigan during a January Arctic Outbreak," *Mon. Weather Rev.*, *100*, 750 (1972).

775 Levengood, W. C., "Instability Effects in Vortex Rings Produced with Liquids," *Nature*, *181*, 1680 (1958).

776 Batchelor, G. K., *An Introduction to Fluid Dynamics*, Cambridge Univ. Press, Cambridge (1967).

777 Reynolds, O., "On the Action of Rain to Calm the Sea" in Ref. 594, pp. 86-88.

778 Sibulkin, M., "Unsteady, Viscous, Circular Flow. Part 3. Application to the Ranque-Hilsch Tube," *J. Fluid Mech.*, *12*, 269 (1962).

779 Hilsch, R., "The Use of the Expansion of Gases in a Centrifugal Field as Cooling Process," *Rev. Scientific Instruments*, *18*, 108 (1947).

780 Fulton, C. D., "Comments on the Vortex Tube," *Refrigerating Engineering*, *59*, 984 (1951).

781 Hartnett, J. P., and E. R. G. Eckert, "Experimental Study of the Velocity and Temperature Distribution in a High-Velocity Vortex-Type Flow," *Trans. Am. Soc. Mechanical Engineers*, *79*, 751 (1957).

782 Martynovskii, V. S., and V. P. Alekseev, "Investigation of the Vortex Thermal Separation Effect for Gases and Vapors," *Soviet Phys. Tech. Papers*, *1*, 2233 (1956).

783 Pengelly, C. D., "Flow in a Viscous Vortex," *J. Appl. Phys.*, *28*, 86 (1957).

784 Van Deemter, J. J., "On the Theory of the Ranque-Hilsch Cooling Effect," *Appl. Scientific Res.*, *A3*, 174 (1952).

785 Cooney, D. O., "Transient Phenomena Observed During Operation of a Ranque-Hilsch Vortex Tube," *Industrial Engineering Chem. Fundamentals*, *10*, 308 (1971).

786 Smith, G. O., "The 'Hilsch' Vortex Tube" in *The Amateur Scientist* by C. L. Stong, Simon and Schuster, New York (1960), pp. 514–520.

787 Scheper, G. W., Jr., "The Vortex Tube—Internal Flow Data and a Heat Transfer Theory," *Refrigerating Engineering*, *59*, 985 (1951).

788 Willmarth, W. W., N. E. Hawk, and R. L. Harvey, "Steady and Unsteady Motions and Wakes of Freely Falling Disks," *Phys. of Fluids*, *7*, 197 (1964).

789 Jayaweera, K. O. L. F., and B. J. Mason, "The Behavior of Freely Falling Cylinders and Cones in a Viscous Fluid," *J. Fluid Mech.*, *22*, 709 (1965).

790 Jayaweera, K. O. L. F., and B. J. Mason, "The Falling Motions of Loaded Cylinders and Discs Simulating Snow Crystals," *Quart. J. Roy. Met. Soc.*, *92*, 151 (1966).

791 Jayaweera, K. O. L. F., B. J. Mason, and G. W. Slack, "The Behaviour of Clusters of Spheres Falling in a Viscous Fluid. Part 1. Experimental," *J. Fluid Mech.*, *20*, 121 (1964).

792 Hocking, L. M., "The Behaviour of Clusters of Spheres Falling in a Viscous Fluid. Part 2. Slow Motion Theory," *J. Fluid Mech.*, *20*, 129 (1964).

793 Bretherton, F. P., "Inertial Effects on Clusters of Spheres Falling in a Viscous Fluid," *J. Fluid Mech.*, *20*, 401 (1964).

794 Lissaman, P. B. S., and C. A. Shollenberger, "Formation Flight of Birds," *Science*, *168*, 1003 (1970).

795 King, R. E., "The Inverted Pendulum," *Am. J. Phys.*, *33*, 855 (1965).

796 Wegener, P. P., and J.-Y. Parlange, "Spherical-Cap Bubbles, in *Annual Review of Fluid Mechanics*, Vol. 5, Annual Reviews, Inc., Pal Alto, Calif. (1973), pp. 79–100.

797 Saffman, P. G., "On the Rise of Small Air Bubbles in Water," *J. Fluid Mech.*, *1*, 249 (1956).

798 Hartunian, R. A., and W. R. Sears, "On the Instability of Small Gas Bubbles Moving Uniformly in Various Liquids," *J. Fluid Mech.*, *3*, 27 (1957).

799 Moore, D. W., "The Rise of a Gas Bubble in a Viscous Liquid," *J. Fluid Mech.*, *6*, 113 (1959).

800 Magarvey, R. H., and P. B. Corkum, "The Wake of a Rising Bubble," *Nature*, *200*, 354 (1963).

801 Parlange, J.-Y., "Spherical Cap Bubbles with Laminar Wakes," *J. Fluid Mech.*, *37*, 257 (1969).

802 Goller, R. R., "The Legacy of 'Galloping Gertie', 25 Years After," *Trans. Am. Soc. Civil Engineers*, *131*, 704 (1966).

803 Shepherd, D. G., *Elements of Fluid Mechanics*, Harcourt, Brace & World, New York (1965), pp. 348–353.

804 Pugsley, A., *The Theory of Suspension Bridges,* Edward Arnold, Ltd., London (1957), pp. 120–127.

805 Kerensky, O. A., "Bridges and Other Large Structures," *Phil. Trans. Roy. Soc. Lond.*, *A269*, 343 (1971).

806 Parkinson, G. V., "Wind-Induced Instability of Structures," *Phil. Trans. Roy. Soc. Lond.*, *A269*, 395 (1971).

807 Scruton, C., and E. W. E. Rogers, "Steady and Unsteady Wind Loading of Buildings and Structures," *Phil. Trans. Roy. Soc. Lond.*, *A269*, 353 (1971).

808 Goller, R. R., "The Legacy of 'Galloping Gertie' 25 Years Later," *Civil Engineering*, *35*, 50 (Oct. 1965).

809 Finch, J. K., "Wind Failures of Suspension Bridges or Evolution and Decay of the Stiffening Truss," *Engineering News-Record*, *126*, 402 (Mar. 13, 1941).

810 "Narrows Nightmare," *Time* (Nov. 18, 1940), p. 21.

811 Steinman, D. B., *The Builders of the Bridge*, Harcourt, Brace and Co., New York (1945); see bibliography pp. 435–438.

812 Steinman, D. B., "Suspension Bridges: The Aerodynamic Problem and Its Solution" in *Science in Progress*, 9th Ser., G. A. Baitsell, ed., Yale Univ. Press, New Haven (1955), pp. 241-291.

813 Schwartz, H. I., "Edgetones and Nappe Oscillation," *J. Acoust. Soc. Am.*, *39*, 579 (1966).

814 Schwarts, H. I., "Nappe Oscillation," *J. Hydraulics Division, Am. Soc. Civil Engineers*, *90*, 129 (Nov. 1964, Part 1).

815 Naudascher, E., Discussion of Ref. 814, *J. Hydraulics Division, Am. Soc. Civil Engineers*, *91* (HY3), 389 (May 1965, Part 1).

816 Petrikat, K., and T. E. Unny, Discussion of Ref. 814, *J. Hydraulics Division, Am. Soc. Civil Engineers*, *91* (HY 5), 223 (Sept. 1965, Part 1).

817 McCarty, G., "Parachutes," *Internat. Science Technology*, (Oct. 1966), pp. 60–71.

818 Brown, W. D., *Parachutes*, Pitman & Sons, London (1951).

819 Dutton, J. A., and H. A. Panofsky, "Clear Air Turbulence: A Mystery May be Unfolding," *Science*, *167*, 937 (1970).

820 Dutton, J. A., "Clear-Air Turbulence, Aviation, and Atmospheric Science," *Rev. Geophys. Space Phys.*, *9*, 613 (1971).

821 Panofsky, H. A., "Up and Down," *Weatherwise*, *25*, 77 (1972).

822 Scorer, R. S., "Clear Air Turbulence in the Jet Stream," *Weather*, *12*, 275 (1957).

823 Raine, A., "Aerodynamics of Skiing," *Science J.*, *6*, 26 (Mar. 1970).

824 Francis, J. R. D., "The Speed of Drifting Bodies in a Stream," *J. Fluid Mech.*, *1*, 517 (1956).

825 Francis, J. R. D., "A Further Note on the Speed of Floating Bodies in a Stream," *J. Fluid Mech.*, *10*, 48 (1961).

826 Pedgley, D. E., "The Shapes of Snowdrifts," *Weather*, *22*, 42 (1967).

827 Shapiro, A. H., *Shape and Flow: The Fluid Dynamics of Drag*, Doubleday Anchor, Science Study Series, Garden City, New York (1961).

828 "Golf," *Encyclopaedia Britannica*, William Benton, Chicago (1970), Vol. 10, pp. 553-554.

829 Hart, C., *Kites: An Historical Survey*, Faber and Faber, London (1967).

830 Schaefer, V. J., "Observations of an Early Morning Cup of Coffee," *Am. Scientist*, *59*, 534 (1971).

831 Whitehead, J. A., Jr., "Cellular Convection," *Am. Scientist*, *59*, 444 (1971).

832 Scriven, L. E., and C. V. Sternling, "The Marangoni Effects," *Nature*, *187*, 186 (1960).

833 Berg, J. C., M. Boudart, and A. Acrivos, "Natural Convection in Pools of Evaporating Liquids," *J. Fluid Mech.*, *24*, 721 (1966).

834 Palm, E., "On the Tendency towards Hexagonal Cells in Steady Convection," *J. Fluid Mech.*, *8*, 183 (1960).

835 Koschmieder, E. L., "On Convection under an Air Surface," *J. Fluid Mech.*, *30*, 9 (1967).

836 Stuart, J. T., "On the Cellular Patterns in Thermal Convection," *J. Fluid Mech.*, *18*, 481 (1964).

837 Roberts, P. H., "Convection in Horizontal Layers with Internal Heat Generation. Theory," *J. Fluid Mech.*, *30*, 33 (1967).

838 Malkus, W. V. R., and G. Veronis, "Finite Amplitude Cellular Convection," *J. Fluid Mech.*, *4*, 225 (1958).

839 Cabelli, A., and G. De Vahl Davis, "A Numerical Study of the Benard Cell," *J. Fluid Mech.*, *45*, 805 (1971).

840 Rossby, H. T., "A Study of Benard Convection with and without Rotation," *J. Fluid Mech.*, *36*, 309 (1969).

841 Pearson, J. R. A., "On Convection Cells Induced by Surface Tension," *J. Fluid Mech.*, *4*, 489 (1958).

842 Scanlon, J. W., and L. A. Segel, "Finite Amplitude Cellular Convection Induced by Surface Tension," *J. Fluid Mech.*, *31*, 1 (1968).

843 Thirlby, R., "Convection in an Internally Heated Layer," *J. Fluid Mech.*, *44*, 673 (1970).

844 Block, M. J., "Surface Tension as the Cause of Benard Cells and Surface Deformation in a Liquid Film," *Nature*, *178*, 650 (1956).

845 Levengood, W. C., "Evidence of Rupture in Droplet Layers on Heated Liquid Surfaces," *Am. J. Phys.*, *26*, 35 (1958).

846 Grodzka, P. G., and T. C. Bannister, "Heat Flow and Convection Demonstration Experiments aboard Apollo 14," *Science*, *176*, 506 (1972).

847 Zern, R. W., and W. C. Reynolds, "Thermal Instabilities in Two-Fluid Horizontal Layers," *J. Fluid Mech.*, *53*, 305 (1972).

848 Loewenthal, M., "Tears of Strong Wine," *Philosophical Mag.* (Series 7), *12*, 462 (1931).

849 Thomson, J., "On Certain Curious Motions Observable at the Surfaces of Wine and Other Alcoholic Liquors," *Philosophical*

Mag. (Series 4), *10*, 330 (1855).

850 Maxworthy, T., "The Structure and Stability of Vortex Rings," *J. Fluid Mech.*, *51*, 15 (1972).

851 Stong, C. L., ed., "The Amateur Scientist," *Sci. Amer.*, *212*, 120 (Jan. 1965).

852 Riehl, H., *Tropical Meteorology*, McGraw-Hill, New York (1954).

853 Woodcock, A. H., "Convection and Soaring over the Open Sea," *J. Marine Res.*, *3*, 248 (1940).

854 Cone, C. D., Jr., "The Soaring Flight of Birds," *Sci. Amer.*, *206*, 130 (Apr. 1962).

855 Strutt, J. W. (Lord Rayleigh), "The Soaring of Birds," in *Scientific Papers, Vol. II, 1881-1887*, Cambridge Univ. Press, Cambridge (1900), pp. 194–197.

856 Rayleigh, Lord, "Flight" in Ref. 129, Vol. 5, pp. 294–295.

857 Wolters, R. A., *The Art and Technique of Soaring*, McGraw-Hill, New York (1971).

858 Tucker, V. A., and G. C. Parrott, "Aerodynamics of Gliding Flight in a Falcon and Other Birds," *J. Exp. Bio.*, *52*, 345 (1970).

859 Cone, C. D., "The Theory of Soaring Flight in Vortex Shells," *Soaring*, *25* (1961): Part 1, p. 8 (April); Part 2, p. 8 (May); Part 3, p. 6 (June).

860 Slater, A. E., "The 'Mystery' of Soaring Flight," *Weather*, *10*, 298 (1955).

861 Bell, G. J., "Some Meteorological Aspects of Soaring Flight," *Weather*, *5*, 8 (1950).

862 Hanna, S. R., "The Formation of Longitudinal Sand Dunes by Large Helical Eddies in the Atmosphere," *J. Appl. Met.*, *8*, 874 (1969).

863 King, W. J. H., "Study of a Dune Belt," *Geography J.*, *51*, 16 (1918).

864 Hastings, J. D., "Sand Streets," *Met. Mag.*, *100*, 155 (1971).

865 Langmuir, I., "Surface Motion of Water Induced by Wind," *Science*, *87*, 119 (1938).

866 Faller, A. J., "The Angle of Windows in the Ocean," *Tellus*, *16*, 363 (1964).

867 Craik, A. D. D., "A Wave-Interaction Model for the Generation of Windrows," *J. Fluid Mech.*, *41*, 801 (1970).

868 Faller, A. J., and A. H. Woodcock, "The Spacing of Windrows of *Sargassum* in the Ocean," *J. Marine Res.*, *22*, 22 (1964).

869 Wilson, I., "Sand Waves," *New Scientist*, *53*, 634 (1972).

870 Harms, J. C., "Hydraulic Significance of Some Sand Ripples," *Geol. Soc. Am. Bul.*, *80*, 363 (1969).

871 Kennedy, J. F., "The Mechanics of Dunes and Antidunes in Erodible-Bed Channels," *J. Fluid Mech.*, *16*, 521 (1963).

872 Raudkivi, A. J., "Bed Flows in Alluvial Channels," *J. Fluid Mech.*, *26*, 507 (1966).

873 Reynolds, A. J., "Waves on the Erodible Bed of an Open Channel," *J. Fluid Mech.*, *22*, 113 (1965).

874 Allen, J. R. L., "Asymmetrical Ripple Marks and the Origin of Cross-Stratification," *Nature*, *194*, 167 (1962).

875 Potter, A., and F. H. Barnes, "The Siphon," *Phys. Ed.*, *6*, 362 (1971).

876 Nokes, M. C., "The Siphon," *School Science Rev.* (Gr. Brit.), *29*, 233 (1948); reviewed in *Am. J. Phys.* *16*, 254 (1948).

877 Hero, of Alexandria, *The Pneumatics of Hero of Alexandria*, American Elsevier, New York (1971).

878 Reyburn, W., *Flushed with Pride: The Story of Thomas Crapper*, Prentice-Hall, New Jersey (1969).

879 *The Way Things Work*, Simon & Schuster, New York (1967).

880 Hobbs, P. V., and A. J. Kezweeny, "Splashing of a Water Drop," *Science*, *155*, 1112 (1967).

881 Harlow, F. H., and J. P. Shannon, "Distortion of a Splashing Liquid Drop," *Science*, *157*, 547 (1967).

882 Macklin, W. C., and P. V. Hobbs, "Subsurface Phenomena and the Splashing of Drops on Shallow Liquids," *Science*, *166*, 107 (1969).

883 Hobbs, P. V., and T. Osheroff, "Splashing of Drops on Shallow Liquids," *Science*, *158*, 1184 (1967).

884 Engel, O. G., "Crater Depth in Fluid Impacts," *J. Appl. Phys.*, *37*, 1798 (1966).

885 Engel, O. G., "Initial Pressure, Initial Flow Velocity, and the Time Dependence of Crater Depth in Fluid Impacts," *J. Appl. Phys.*, *38*, 3935 (1967).

886 Rambant, P. C., C. T. Bourland, N. D. Heidelbaugh, C. S. Huber, and M. C. Smith, Jr., "Some Flow Properties of Foods in Null Gravity," *Food Tech.*, *26*, 58 (1972).

887 Lane, W. R., and H. L. Green, "The Mechanics of Drops and Bubbles," in *Surveys in Mechanics*, G. K. Batchelor and R. M. Davies, eds., Cambridge

Univ. Press, Cambridge (1956), pp. 162–215.

888 Davies, D. P., "Road Marks," *New Scientist*, *47*, 42 (1970).

889 Durrell, L., *Prospero's Cell*, Dutton, New York (1962), pp. 38–39.

890 Lamb, Sir H., *Hydrodynamics*, 6th ed., Dover, New York (1945).

891 Rayleigh, Lord, "Foam" in Ref. 129, Vol. 4, pp. 26–38.

892 Hardy, W. B., "Films" in Ref. 129, Vol. 9, pp. 109–113.

893 McCutchen, C. W., "Surface Films Compacted by Moving Water: Demarcation Lines Reveal Film Edges," *Science*, *170*, 61 (1970).

894 Mockros, L. F., and R. B., Krone, "Hydrodynamic Effects on an Interfacial Film," *Science*, *161*, 361 (1968).

895 Dietz, R. S., and E. C. Lafond, "Natural Slicks," *J. Marine Res.*, *9*, 69 (1950).

896 Ewing, G., "Slicks, Surface Films, and Internal Waves," *J. Marine Res.*, *9*, 161 (1950).

897 Stommel, H., "Streaks on Natural Water Surfaces," *Weather*, *6*, 72 (1951).

898 Totton, A. K., "Calm Lanes in Ruffled Water," *Weather*, *5*, 289 (1950).

899 Taylor, G. I., "The Dynamics of Thin Sheets of Fluid. I. Water Bells," *Proc. Roy. Soc. Lond.*, *A253*, 289 (1959); also in Ref. 255, Vol. IV, pp. 344–350.

900 Hopwood, F. L., "Water Bells," *Proc. Physical Soc.* (London), *B65*, 2 (1952).

901 Lance, G. N., and R. L. Perry, "Water Bells," *Proc. Physical Soc.*, (London), *B66*, 1067 (1953).

902 Parlange, J.-Y., "A Theory of Water Bells," *J. Fluid Mech.*, *29*, 361 (1967).

903 Huang, C. P., and J. H. Lienhard, "The Influence of Gravity upon the Shape of Water Bells," *J. Appl. Mech.*, *Trans. Am. Soc. Mechanical Engineers*, Part E, *33*, 457 (1966).

904 Tyndall, J., "On Some Phenomena Connected with the Motion of Liquids" in Ref. 129, Vol. 1, pp. 131–133.

905 Taylor, G. I., "The Dynamics of Thin Sheets of Fluid. II. Waves on Fluid Sheets," *Proc. Roy. Soc. Lond.*, *A253*, 296 (1959); also in Ref. 255, Vol. IV, p. 351.

906 Taylor, G. I., "The Dynamics of Thin Sheets of Fluid. III. Disintegration of Fluid Sheets," *Proc. Roy. Soc. Lond.*, *A253*, 313 (1959); also in Ref. 255, Vol. IV, p. 368.

907 Taylor, G. I., "Formation of Thin Flat Sheets of Water," *Proc. Roy. Soc. Lond.*, *A259*, 1 (1960); also in Ref. 255, Vol. IV, p. 378.

908 Bond, W. N., "The Surface Tension of a Moving Water Sheet," *Proc. Physical Soc.*, (London) *47*, 549 (1935).

909 Huang, J. C. P., "The Break-Up of Axisymmetric Liquid Sheets," *J. Fluid Mech.*, *43*, 305 (1970).

910 Huang, J. C., "Dynamics of Free Axisymmetric Liquid Sheets," Bulletin 306, Research Division, College of Engineering, Washington State Univ., Pullman, Wash. (Aug. 1967).

911 Reiner, M., "The Teapot Effect. . .A Problem," *Phys. Today*, *9*, 16 (Sept. 1956).

912 Keller, J. B., "Teapot Effect," *J. Appl. Phys.*, *28*, 859 (1957).

913 French, T., "Pneumatic Tyres," *Science J.*, *5A*, 35 (Nov. 1969).

914 Reynolds, O., "On the Floating of Drops on the Surface of Water Depending Only on the Purity of the Surface" in Ref. 594, pp. 413–414.

915 Benedicks, C., and P. Sederholm, "Adsorption as the Cause of the Phenomenon of the 'Floating Drop,' and Foam Consisting Solely of Liquids," *Nature*, *153*, 80 (1944).

916 Schol, G., in "The Amateur Scientist," C. L. Stong, ed., *Sci. Amer.*, *229*, 104 (Aug. 1973).

917 Collyer, A. A., "Demonstrations with Viscoelastic Liquids," *Phys. Ed.*, *8*, 111 (1973).

918 Barnes, G., and R. Woodcock, "Liquid Rope-Coil Effect," *Am. J. Phys.*, *26*, 205 (1958).

919 Barnes, G., and J. MacKenzie, "Height of Fall Versus Frequency in Liquid Rope-Coil Effect," *Am. J. Phys.*, *27*, 112 (1959).

920 Taylor, G. I., "Instability of Jets, Threads, and Sheets of Viscous Fluids" in Ref. 255, Vol. IV, pp. 543–546.

921 Trevena, D. H., "Elastic Liquids," *Sources of Physics Teaching*, Part 5, Taylor and Francis, London (1970), pp. 57-68.

922 Kaye, A., "A Bouncing Liquid Stream," *Nature*, *197*, 1001 (1963).

923 Lodge, A. S., *Elastic Liquids*, Academic Press, New York (1964).

924 Longwell, P. A., *Mechanics of Fluid Flow*, McGraw-Hill, New York (1966).

925 Fredrickson, A. G., *Principles and Applications of Rheology*, Prentice-Hall, New Jersey (1964).

926 Reiner, M., "Phenomenological Macrorheology," in *Rheology Theory and Applications*, Vol. 1, F. R. Eirick, ed., Academic Press,

New York (1956), pp. 9-62.
927 Oldroyd, J. G., "Non-Newtonian Flow of Liquids and Solids," in *Rheology Theory and Applications*, Vol. I, F. R. Eirick, ed., Academic Press, New York (1956), pp. 653-682.
928 Jobling, A., and J. E. Roberts, "Goniometry of Flow and Rupture," in *Rheology Theory and Applications*, Vol. II, F. R. Eirick, ed., Academic Press, New York (1958), pp. 503-535.
929 Muller, H. G., "Weissenberg Effect in the Thick White of the Hen's Egg," *Nature*, *189*, 213 (1961).
930 Sharman, R. V., "Non-Newtonian Liquids," *Phys. Ed.*, *4*, 375 (1969).
931 Saville, D. A., and D. W. Thompson, "Secondary Flows Associated with the Weissenberg Effect," *Nature, 223,* 391 (1969).
932 Stong, C. L., ed., "The Amateur Scientist," *Sci. Amer.*, *212*, 118 (Jan. 1965).
933 Wiegand, J. H., "Deomonstrating the Weissenberg Effect with Gelatin," *J. Chem. Ed.*, *40*, 475 (1963).
934 Pryce-Jones, J., "The Rheology of Honey," in *Foodstuffs. Their Plasticity, Fluidity and Consistency*, G. W. Scott Blair, ed., Wiley-Interscience, New York (1953), pp. 148-176.
935 Bosworth, R. C. L., *Physics in Chemical Industry*, Macmillan, London (1950).
936 Stanley, R. C., "Non Newtonain Viscosity and Some Aspects of Lubrication," *Phys. Ed.*, *7*,193 (1972).
937 Bauer, W. H., and E. A. Collins, "Thixotropy and Dilatancy," Chapter 8 of *Rheology Theory and Applications*. Vol. 4, F. R. Eirich, ed., Academic Press, New York (1967), pp. 423-459.
938 Collyer, A. A., "Time Independent Fluids," *Phys. Ed.*, *8*, 333 (1973).
939 Busse, W. F., and F. C. Starr, "Change of a Viscoelastic Sphere to a Torus by Random Impacts," *Am. J. Phys.*, *28*, 19 (1960).
940 James, D. F., "Open Channel Siphon with Viscoelastic Fluids," *Nature*, *212*, 754 (1966).
941 Matthes, G. H., "Quicksand," *Sci. Amer.*, *188*, 97 (June 1953).
942 Goldsmith, H. L., and S. G. Mason, "The Microrheology of Dispersions" in *Rheology Theory and Applications*, Vol. 4, F. R. Eirich, ed., Academic, Press, New York (1967), pp. 191-194.
943 Heller, J. P., "An Unmixing Demonstration," *Am. J. Phys.*, *28*, 348 (1960).
944 Cornish, R. E., "Improving Underwater Vision of Lifeguards and Naked Divers," *J. Opt. Soc. Am.*, *23*, 430 (1933).
945 Baddeley, A. D., "Diver Performance" in *Underwater Science*, J. D. Woods and J. N. Lythgoe, eds., Oxford Univ. Press, New York (1971), pp. 47-50.
946 Laird, E. R., "The Position of the Image of an Object under Water," *Am. J. Phys.*, *6*, 40 (1938).
947 Reese, H. M., "Where is a Fish Seen?", *Am. J. Phys.*, *6*, 163 (1938); with reply by E. R. Laird.
948 Arvidsson, G., "Image of an Object under Water," *Am. J. Phys.*, *6*, 164 (1938).
949 Kinsler, L. E., "Imaging of Underwater Objects," *Am. J. Phys.*, *13*, 255 (1945).
950 Bruyne, N. A. de, "The Action of Adhesives," *Sci. Amer.*, *206*, 114 (Apr. 1962).
951 Adler, C. "Shadow-Sausage Effect," *Am. J. Phys.*, *35*, 774 (1967).
952 Smith, M. J., "Comment on: Shadow-Sausage Effect," *Am. J. Phys.*, *36*, 912 (1968).
953 Swindell, W., "Effect of Environmental Changes on the Ghosting of Distant Objects in Twin-Glazed Windows," *Appl. Optics*, *11*, 2033 (1972).
954 Minnaert, M., *Light and Colour in the Open Air*, Dover, New York (1954).
955 Ives, R. L., "Meteorological Conditions Accompanying Mirages in the Salt Lake Desert," *J. Franklin Institute*, *245*, 457 (1948).
956 Ashmore, S. E., "A North Wales Road-Mirage," *Weather*, *10*, 336 (1955).
957 Paton, J., "The Optical Properties of the Atmosphere," *Weather*, *3*, 243 (1948).
958 Ives, R. L., "Recurrent Mirages at Puerto Penasco, Sonora," *J. Franklin Institute*, *252*, 285 (1951).
959 Ives, R. L., "The Mirages of La Encantada," *Weather*, *23*, 55 (1968).
960 Vollprecht, R., "The 'Cold Mirage' in Western Australia," *Weather*, *2*, 174 (1947).
961 Botley, C. M., "Folk-Lore in Meteorology," *Weather*, *21*, 263 (1966).
962 Botley, C. M., "Mirages—What's in a Name?", *Weather*, *20*, 22 (1965).
963 Strouse, W. M., "Bouncing Light Beam," *Am. J. Phys.*, *40*, 913 (1972).
964 Cameron, W. S., J. H. Glenn,

M. S. Carpenter, and J. A. O'Keefe, "Effect of Refraction on the Setting Sun as Seen from Space in Theory and Observation," *Astron. J.*, *68*, 348 (1963).

965 Photos, *National Geographic Mag.*, *135*, 370–371 (1969); *128*, 682–683 (1965).

966 O'Connell, D. J. K., *The Green Flash and Other Low Sun Phenomena,* North Holland Publishing Co., Amesterdam (1958); order from the Vatican Observatory, Castel Gandolfo, Italy.

967 O' Connell, D. J. K., "The Green Flash," *Sci. Amer.*, *202*, 112 (Jan. 1960).

968 O'Connell, D. J. K., "The Green Flash and Kindred Phenomena," *Endeavor*, *20*, 131 (1961).

969 Taylor, J. H., and B. T. Matthias, "Green Flash from High Altitude," *Nature*, *222*, 157 (1969).

970 Seebold, R. E., "Green Flash," *J. Opt. Soc. Am.*, *51*, 237 (1961).

971 Koblents, Ya. P., "Conditions Attending a Green Flash in the Antarctic," *Sov. Antarctic Expedition Information Bul.* (transl. from Russian), no. 73, 57 (1969).

972 Thackeray, A. D., "An Unusual View of the Green Flash," *Mon. Notes Astron. Soc. South Africa*, *27*, 131 (1968).

973 Feibelman, W. A., "Low Sun Phenomena," *Appl. Optics*, *2*, 199 (1963).

974 Kirkpatrick, P., "Green Flash," *Am. J. Phys.*, *24*, 532 (1956).

975 Lovell, D. J., "Green Flash at Sunset," *Am. J. Phys.*, *25*, 206 (1957).

976 Ellis, J. W., "Green Flash from a Looming Setting Sun," *Am. J. Phys.*, *25*, 387 (1957).

977 Gorton, H. C., "Method to Facilitate Observation of the Green Flash," *Am. J. Phys.*, *25*, 586 (1957).

978 Jacobsen, T. S., "The Green Flash at Sunset and at Sunrise," *Sky Telescope*, *12*, 233 (July 1953).

979 Ashmore, S. E., "A Note on the Green Ray," *Quart. J. Roy. Met. Soc.*, *71*, 383 (1945).

980 Hulburt, E. O., "The Green Segment Seen from an Airplane," *J. Opt. Soc. Am.*, *39*, 409 (1949).

981 Wilson, R. H., Jr., letter on green flash, *Sky Telescope*, *42*, 327 (Dec. 1971).

982 Feibelman, W. A., in "The Amateur Scientist," C. L. Stong, ed., *Sci. Amer.*, *204*, 177 (Jan. 1961).

983 Tricker, R. A. R., *Introduction to Meteorological Optics,* American Elsevier, New York (1970).

984 Fitch, J. M., "The Control of the Luminous Environment," *Sci. Amer.*, *219*, 190 (Sep. 1968).

985 Hull, N., "Simple Visual Aid to Understanding Plane Mirrors at an Angle Theta," *Am. J. Phys.*, *27*, 610 (1959).

986 Kulkarni, V. M., "Number of Images Produced by Multiple Reflection," *Am. J. Phys.*, *28*, 317 (1960).

987 Liu, C.-H., "Number of Images Produced by Multiple Reflection," *Am. J. Phys.*, *30*, 380 (1962).

988 Brown, F. L., "Multiple Reflections from Plane Mirrors," *Am. J. Phys.*, *13*, 278 (1945).

989 Chai, A.-T., "The Number of Images of an Object between Two Plane Mirrors," *Am. J. Phys.*, *39*, 1390 (1971).

990 Goodell, J. B., "On the Appearance of the Sea Reflected Sky," *Appl. Optics*, *10*, 223 (1971).

991 Minnaert, M., "Unusual or Neglected Optical Phenomena in the Landscape," *J. Opt. Soc. Am.*, *58*, 297 (1968).

992 Jerlov, N. G., *Optical Oceanography*, Elsevier, New York (1968).

993 Hirsh, F. R., Jr., and E. M. Thorndike, "On the Pinhead Shadow Inversion Phenomenon," *Am. J. Phys.*, *12*, 164 (1944).

994 Young M., "Pinhole Imagery," *Am. J. Phys.*, *40*, 715 (1972).

995 Young, M., "Pinhole Optics," *Appl. Optics*, *10*, 2763 (1971).

996 Baez, A. V., "Pinhole-Camera Experiment for the Introductory Physics Course," *Am. J. Phys.*, *25*, 636 (1957).

997 Turner, L. A., "Resolving Power and the Theory of the Pinhole Camera," *Am. J. Phys.*, *8*, 112 (1940).

998 Turner, L. A., "Best Definition with the Pinhole Camera," *Am. J. Phys.*, *8*, 365 (1940).

999 Arakawa, H., "Crescent-shaped Shadows during a Partial Eclipse of the Sun," *Weather*, *16*, 254 (1961).

1000 Howard, J. A., "Increased Luminance in the Direction of Reflex Reflexion—A Recently Observed Natural Phenomenon," *Nature*, *224*, 1102 (1969).

1001 Minnaert, M., "Retro-reflection," *Nature*, *225*, 718 (1970).

1002 Preston, J. S., "Retro-reflexion by Diffusing Surfaces," *Nature*, *213*, 1007 (1967).

1003 "'Hot-Spot' in Aerial Photograph," *Weather*, *21*, 288 (1966).

1004 Butler, C. P., "Heiligenschein Seen from an Airplane,"

J. Opt. Soc. Am., *45*, 328 (1955).

1005 Wildey, R. L., and H. A. Pohn, "The Normal Albedo of the Apollo 11 Landing Site and Intrinsic Dispersion in the Lunar Heiligenschein," *Astrophys. J.*, *158*, L 129 (1969).

1006 Pohn, H. A., H. W. Radin, and R. L. Wildey, "The Moon's Photometric Function Near Zero Phase Angle from Apollo 8 Photography," *Astrophys. J.*, *157*, L 193 (1969).

1007 Mattsson, J. O., and C. Cavallin, "Retroreflection of Light from Dew-Covered Surfaces and an Image-Producing Device for Registration of this Light," *Oikos*, *23*, 285 (1972).

1008 Monteith, J. L., "Refraction and the Spider," *Weather*, *9*, 140 (1954).

1009 Jacobs, S. F., "Self-Centered Shadow," *Am. J. Phys.*, *21*, 234 (1953).

1010 Wanta, R. C., "The Self-Centered Shadow," *Am. J. Phys.*, *21*, 578 (1953).

1011 Van Lear, G. A., Jr., "Reflectors Used in Highway Signs and Warning Signals. Parts I, II, and III," *J. Opt. Soc. Am.*, *30*, 462 (1940).

1012 Pirie, A., "The Biochemistry of the Eye," *Nature*, *186*, 352 (1960).

1013 Vonnegut, B., and C. B. Moore, "Visual Analogue of Radar Bright Band Phenomenon," *Weather*, *15*, 277 (1960).

1014 Van de Hulst, H. C., *Light Scattering by Small Particles*, Wiley, New York (1957).

1015 Boyer, C. B., "Kepler's Explanation of the Rainbow," *Am. J. Phys.*, *18*, 360 (1950).

1016 Querfeld, C. W., "Mie Atmospheric Optics," *J. Opt. Soc. Am.*, *55*, 105 (1965).

1017 Nussenzveig, H. N., "High-Frequency Scattering by a Transparent Sphere. II. Theory of the Rainbow and the Glory," *J. Math. Phys.*, *10*, 125 (1969).

1018 McDonald, J. E., "Caustic of the Primary Rainbow," *Am. J. Phys.*, *31*, 282 (1963).

1019 Mason, E. A., R. J. Munn, and F. J. Smith, "Rainbows and Glories in Molecular Scattering," *Endeavor*, *30*, 91 (1971).

1020 Boyer, C. B., *The Rainbow from Myth to Mathematics*, Thomas Yoseloff, New York (1959).

1021 Malkus, W. V. R., "Rainbows and Cloudbows," *Weather*, *10*, 331 (1955).

1022 Fraser, A. B., "Inhomogeneities in the Color and Intensity of the Rainbow," *J. Atmos. Sci.*, *29*, 211 (1972).

1023 Greenler, R. G., "Infrared Rainbow," *Science*, *173*, 1231 (1971).

1024 Saunders, P. M., "Infra-Red Rainbow," *Weather*, *13*, 352 (1958).

1025 Walker, D., "A Rainbow and Supernumeraries with Graduated Separations," *Weather*, *5*, 324 (1950).

1026 Humphreys, W. J., "Why We Seldom See A Lunar Rainbow," *Science*, *88*, 496 (1938).

1027 Wentworth, C. K., "Frequency of Lunar Rainbows," *Science*, *88*, 498 (1938).

1028 McIntosh, D. H., ed., *Meteorological Glossary*, Her Majesty's Stationary Office, London (1963).

1029 Bull, G. A., "Reflection Rainbow," *Weather*, *16*, 267 (1961).

1030 Mattsson, J. O., S. Nordbeck, and B. Rystedt, "Dewbows and Fogbows in Divergent Light," Lund Studies in Geography, Ser. C. General, Mathematical and Regional Geography, No. 11 (1971); order from authors, Dept. of Geography, Lunds Universitet, Soelvegaton 13, S-223 62 Lund, Sweden.

1031 McDonald, J. E., "A Gigantic Horizontal Cloudbow," *Weather*, *17*, 243 (1962).

1032 Palmer, F., "Unusual Rainbows," *Am. J. Phys.*, *13*, 203 (1945).

1033 Ohtake, T., and K. O. L. F. Jayaweera, "Ice Crystal Displays from Power Plants," *Weather*, *27*, 271 (1972).

1034 Greenler, R. G., and A. J. Mallmann, "Circumscribed Halos," *Science*, *176*, 128 (1972).

1035 Botley, C. M., "Halos and Coronae," *Weather*, *1*, 85 (1946).

1036 Lacy, R. E., "The Halo Display of 2 March 1954," *Weather*, *9*, 206 (1954).

1037 Photographs, *Weather*, *27*, 240 (1972).

1038 Evans, W. F. J., and R. A. R. Tricker, "Unusual Arcs in the Saskatoon Halo Display," *Weather*, *27*, 234 (1972).

1039 Brain, J. P., "Halo Phenomena—An Investigation," *Weather*, *27*, 409 (1972).

1040 Jacobowitz, H., "A Method for Computing the Transfer of Solar Radiation through Clouds of Hexagonal Ice Crystals," *J. Quantative Spectrosc. Radiat. Transfer*, *11*, 691 (1971).

1041 Tricker, R. A. R., "A Note on J. R. Blake's 'Circumscribed Halo,'" *Weather*, *28*, 159 (1973).

1042 Mattsson, J. O., "Experimental Optical Phenomena," *Weather*, *21*, 14 (1966).

1043 Neuberger, H., "Forcasting Significance of Halos in Proverb and Statistics," *Bul. Am. Met. Soc.*, *22*, 105 (1941).

1044 Davies, P. W., "Upper Tangent Arc to the 22° Solar Halo," *Weather*, *21*, 138 (1966).

1045 Jones, G. A., and K. J. Wiggins, "Halo Phenomena at Odiham," *Weather*, *19*, 289 (1964).

1046 Goldie, E. C. W., and J. M. Heighes, "The Berkshire Halo Display of 11 May 1965," *Weather*, *23*, 61 (1968).

1047 Winstanley, D., "Halo Phenomena over Oxford on 12 June 1969," *Weather*, *25*, 131 (1970).

1048 Goldie, E. C. W., "A Graphical Guide to Haloes," *Weather*, *26*, 391 (1971).

1049 Ripley, E. A., and B. Saugier, "Photometeors at Saskatoon on 3 December 1970," *Weather*, *26*, 150 (1971).

1050 Stevens, G. C., and S. Fritz, "Two Halo Displays over Eastern U. S. in December 1948," *Bul. Am. Met. Soc.*, *31*, 318 (1950).

1051 Minnaert, M., "An Exceptional Phenomenon of Atmospheric Optics: A One-Sided Mocksun at 3°30'," *Weather*, *21*, 250 (1966).

1052 Jacquinot, P., and C. Squire, "Note on Reflection and Diffraction from Ice Crystals in the Sky," *J. Opt. Soc. Am.*, *43*, 318 (1953).

1053 Squire, C. F., "Note on Reflection and Dfffraction from Ice Crystals in the Sky," *J. Opt. Soc. Am.*, *42*, 782 (1952).

1054 Verschure, P. P. H., "Rare Halo Displays in Amsterdam," *Weather*, *26*, 532 (1971).

1055 Rott, H., "Sub-Sun and Sub-Parhelion," *Weather*, *28*, 65 (1973).

1056 Botley, C. M. "Parry of the Parry Arc," *Weather*, *10*, 343 (1955).

1057 Goldie, E. C. W. "Observation of a Rare Halo," *Weather*, *19*, 328 (1964).

1058 Tricker, R. A. R., "Observations on Certain Features to be Seen in a Photograph of Haloes Taken by Dr. Emil Schulthess in Antarctica," *Quart. J. Roy. Met. Soc.*, *98*, 542 (1972).

1059 Georgi, J., "A 22° Halo Represented by the Photographic Sky-Mirror," *Weather*, *18*, 135 (1963).

1060 Scorer, R. S., "Rare Halo in 'Arctic Smoke,'" *Weather*, *18*, 319 (1963).

1061 Neuberger, H., "When a Rare Halo is Photographed, a Jet Contrail Is Not Important," *Bul. Am. Met. Soc.*, *49*, 1060 (1968).

1062 Deirmendjian, D., "Sun's Image on Airborne Ice Crystals?", *Appl. Optics*, *7*, 556 (1968).

1063 Barmore, F. E., "Comments on Sun's Image on Airborne Ice Crystals," *Appl. Optics*, *7*, 1654 (1968).

1064 Jayaweera, K. O. L. F., and G. Wendler, "Lower Parry Arc of the Sun," *Weather*, *27*, 50 (1972).

1065 Greenler, R. G., M. Drinkwine, A. J. Mallmann, and G. Blumenthal, "The Origin of Sun Pillars," *Am. Scientist*, *60*, 292 (1972).

1066 Mattsson, J. O., "'Sub-Sun' and Light-Pillars of Street Lamps," *Weather*, *28*, 66 (1973).

1067 Gall, J. C., and M. E. Graves, "Possible Newly Recognized Meteorological Phenomenon Called Crown Flash," *Nature*, *229*, 184 (1971).

1068 Graves, M. E., J. C. Gall, and B. Vonnegut, "Meteorological Phenomenon Called Crown Flash," *Nature*, *231*, 258 (1971).

1069 Vonnegut, B., "Orientation of Ice Crystals in the Electric Field of a Thunderstorm," *Weather*, *20*, 310 (1965).

1070 Shurcliff, W. A., and S. S. Ballard, *Polarized Light*, D. Van Nostrand, New Jersey (1964).

1071 Grabau, M., "Polarized Light Enters the World of Everyday Life," *J. Appl. Phys.*, *9*, 215 (1938).

1072 Land, E. H., "Polaroid and the Headlight Problem," *J. Franklin Institute*, *224*, 269 (1937).

1073 Meyer-Arendt, J. R., T. M. Alexander, C. M. Landes, and K. Wilder, "Gradient Density Glasses for Night Driving," *Appl. Optics*, *9*, 2176 (1970).

1074 Land, E. H., "Some Aspects of the Development of Sheet Polarizers," *J. Opt. Soc. Am.*, *41*, 957 (1951); also in *Polarized Light*, published by the Am. Institute of Physics (1963).

1075 Rozenberg, G. V., *Twilight: A Study in Atmospheric Optics*, Plenum Press, New York (1966).

1076 Hallden, U., "A Simple Device for Observing the Polarization of Light from the Sky," *Nature*, *182*, 333 (1958).

1077 Baez, A. V., "Photoelastic Patterns without Crossed Polarizer and Analyzer," *Am. J. Phys.*, *22*, 39 (1954).

1078 Ferguson, I. F., "Photoelastic Patterns without Crossed Polarizer and Analyzer," *Am. J. Phys.*, *22*, 495 (1954).

1079 Rayleigh, Lord, "The Blue Sky and the Optical Properties of Air" in Ref. 129, Vol. 8, pp. 309-317.

1080 Cornford, S. G., "An Effect of Polarized Sky Light," *Weather*, *23*, 39 (1968).

1081 Wilson, R. M., E. J. Gardner, R. H. Squire,"'The Absorption of Light by Oriented Molecules," *J. Chem. Ed.*, *50*, 94 (1973).

1082 Wood, E. A., *Crystals and Light*, Van Nostrand, Momentum, New Jersey (1964).

1083 Mason, S. F., "Optical Activity and Molecular Dissymmetry," *Contemp. Phys.*, *9*, 239 (1968).

1084 Nye, J. F., *Physical Properties of Crystals*, Oxford, London (1957), pp. 260–274.

1085 Lockley, R. M., *Animal Navigation*, Hart, New York (1967).

1086 Kennedy, D., and E. R. Baylor, "Analysis of Polarized Light by the Bee's Eye," *Nature*, *191*, 34 (1961).

1087 Kalmus, H., "Orientation of Animals to Polarized Light," *Nature*, *184*, 228 (1959).

1088 Waterman, T., "Polarized Light and Animal Navigation," *Sci. Amer. 193*, 88 (July 1955); also in *Polarized Light* published by the Am. Institute of Physics (1963).

1089 Snyder, A. W., and C. Pask, "How Bees Navigate," *Nature*, *239*, 48 (1972).

1090 Seliger, H. H., and W. D. McElroy, *Light*: *Physical and Biological Action*, Academic Press, New York (1965).

1091 Helmholtz, H. von, *Physiological Optics*, J. P. C. Southall, ed., Dover, New York (1962).

1092 Helmholtz, H. von, pp. 304-307 of Ref. 1091; also in *Polarized Light*, Published by the Am. Institute of Physics (1963).

1093 Shurcliff, W. A., "Haidinger's Brushes and Circularly Polarized Light," *J. Opt. Soc. Am.*, *45*, 399 (1955); also in *Polarized Light*, published by the Am. Institute of Physics (1963).

1094 Summers, D. M., G. B. Friedmann, and R. M. Clements, "Physical Model for Haidinger's Brush," *J. Opt. Soc. Am.*, *60*, 271 (1970).

1095 Weihs, D., "Hydromechanics of Fish Schooling," *Nature*, *241*, 290 (1973).

1096 Fahy, E. F., and M. A. MacConaill, "Optical Properties of 'Cellophane'," *Nature*, *178*, 1072 (1956).

1097 Winans, J. G., "Demonstrations with Half-Wave Plates," *Am. J. Phys.*, *21*, 170 (1953).

1098 Panofsky, W. K. H., and M. Phillips, *Classical Electricity and Magnetism*, Addison-Wesley, Mass, (1962), Sec. 22-8 and pp. 414 ff.

1099 Lorentz, H. A., *Problems of Modern Physics*, Dover, New York (1967), pp. 52 ff.

1100 Reif, F., *Statistical Physics*, Berkeley Physics Course, Vol. 5, McGraw-Hill, New York (1964), p. 33.

1101 La Mer, V. K., and M. Kerker, "Light Scattered by Particles," *Sci. Amer.*, *188*, 69 (Feb. 1953).

1102 Dave, J. V., and C. L. Mateer, "The Effect of Stratospheric Dust on the Color of the Twilight Sky," *J. Geophys. Res.*, *73*, 6897 (1968).

1103 Ives, R. L., "Apparent Relation of 'Stepped' Sunset Red to Inversions," *Bul. Am. Met. Soc.*, *40*, 311 (1959).

1104 Volz, F. E., "Twilights and Stratospheric Dust before and after the Agung Eruption," *Appl. Optics*, *8*, 2505 (1969).

1105 Shah, G. M., "Enhanced Twilight Glow Caused by the Volcanic Eruption on Bali Island in March and September 1963," *Tellus*, *21*, 636 (1969).

1106 Volz, F. E., "Twilight Phenomena Caused by the Druption of Agung Volcano," *Science*, *144*, 1121 (1964).

1107 Meinel, M. P., and A. B. Meinel, "Late Twilight Glow of the Ash Stratum from the Eruption of Agung Volcano," *Science*, *142*, 582 (1963).

1108 Burdecki, F., "Meteorological Phenomena after Volcanic Eruptions," *Weather*, *19*, 113 (1964).

1109 Lamb, H. H., "Volcanic Dust in the Atmosphere; With a Chronology and Assessment of its Meteorological Significance," *Phil. Trans. Roy. Soc. Lond.*, *A266*, 425 (1970).

1110 Deacon, E. L., "The Second Purple Light," *Nature*, *178*, 688 (1956).

1111 Middleton, W. E. K., *Vision through the Atmosphere*, Univ. Toronto Press, Toronto (1968).

1112 Went, F. W., "Air Pollution," *Sci. Amer.*, *192*, 62 (May 1955).

1113 Stefansson, V., *The Friendly Arctic*, Macmillan, New York (1944).

1114 Smith, A. G., "Daylight Visibility of Stars from a Long Shaft," *J. Opt. Soc. Am.*, *45*, 482 (1955).

1115 Hynek, J. A., "Photographing Stars in the Daytime," *Sky Telescope*, *10*, 61 (1951).

1116 Tyler, J. E., "Colour of 'Pure Water'," *Nature*, *208*, 549 (1965).

1117 Stamm, G. L., and R. A. Langel, "Some Spectral Irradiance Measurements of Upwelling Natural Light off the East Coast of the United States," *J. Opt. Soc. Am.*, *51*, 1090 (1961).

1118 Rayleigh, Lord, "Colours of Sea and Sky" in Ref. 129, Vol. 7, pp. 93–99.

1119 Middleton, W. E. K., "The Color of the Overcast Sky," *J. Opt. Soc. Am.*, *44*, 793 (1954).

1120 Scoresby, W. A., *An Account of the Arctic Regions*, Vol. 1, Archibald Constable & Co., Edinburgh (1820), pp. 299–300.

1121 Moeller, F., "On the Backscattering of Global Radiation by the Sky," *Tellus*, *17*, 350 (1965).

1122 Catchpole, A. J. W., and D. W. Moodie, "Multiple Reflection in Arctic Regions," *Weather*, *26*, 157 (1971).

1123 Weisskopf, V. F., "How Light Interacts with Matter," *Sci. Amer.*, *219*, 60 (Sep. 1968).

1124 Hallett, J., and R. E. J. Lewis, "Mother-of-Pearl Clouds," *Weather*, *22*, 56 (1967).

1125 Scorer, R. S., "Mother-of-Pearl Clouds," *Weather*, *19*, 115 (1964).

1126 McIntosh, D. H., "Mother-of-Pearl Cloud over Scotland," *Weather*, *27*, 14 (1972).

1127 Williams, G. C., "Nacreous Clouds Observed in Southeastern Alaska January 24, 1950," *Bul. Am. Met. Soc.*, *31*, 322 (1950).

1128 Sharpe, J. M., Jr., "Nacreous Clouds at White Sands Missile Range," *Bul. Am. Met. Soc.*, *51*, 1148 (1970).

1129 Stormer, C., "Mother-of-Pearl Clouds," *Weather*, *3*, 13 (1948).

1130 Pohl, R. W., "Discovery of Interference by Thomas Young," *Am. J. Phys.*, *28*, 530 (1960).

1131 De Witte, A. J., "Interference in Scattered Light," *Am. J. Phys.*, *35*, 301 (1967).

1132 Fergason, J. L., "Liquid Crystals," *Sci. Amer.*, *211*, 76 (Aug. 1964).

1133 Jeppesen, M. A., and W. T. Hughes, "Liquid Crystals and Newton's Rings," *Am. J. Phys.*, *38*, 199 (1970).

1134 Gray, G. W., *Molecular Structure and the Properties of Liquid Crystals*, Academic Press, London (1962).

1135 Brown, G. H., G. T. Dienes, and M. M. Labes, *Liquid Crystals*, Gordon and Breach, New York (1967).

1136 Brown, G. H., "Liquid Crystals and Their Roles in Inanimate and Animate Systems," *Am. Scientist*, *60*, 64 (1972).

1137 Luckhurst, G. R., "Liquid Crystals," *Phys. Bul.*, *23*, 279 (1972).

1138 Wright, W. D., "'The Rays Are Not Coloured'," *Nature*, *198*, 1239 (1963).

1139 Ghiradella, H. D. Aneshansley, T. Eisner, R. E. Silberglied, and H. E. Hinton, "Ultraviolet Reflection of a male Butterfly: Interference Color Caused by Thin-Layer Elaboration of Wing Scales," *Science*, *178*, 1214 (1972).

1140 Anderson, T. F., and A. G. Richards, Jr., "An Electron Microscope Study of Some Structural Colors of Insects," *J. Appl. Phys.*, *13*, 748 (1942).

1141 Merritt, E., "A Spectrophotometric Study of Certain Cases of Structural Colors," *J. Opt. Soc. Am.*, *11*, 93 (1925).

1142 Mason, C. W., "Structural Colors in Feathers," *J. Phys. Chem.*, *27*, 201, 401 (1923).

1143 Roosen, R. G., "The Gegenschein," *Rev. Geophys. Space Phys.*, *9*, 275 (1971).

1144 Roosen, R. G., "A Photographic Investigation of the Gegenschein and the Earth-Moon Libration Point L_5," *Icarus*, *9*, 429 (1968); also see photo in errata, *Icarus*, *10*, 352 (1969).

1145 Roosen, R. G., "The Gegenschein and Interplanetary Dust outside the Earth's Orbit," *Icarus*, *13*, 184 (1970).

1146 Blackwell, D. E., "The Zodiacal Light," *Sci. Amer.*, *203*, 54 (July 1960).

1147 Hulburt, E. O., "Optics of Searchlight Illumination," *J. Opt. Soc. Am.*, *36*, 483 (1946).

1148 Kirkpatrick, P., "A Binocular Illusion," *Am. J. Phys.*, *22*, 493 (1954).

1149 van de Hulst, H. C., "A Theory of the Anti-Coronae," *J. Opt. Soc. Am.*, *37*, 16 (1947).

1150 Naik, Y. G., and R. M. Joshi "Anti-Coronas or Brocken Bows," *J. Opt. Soc. Am.*, *45*, 733 (1955).

1151 Bryant, H. C., and A. J. Cox, "Mie Theory and the Glory," *J. Opt. Soc. Am.*, *56*, 1529 (1966).

1152 Brandt, J. C., "An Unusual Observation of the 'Glory'," *Pub. Astron. Soc. Pacific*, *80*, 25 (1968).

1153 Saunders, M. J., "Near-Field Backscattering Measurements from a Microscopic Water Droplet," *J. Opt. Soc. Am.*, *60*, 1359 (1970).

1154 Kang, P., E. A. Mason, and R. J. Munn, " 'Glorified Shadows' in Molecular Scattering: Some Optical Analogies," *Am. J. Phys.*, *38*, 294 (1970).

1155 Fahlen, T. S., and H. C. Bryant, "Direct Observation of Surface Waves on Water Droplets,"

J. Opt. Soc. Am., *56*, 1635 (1966).

1156 Fahlen, T. S., and H. C. Bryant, "Optical Back Scattering from Single Water Droplets," *J. Opt. Soc. Am.*, *58*, 304 (1968).

1157 Mellerio, J., and D. A. Palmer, "Entopic Halos," *Vision Res.*, *10*, 595 (1970).

1158 Mellerio, J., and D. A. Palmer, "Entopic Halos and Glare," *Vision Res.*, *12*, 141 (1972).

1159 Wilson, R., "The Blue Sun of 1950 September," *Mon. Notices Roy. Astron. Soc.*, *111*, 478 (1951); this footnoted in *Principles of Optics* by M. Born and E. Wolf, 3rd ed., Pergamon Press, New York (1965), p. 661.

1160 Paul, W., and R. V. Jones, "Blue Sun and Moon," *Nature*, *168*, 554 (1951).

1161 Lothian, G. F., "Blue Sun and Moon," *Nature*, *168*, 1086 (1951).

1162 Porch, W. M., D. S. Ensor, R. J. Charlson, and J. Heintzenberg, "Blue Moon: Is This a Property of Background Aerosol?" , *Appl. Optics*, *12*, 34 (1973).

1163 Horvath, H., "On the Brown Colour of Atmospheric Haze," *Atmos. Environment*, *5*, 333 (1971); see also *6*, 143 (1972).

1164 Charlson, R. J., and N. C. Ahlquist, "Brown Haze: NO_2 or Aerosol?", *Atmos. Environment*, *3*, 653 (1969).

1165 Fish, B. R., "Electrical Generation of Natural Aerosols from Vegetation," *Science*, *175*, 1239 (1972).

1166 Went, F. W., "Blue Hazes in the Atmosphere," *Nature*, *187*, 641 (1960).

1167 White, H. E., and P. Levatin, " 'Floaters' in the Eye," *Sci. Amer.*, *206*, 119 (June 1962).

1168 Fireman, R. A., E. Prenner, and H. Norden, in "The Amateur Scientist," C. L. Stong, ed., *Sci. Amer.*, *198*, 122 (June 1958).

1169 Meyer-Arendt, J. R., *Introduction to Classical and Modern Optics*, Prentice-Hall, New Jersey (1972).

1170 Jenkins, F. A., and H. E. White, *Fundamentals of Optics*, McGraw-Hill, New York (1957).

1171 Hults, M. E., R. D. Burgess, D. A. Mitchell, and D. W. Warn, "Visual, Photographic and Photoelectric Detection of Shadow Bands at the March 7, 1970, Solar Eclipse," *Nature*, *231*, 255 (1971).

1172 Susel, F. M., "Recording the Eclipse Shadow Bands on Magnetic Tape," *J. Roy. Astron. Soc. Canada*, *65*, 273 (1971).

1173 Burgess, R. D., and M. E. Hults, "A Shadow-Band Experiment," *Sky Telescope*, *38*, 95 (Aug. 1969).

1174 Young, A. T., *Sky Telescope*, *38*, 309 (Nov. 1969).

1175 Paulton, E. M., "Recording Shadow Bands at the March Eclipse," *Sky Telescope*, *39*, 132 (Feb. 1970).

1176 Hults, M., *Sky Telescope*, *33*, 147 (Mar. 1967).

1177 Paulton, E. M., "Eclipse Shadow Band Motion—An illusion?", *Sky Telescope, 25,* 328 (June 1963).

1178 Paulton, E. M., "Observing and Reporting Shadow Bands," *Sky Telescope, 18,* 627 (Sep. 1959).

1179 Quann, J. J., and C. J. Daly, "The Shadow Band Phenomenon," *J. Atmos. Terrestrial Phys.*, *34*, 577 (1972).

1180 Kerr, D. E., G. G. Sivjee, W. McKinney, P. Takacs, and W. G. Fastie, "Brightness of Forbidden OI Lines and Properties of Shadow Bands During the Eclipse of 7 March 1970," *J. Atmos. Terrestrial Phys.*, *34*, 585 (1972).

1181 Young, A. T., "The Problem of Shadow Band Observations," *Sky Telescope*, *43*, 291 (1972).

1182 Ives, R. L., "Sunset Shadow Bands," *J. Opt. Soc. Am.*, *35*, 736 (1945).

1183 Burke, J. J., "Observations of the Wavelength Dependence of Stellar Scintillation," *J. Opt. Soc. Am.*, *60*, 1262 (1970).

1184 Young, A., "Saturation of Scintillation," *J. Opt. Soc. Am.*, *60*, 1495 (1970).

1185 Riggs, L. A., C. G. Mueller, C. H. Graham, and F. A. Mote, "Photographic Measurements of Atmospheric Boil," *J. Opt. Soc. Am.*, *37*, 415 (1947).

1186 Lawrence, E. N., "Large Air-Temperature Lapse-Rates near the Ground," *Weather*, *27*, 27 (1972).

1187 Mikesell, A. H., A. A. Hoag, and J. S. Hall, "The Scintillation of Starlight," *J. Opt. Soc. Am.*, *41*, 689 (1951).

1188 Gifford, F., and A. H. Mikesell, "Atmospheric Turbulence and the Scintillation of Starlight," *Weather*, *8*, 195 (1953).

1189 Ashkin, A., "Acceleration and Trapping of Particles by Radiation Pressure," *Phys. Rev. Letters*, *24*, 156 (1970).

1190 Ashkin, A., and J. M. Dziedzic, "Optical Levitation by Radiation Pressure," *Appl. Phys. Letters*, *19*, 283 (1971).

1191 Ashkin, A., "The Pressure of Laser Light," *Sci. Amer.*, *226*, 62 (Feb. 1972).

1192 Vonnegut, B., and J. R. Weyer, "Luminous Phenomena Accompanying Tornadoes," *Weatherwise, 19,* 66 (1966).

1193 Vonnegut, B., and J. R. Weyer, "Luminous Phenomena in Nocturnal Tornadoes," *Science*, *153*, 1213 (9 September 1966).

1194 Harvey, E. N., *A History of Luminescence*, American Philosophical Society, Philadelphia (1957).

1195 Alzetta, G., I. Chudacek, and R. Scarmozzino, "Excitation of Triboluminescence by Deformation of Single Crystal," *Physica Status Solidi (A)*, *1*, 775 (1970).

1196 Smith, H. M., "Synchronous Flashing of Fireflies," *Science*, *82*, 151 (16 August 1935).

1197 Buck, J., and E. Buck, "Biology of Synchronous Flashing of Fireflies," *Nature*, *211*, 562 (1966).

1198 Buck, J., and E. Buck, "Mechanism of Rhythmic Synchronous Flashing of Fireflies," *Science*, *159*, 1319 (1968).

1199 Hanson, F. E., J. F. Case, E. Buck, and J. Buck, "Synchrony and Flash Entrainment in a New Guinea Firefly," *Science*, *174*, 161 (1971).

1200 McElroy, W. D., and H. H. Seliger, "Biological Luminescence," *Sci. Amer.*, *207*, 76 (Dec. 1962).

1201 Bowen, E. J., and G. F. J. Garlick, "Luminescence," *Intern. Sci. Tech.*, (Aug. 1966), p. 18.

1202 Zahl, P. A., "Sailing a Sea of Fire," *Nat. Geographic Mag.*, *118*, 120 (July 1960).

1203 Daniels, F., Jr., J. C. van der Leun, and B. E. Johnson, "Sunburn," *Sci. Amer.*, *219*, 38 (July 1968).

1204 Greene, C. H., "Glass," *Sci. Amer.*, *204*, 92 (Jan. 1961).

1205 Stevens, W. R., *Building Physics: Lightning*, Pergamon Press, New York (1969).

1206 Hecht, E., "Speckle Patterns in Unfiltered Sunlight," *Am. J. Phys.*, *40*, 207 (1972).

1207 Rigden, J. D., and E. I. Gordon, "The Granularity of Scattered Optical Maser Light," *Proc. Institute Radio Engineers* (now: *IEEE Proc.*), *50*, 2367 (1962).

1208 Sinclair, D. C., "Demonstration of Chromatic Aberration in the Eye Using Coherent Light," *J. Opt. Soc. Am.*, *55*, 575 (1965).

1209 Mallette, V., "Comment on 'Speckle Patterns in Unfiltered Sunlight'," *Am. J. Phys.*, *41*, 844 (1973).

1210 Rushton, W. A. H., "Effect of Humming on Vision," *Nature*, *216*, 1173 (1967).

1211 Williams, P. C., and T. P. Williams, "Effect of Humming on Watching Television," *Nature*, *239*, 407 (1972).

1212 Enright, J. T., "Distortions of Apparent Velocity: A New Optical Illusion," *Science*, *168*, 464 (1970).

1213 Enright, J. T., "Stereopsis, Visual Latency, and Three-dimensional Moving Pictures," *Am. Scientist*, *58*, 536 (1970).

1214 Julesz, B., and B. White, "Short Term Visual Memory and the Pulfrich Phenomenon," *Nature*, *222*, 639 (1969).

1215 Hansteen, R. W., "Visual Latency as a Function of Stimulus Onset, Offset, and Background Luminance," *J. Opt. Soc. Am.*, *61*, 1190 (1971).

1216 Ives, R. L., "The Sequence Illusion," *J. Franklin Institute*, *230*, 755 (1940).

1217 Katz, M. S., and I. Schwartz, "New Observation of the Pulfrich Effect," *J. Opt. Soc. Am.*, *45*, 523 (1955).

1218 Rogers, B. J., and S. M. Anstis, "Intensity Versus Adaptation and the Pulfrich Stereophenomenon," *Vision Res.*, *12*, 909 (1972).

1219 Wilson, J. A., and S. M. Anstis, "Visual Delay as a Function of Luminance," *Am. J. Psychology*, *82*, 350 (1969).

1220 Prestrude, A. M., and H. D. Baker, "Light Adaptation and Visual Latency," *Vision Res.*, *11*, 363 (1971).

1221 Prestrude, A. M., "Visual Latencies at Photopic Levels of Retinal Illuminance," *Vision Res.*, *11*, 351 (1971).

1222 Prestrude, A. M., and H. D. Baker, "New Method of Measuring Visual-Perceptual Latency Differences," *Perception and Psychophysics*, *4*, 152 (1968).

1223 Oster, G., "Phosphenes," *Sci. Amer.*, *222*, 82 (Feb. 1970).

1224 Moreland, J. D., "On Demonstrating the Blue Arcs Phenomenon," *Vision Res.*, *8*, 99 (1968).

1225 Moreland, J. D., "Threshold Measurements of the Blue Arcs Phenomenon," *Vision Res.*, *8*, 1093 (1968).

1226 Moreland, J. D., "Possible Mechanisms of the Blue Arcs of the Retina," *Proc. Physiological Soc.*, Jan. 10–11, 1969, *J. Physiology*, *201*, 60P (Apr. 1969).

1227 Moreland, J. D., "Retinal Topography and the Blue-Arcs Phenomenon," *Vision Res.*, *9*, 965 (1969).

1228 Ratliff, F., *Mach Bands: Quantatitive Studies on Neural Networks in the Retina*, Holden-Day, Inc., San Francisco (1965).

1229 Ratliff, F., "Contour and Contrast," *Sci. Amer.*, *226*, 90 (June 1972).

1230 Green, D. G., and M. B. Fast, "On the Appearance of Mach Bands in Gradients of Varying

Color," *Vision Res.*, *11*, 1147 (1971).

1231 Jacobson, J. Z., and G. E. MacKinnon, "Coloured Mach Bands," *Canadian J. Psychology and Review Canadian Psychology*, *23*, 56 (1969).

1232 Welford, W. T., "The Visual Mach Effect," *Phys. Ed.*, *3*, 83 (1968).

1233 Cornsweet, T. N., *Visual Perception*, Academic Press, New York (1970).

1234 Remole, A., "Subjective Patterns in a Flickering Field: Binocular vs. Monocular Observation," *J. Opt. Soc. Am.*, *63*, 745 (1973).

1235 Hayward, R., in "Amateur Scientist," C. L. Stong, ed., *Sci. Amer.*, *198*, 100 (Jan. 1958).

1236 Land, E. H., "Experiments in Color Vision," *Sci. Amer.*, *200*, 84 (May 1959).

1237 Rushton, W. A. H., "The Eye, the Brain and Land's Two-Colour Projections," *Nature*, *189*, 440 (1961).

1238 Land, E. H., "Color Vision and the Natural Image," *Proc. Nat. Academy Sci.*, *45*, 115, 636 (1959).

1239 McCann, J. J., "Rod-Cone Interactions: Different Color Sensations from Identical Stimuli," *Science*, *176*, 1255 (1972).

1240 Festinger, L., M. R. Allyn, and C. W. White, "The Perception of Color with Achromatic Stimulation," *Vision Res.*, *11*, 591 (1971).

1241 Campenhausen, C. von, "The Colors of Benham's Top under Metameric Illuminations," *Vision Res.*, *9*, 677 (1969).

1242 Vatsa, L. P. S., "Prevost-Fetchner-Benham Effect," *Am. J. Phys.*, *40*, 914 (1972).

1243 Hammond, J. A., "Stroboscopic Effect with Fluorescent Lighting," *Am. J. Phys.*, *33*, 506 (1965).

1244 Baker, D. J., Jr., "Time Dependence of Fluorescent Lamp Emission—A Simple Demonstration," *Am. J. Phys.*, *34*, 627 (1966).

1245 Hauver, G. E., "Color Effect of Fluorescent Lighting," *Am. J. Phys.*, *17*, 446 (1949).

1246 Ficken, G. W., Jr., "Melde Experiment Viewed with Fluorescent Lights," *Am. J. Phys.*, *36*, 63 (1968).

1247 Crookes, T. G., "Television Images," *Nature*, *179*, 1024 (1957).

1248 Dadourian, H. M., "The Moon Illusion," *Am. J. Phys.*, *14*, 65 (1946).

1249 Boring, E. G., "The Perception of Objects," *Am. J. Phys.*, *14*, 99 (1946).

1250 Boring, E., "The Moon Illusion," *Am. J. Phys.*, *11*, 55 (1943).

1251 Kaufman, L, and I. Rock, "The Moon Illusion," *Science*, *136*, 953, 1023 (1962).

1252 Kaufman, L., and I. Rock, "The Moon Illusion," *Sci. Amer.*, *207*, 120 (July 1962).

1253 Restle, F., "Moon Illusion Explained on the Basis of Relative Size," *Science*, *167*, 1092 (1970).

1254 Miller, A., and H. Neuberger, "Investigations into the Apparent Shape of the Sky," *Bul. Am. Met. Soc.*, *26*, 212 (1945).

1255 Sundet, J. M., "The Effect of Pupil Size Variations on the Colour Stereoscopic Phenomenon," *Vision Res.*, *12*, 1027 (1972).

1256 Kishto, B. N., "The Colour Stereoscopic Effect," *Vision Res.*, *5*, 313 (1965).

1257 Vos, J. J., "Some New Aspects of Color Stereoscopy," *J. Opt. Soc. Am.*, *50*, 785 (1960).

1258 Vos., J. J., "An Antagonistic Effect in Colour Stereoscopy," *Ophthalmologica*, *142*, 442 (1963).

1259 Vos, J. J., "The Colour Stereoscopic Effect," *Vision Res.*, *6*, 105 (1966).

1260 Greenslade, T. B., and M. W. Green, "Experiments with Stereoscopic Images," *Phys. Teacher*, *11*, 215 (1973).

1261 Bugelski, B. R., "Traffic Signals and Depth Perception," *Science*, *157*, 1464 (1967).

1262 Fritz, S., "The 'Polar Whiteout'," *Weather*, *12*, 345 (1957).

1263 Kilston, S., C. Sagan, and R. Drummond, "A Search for Life on Earth At Kilometer Resolution," *Icarus*, *5*, 79 (1966).

1264 Boeke, K., *Cosmic View, the Universe in 40 Jumps*, John Day Co., Inc., New York (1957).

1265 Parmenter, F. C., "Trans-Canada Highway," *Mon. Weather Rev.*, *98*, 252 (1970).

1266 Berry, M. V., "Reflections on a Christmas-tree Bauble," *Phys. Ed.*, *7*, 1 (1972).

1267 Woods, E. A., "Moire Patterns—A Demonstration," *Am. J. Phys.*, *30*, 381 (1962).

1268 Stecher, M., "The Moiré Phenomenon," *Am. J. Phys.*, *32*, 247 (1964).

1269 Oster, G., and Y. Nishijima, "Moiré Patterns," *Sci. Amer.*, *208*, 54 (May 1963).

1270 Stong, C. L., "Moiré Patterns Provide both Recreation and Some Analogues for Solving Problems" in "The Amateur Scientist," *Sci. Amer.*, *211*, 134 (Nov. 1964).

1271 Mawdsley, J., "Demonstrating Phase Velocity and Group Velocity," *Am. J. Phys., 37*, 842 (1969).

1272 Chiang, C., "Stereoscopic Moiré Patterns," *J. Opt. Soc. Am., 57*, 1088 (1967).

1273 Dalziel, C. F., "Electric Shock Hazard," *IEEE Spectrum, 9*, 41 (Feb. 1972).

1274 Friedlander, G. D., "Electricity in Hospitals: Elimination of Lethal Hazards," *IEEE Spectrum, 8*, 40 (Sep. 1971).

1275 Galambos, R., *Nerves and Muscles*, Anchor Books, Doubleday, New York (1962).

1276 Wooldridge, D. E., *The Machinery of the Brain*, McGraw-Hill, New York (1963), pp. 16-17.

1277 Cox, R. T., "Electric Fish," *Am. J. Phys., 11*, 13 (1943).

1278 Grundfest, H., "Electric Fishes," *Sci. Amer., 203*, 115 (Oct. 1960).

1279 Lissmann, H. W., "Electric Location by Fishes," *Sci. Amer., 208*, 50 (Mar. 1963).

1280 Rommel, S. A., Jr., and J. D. McCleave, "Oceanic Electric Fields: Perception by American Eels?", *Science, 176*, 1233 (1972).

1281 Bullock, T. H., "Seeing the World through a New Sense: Electroreception in Fish," *Am. Scientist, 61*, 316 (1973).

1282 Peckover, R. S., "Oceanic Electric Currents Induced by Fluid Convection," *Phys. Earth Planetary Interiors, 7*, 137 (1973).

1283 Baez, A. V., "Some Observations on the Electrostatic Attraction of a Stream of Water," *Am. J. Phys., 20*, 520 (1952).

1284 Magarvey, R. H., and L. E. Outhouse, "Note on the Break-up of a Charged Liquid Jet," *J. Fluid Mech., 13*, 151 (1962).

1285 Huebner, A. L., and H. N. Chu, "Instability and Breakup of Charged Liquid Jets," *J. Fluid Mech., 49*, 361 (1971).

1286 Huebner, A. L., "Disintegration of Charged Liquid Jets," *J. Fluid Mech., 38*, 679 (1969).

1287 Goedde, E. F., and M. C. Yuen, "Experiments on Liquid Jet Instability," *J. Fluid Mech., 40*, 495 (1970).

1288 Plumb, R. C., "Triboelectricity," in "Chemical Principles Exemplified," *J. Chem. Ed., 48*, 524 (1971).

1289 Harper, W. R., "The Generation of Static Charge," *Advances in Physics, 6*, 365 (1957).

1290 Kornfel'd, M. I., "Nature of Frictional Electrification," *Soviet Physics—Solid State, 11*, 1306 (1969).

1291 Ainslie, D. S., "What Are the Essential Conditions for Electrification by Rubbing?", *Am. J. Phys., 35*, 535 (1967).

1292 Moore, A. D., "Electrostatics," *Sci. Amer., 226*, 47 (Mar. 1972).

1293 Harper, W. R., "Triboelectrification," *Phys. Ed., 5*, 87 (1970).

1294 Jefimenko, O., "Lecture Demonstrations on Electrification by Contact," *Am. J. Phys., 27*, 604 (1959).

1295 Cunningham, R. G., and D. J. Montgomery, "Demonstration in Static Electricity," *Am. J. Phys., 24*, 54 (1956).

1296 Vonnegut, B., "Atmospheric Electrostatics" in Ref. 540, pp. 390–424.

1297 Hendricks, C. D., "Charging Macroscopic Particles" in Ref. 540, pp. 57–85; see pp. 64–67.

1298 Stow, C. D., "The Generation of Electricity by Blowing Snow," *Weather, 22*, 371 (1967).

1299 Stow, C. D., "Dust and Sand Storm Electrification," *Weather, 24*, 134 (1969).

1300 Kamra, A. K., "Electrification in an Indian Dust Storm," *Weather, 24*, 145 (1969).

1301 Latham, J., and C. D. Stow, "Electrification of Snowstorms," *Nature, 202*, 284 (1964).

1302 Seifert, H. S., "Friction Tape Produces Glow Discharge," *Am. J. Phys., 20*, 380 (1952).

1303 Miller, J. S., "Concerning the Electric Charge on a Moving Vehicle," *Am. J. Phys., 21*, 316 (1953).

1304 Miller, R. F., "Electric Charge on a Moving Vehicle," *Am. J. Phys., 21*, 579 (1953).

1305 Mackeown, S. S., and V. Wouk, "Generation of Electric Charges by Moving Rubber-Tired Vehicles," *Am. Institute Electrical Engineers Trans., 62*, 207 (1943).

1306 Beach, R., "Static Electricity on Rubber-Tired Vehicles," *Electrical Engineering, 60*, 202 (1941).

1307 Pierce, E. T., and A. L. Whitson, "Atmospheric Electricity in a Typical American Bathroom," *Weather, 21*, 449 (1966).

1308 Hughes, J. F., "Electrostatic Hazards in Supertanker Cleaning Operations," *Nature, 235*, 381 (1972).

1309 Smy, P. R., "Charge Production, Supertankers, and Supersonic Aircraft," *Nature, 239*, 269 (1972).

1310 Pierce, E. T., and A. L. Whitson, "Atmospheric Electricity and the Waterfalls of Yosemite Valley," *J. Atmos. Sci., 22*, 314 (1965).

1311 Lyle, A. R., and H. Strawson, "Electrostatic Hazards in Tank Filling Operations," *Phys. Bul.*, *23*, 453 (1972).

1312 Rosser, W. G. V., "What Makes an Electric Current 'Flow'," *Am. J. Phys.*, *31*, 884 (1963).

1313 Weyl, W. A., and W. C. Ormsby, "Atomistic Approach to the Rheology of Sand-Water and of Clay-Water Mixtures" in *Rheology Theory and Applications*, Vol. III, F. R. Eirick, ed., Academic Press, New York (1960), pp. 249-297.

1314 Pilpel, N., "Crumb Formation," *Endeavor*, *30*, 77 (May 1971).

1315 Pilpel, N., "The Cohesiveness of Powders," *Endeavor*, *28*, 73 (May 1969).

1316 Mackay, R. S., "Two Startling Demonstrations with a Magnet," *Am. J. Phys.*, *28*, 678 (1960).

1317 Sumner, D. J., and A. K. Thakkrar, "Experiments with a 'Jumping Ring' Apparatus," *Phys. Ed.*, *7*, 238 (1972).

1318 Laithwaite, E. R., *Propulsion without Wheels*, Hart, New York (1968), Chapter 10.

1319 "Discussion on Electromagnetic Levitation," *Proc. Institution Electrical Engineers* (London), *113*, 1395 (1966).

1320 White, H. E., and H. Weltin, "Electromagnetic Levitator," *Am. J. Phys.*, *31*, 925 (1963).

1321 Fleming, J. A., "Electromagnetic Repulsion" in Ref. 129, Vol. 4, pp. 72-92.

1322 Brigman, G. H., "The Martini Perpetual Motion Machine: A Semiprogrammed Problem," *Am. J. Phys.*, *40*, 1001 (1972).

1323 "Violation of the First Law of Thermodynamics," *Phys. Teacher*, *2*, 383 (1964).

1324 "Perpetual Motion," *Encyclopaedia Britannica*, William Benton, Chicago (1970), Vol. 17, pp. 639-641.

1325 Angrist, S. W., "Perpetual Motion Machines," *Sci. Amer.*, *218*, 114 (Jan. 1968).

1326 Gautier, T. N., "The Ionosphere," *Sci. Amer.*, *193*, 126 (Sep. 1955).

1327 "Measurement of V-H-F Bursts," *Electronics*, *18*, 105 (1945).

1328 Chapman, S., "Sun Storms and the Earth: The Aurora Polaris and the Space around the Earth," *Am. Scientist*, *49*, 249 (1961).

1329 Heikkila, W. J., "Aurora," *EOS, Trans. Am. Geophys. Union*, *54*, 764 (1973).

1330 Storey, L. R. O., "Whistlers," *Sci. Amer.*, *194*, 34 (Jan. 1956).

1331 Croom, D. L., "Whistlers," *Weather*, *18*, 258, 296 (1963).

1332 Hill, R. D., "Thunderbolts," *Endeavor*, *31*, 3 (1972).

1333 Orville, R. E., "The Colour Spectrum of Lightning," *Weather*, *21*, 198 (1966).

1334 Orville, R. E., "Lightning Photography," *Phys. Teacher*, *9*, 333 (1971).

1335 "First Photograph of Lightning," *J. Franklin Institute*, *253*, xxii (1952).

1336 Wood, E. A., "Physics for Automobile Passengers," *Phys. Teacher*, *11*, 239 (1973).

1337 Vonnegut, B., "Some Facts and Speculations Concerning the Origin and Role of Thunderstorm Electricity," *Met. Monographs*, *5*, 224 (1963).

1338 Jefimenko, O., "Operation of Electric Motors from the Atmospheric Electric Fields," *Am. J. Phys.*, *39*, 776 (1971).

1339 Jefimenko, O. D., "Electrostatics Motors" in Ref. 540, pp. 131-147.

1340 Matthias, B. T., and S. J. Buchsbaum, "Pinched Lightning," *Nature*, *194*, 327 (1962); photo also in Ref. 1351.

1341 Uman, M. A., "Bead Lightning and the Pinch Effect," *J. Atmos. Terrestrial Phys.*, *24*, 43 (1962).

1342 Blanchard, D. C., "Charge Separation from Saline Drops on Hot Surfaces," *Nature*, *201*, 1164 (1964).

1343 Anderson, R., S. Bjoernsson, D. C. Blanchard, S. Gathman, J. Hughes, S. Jonasson, C. B. Moore, H. J. Survilas, and B. Vonnegut, "Electricity in Volcanic Clouds," *Science*, *148*, 1179 (1965).

1344 Blanchard, D. C., "Volcanic Electricity," *Oceanus*, *13*, 2 (Nov. 1966).

1345 Brook, M., and C. B. Moore, and T. Sigurgeirsson, "Lightning in Volcanic Clouds," *J. Geophys. Res.*, *79*, 472 (1974).

1346 Photo, *Nat. Geographic Mag.*, *113*, 755 (1958).

1347 Uman, M. A., D. F. Seacord, G. H. Price, and E. T. Pierce, "Lightning Induced by Thermonuclear Detonations," *J. Geophys. Res.*, *77*, 1591 (1972).

1348 Hill, R. D., "Lightning Induced by Nuclear Bursts," *J. Geophys. Res.*, *78*, 6355 (1973).

1349 Singer, S., *The Nature of Ball Lightning*, Plenum Press, New York (1971).

1350 Powell, J. R., and D. Finkelstein, "Ball Lightning," *Am. Scientist*, *58*, 262 (1970).

1351 Lewis, H. W., "Ball Lightning," *Sci. Amer.*, *208*, 106 (Mar. 1963).

1352 Charman, W. N., "Perceptual Effects and the Reliability of Ball Lightning Reports," *J. Atmos. Terrestrial Phys.*, *33*, 1973 (1971).

1353 Ashby, D. E. T. F., and C. Whitehead, "Is Ball Lightning Caused by Antimatter Meteorites," *Nature*, *230*, 180 (1971).

1354 Uman, M. A., "Some Comments on Ball Lightning," *J. Atmos. Terrestrial Phys.*, *30*, 1245 (1968).

1355 Mills, A. A., "Ball Lightning and Thermoluminescence," *Nature Physical Science*, *233*, 131 (1971).

1356 Powell, J. R., and D. Finkelstein, "Structure of Ball Lightning," *Advances in Geophysics*, Vol. 13, Academic Press, New York (1969), pp. 141–189.

1357 Altschuler, M. D., L. L. House, and E. Hildner, "Is Ball Lightning a Nuclear Phenomenon?", *Nature*, *228*, 545 (1970).

1358 Argyle, E., "Ball Lightning as an Optical Illusion," *Nature*, *230*, 179 (1971).

1359 Wittmann, A., "In Support of a Physical Explanation of Ball Lightning," *Nature*, *232*, 625 (1971).

1360 Jennison, R. C., "Ball Lightning," *Nature*, *224*, 895 (1969).

1361 Jennison, R. C., "Ball Lightning and After-images," *Nature*, *230*, 576 (1971).

1362 Charman, W. N., "After-images and Ball Lightning," *Nature*, *230*, 576 (1971).

1363 Davies, P. C. W., "Ball Lightning or Spots before the Eyes," *Nature*, *230*, 576 (1971).

1364 Wagner, G., "Optical and Acoustic Detection of Ball Lightning," *Nature*, *232*, 187 (1971).

1365 Johnson, P. O., "Ball Lightning and Self-Containing Electromagnetic Fields," *Am. J. Phys.*, *33*, 119 (1965).

1366 Anderson, F. J., and G. D. Freir, "A Report on Ball Lightning," *J. Geophys. Res.*, *77*, 3928 (1972).

1367 Wooding, E. R., "Laser Analogue to Ball Lightning," *Nature*, *239*, 394 (1972).

1368 Crawford, J. F., "Antimatter and Ball Lightning," *Nature*, *239*, 395 (1972).

1369 Davies, D. W., and R. B. Standler, "Ball Lightning," *Nature*, *240*, 144 (1972).

1370 Charman, N., "The Enigma of Ball Lightning," *New Scientist*, *56*, 632 (1972).

1371 Finkelstein, D. and J. Powell, *"Earthquake Lightning," Nature*, *228*, 759 (1970).

1372 Finkelstein, D., R. D. Hill, and J. R. Powell, "The Piezoelectric Theory of Earthquake Lightning," *J. Geophys. Res.*, *78*, 992 (1973).

1373 MacLaren, M., "Early Electrical Discoveries by Benjamin Franklin and his Contemporaries," *J. Franklin Institute*, *240*,1 (1945).

1374 Schonland, B. F. J., "The Work of Benjamin Franklin on Thunderstorms and the Development of the Lightning Rod," *J. Franklin Institute*, *253*, 375 (1952).

1375 Chalmers, J. A., "The Action of a Lightning Conductor," *Weather*, *20*, 183 (1965).

1376 Gillespie, P. J. "Ionizing Radiation: a Potential Lightning Hazard?", *Nature*, *208*, 577 (1965).

1377 Roberts, J. E., "Ionizing Radiation and Lightning Hazards," *Nature*, *210*, 514 (1966).

1378 Gillespie, P. J., "Ionizing Radiation and LIghtning Hazards," *Nature*, *210*, 515 (1966).

1379 McEachron, K. B., "Lightning Protection Since Franklin's Day," *J. Franklin Institute*, *253*, 441 (1952).

1380 Cohen, I. B., "Prejudice against the Introduction of Lighting Rods," *J. Franklin Institute*, *253*, 393 (1952).

1381 Marshall, J. L., *Lightning Protection*, Wiley, New York (1973).

1382 Fuquay, D. M., A. R. Taylor, R. G. Hawe, and C. W. Schmid, Jr., "Lightning Discharges that Caused Forest Fires," *J. Geophys. Res.*, *77*, 2156 (1972).

1383 Taylor, A. R., "Lightning Effects on the Forest Complex," *Annual Tall Timbers Fire Ecology Conf. Proc.*, *9*, 127 (1969).

1384 Fuquay, D. M., R. G. Baugham, A. R. Taylor, and R. G. Hawe, "Characteristics of Seven Lightning Discharges that Caused Forest Fires," *J. Geophys. Res.*, *72*, 6371 (1967).

1385 Taylor, A. R., "Diameter of Lightning as Indicated by Tree Scars," *J. Geophys. Res.*, *70*, 5693 (1965).

1386 Taylor, A. R., "Agent of Change in Forest Ecosystems," *J. Forestry*, *68*, 477 (1971).

1387 Orville, R. E., "Photograph of a Close Lightning Flash," *Science*, *162*, 666 (1968).

1388 Green, F. H. W., and A. Millar, "A Tree Struck by Lightning, July 1970," *Weather*, *26*, 174 (1971).

1389 Shipley, J. F., "Lightning and Trees," *Weather*, *1*, 206 (1946).

1390 Orville, R. E., "Close Lightning," *Weather*, *26*, 394 (1971).

1391 Beck, E., *Lightning Protection for Electric Systems*, McGraw-Hill, New York (1954).

1392 Falconer, R. E., "Lightning Strikes a Parked School Bus," *Weather*, *21*, 280 (1966).

1393 Vonnegut, B., "Effects of a Lightning Discharge on an Aeroplane," *Weather*, *21*, 277 (1966).

1394 "Lightning Strikes an Aircraft—I," *Weather*, *19*, 206 (1964).

1395 Mason, D., "Lightning Strikes an Aircraft—II," *Weather*, *19*, 248 (1964).

1396 Hagenguth, J. H., "Lightning Stroke Damage to Aircraft," *Am. Institute Electrical Engineers Trans.*, *68*, 1036 (1949).

1397 Fitzgerald, D. R., "Probable Aircraft 'Triggering' of Lightning in Certain Thunderstorms," *Mon. Weather Rev.*, *95*, 835 (1967).

1398 Moore, C. B., B. Vonnegut, J. A. Machado, and H. J. Survilas, "Radar Observations of Rain Gushes Following Overhead Lightning Strokes," *J. Geophys. Res.*, *67*, 207 (1962).

1399 Shackford, C. R., "Radar Indications of a Precipitation—Lightning Relationship in New England Thunderstorms," *J. Met.*, *17*, 15 (1960).

1400 Moore, C. B., B. Vonnegut, E. A. Vrablik, and D. A. McCraig, "Gushes of Rain and Hail after Lightning," *J. Atmos. Sci.*, *21*, 646 (1964).

1401 Taussig, H. B., "Death from Lightning and the Possibility of Living Again," *Am. Scientist*, *57*, 306 (1969).

1402 McIntosh, D. H., ed., *Meteorological Glossary*, Her Majesty's Stationary Office, London (1963).

1403 Kamra, A. K., "Visual Observation of Electric Sparks on Gypsum Dunes," *Nature*, *240*, 143 (1972).

1404 Botley, C. M., and W. E. Howell, "Mystery on Mount Adams," *Mt. Washington Observatory News Bul.*, *8* (1), 9 (Mar. 1967).

1405 Markson, R., and R. Nelson, "Mountain-Peak Potential-Gradient Measurements and the Andes Glow," *Weather*, *25*, 350 (1970).

1406 Young, J. R. C., "The Andes Glow," *Weather*, *26*, 39 (1971).

1407 Robinson, M., "A History of the Electric Wind," *Am. J. Phys.*, *30*, 366 (1962).

1408 Cobine, J. D., "Other Electrostatic Effects and Applications" in Ref. 540, pp. 441-455; see pp. 452-453.

1409 Markowitz, W., "The Physics and Metaphysics of Unidentified Flying Objects," *Science*, *157*, 1274 (1967).

1410 Bondi, H., *The Universe at Large*, Anchor Books, Doubleday, New York (1960), Chapter 2.

1411 Sciama, D. W., *The Unity of the Universe*, Doubleday, New York (1959), Chapter 6.

1412 Bondi, H., *Cosmology*, Cambridge Univ. Press, Cambridge (1961), Chapter 3.

1413 Jaki, S. L., *The Paradox of Olbers' Paradox*, Herber and Herber, New York (1969).

1414 Jaki, S. L., "Olbers', Halley's, or Whose Paradox?", *Am. J. Phys.*, *35*, 200 (1967).

1415 Harrison, E. R., "Olbers' Paradox," *Nature*, *204*, 271 (1964).

1416 Layzer, D., "Why Is the Sky Dark at Night?", *Nature*, *209*, 1340 (1966).

1417 Soberman, R. K., "Noctilucent Clouds," *Sci. Amer.*, *208*, 50 (June 1963); also see cover of this issue.

1418 Ludlam, F. H., "Noctilucent Clouds," *Weather*, *20*, 186 (1965).

1419 Hanson, A. M. "Noctilucent Clouds at 76.3° North," *Weather*, *18*, 142 (1963).

1420 Dietze, G., "Zones of the Visibility of a Noctilucent Cloud," *Tellus*, *21*, 436 (1969).

1421 Hallett, J., "Noctilucent Clouds 27-28 June 1966," *Weather*, *22*, 66 (1967).

1422 Fogle, B., "Noctilucent Clouds in the Southern Hemisphere," *Nature*, *204*, 14 (1964).

1423 Grishin, N. I., "Blue Clouds," *Solar System Res.*, *2*, 1 (Jan.-Mar. 1968).

1424 Stone, D., letter to editor, *Geotimes*, *14*, 8 (Oct. 1969).

1425 "Great Leap Downward," *Time*, *94*, 60 (Dec. 19, 1969); also see letters *95*, 4 (Jan. 12, 1970) and *95*, 4 (Jan. 19, 1970).

1426 Truman, J. C., "Wave Propagation in Snow," *Am. J. Phys.*, *41*, 282 (1973).

1427 Spenceley, B., and L. Hastings, "On Wave Propagation in Snow," *Am. J. Phys.*, *41*, 1025 (1973).

1428 Shinbrot, M., "Fixed-Point Theorems," *Sci. Amer.*, *214*, 105 (Jan. 1966).

1429 Hilbert, D., and S. Cohn-Vossen, *Geometry and the Imagination*, Chelsea, New York (1952), pp. 324 ff.

1430 Courant, R., and H. Robbins, *What Is Mathematics? An Elementary Approach to Ideas and Methods*, Oxford Univ. Press, New York (1941), pp. 251-255.

1431 Cobb, V., *Science Experiments You Can Eat*, J. B. Lippincott Co., Philadelphia (1972), pp. 58-63.

1432 de Bruyne, N. A., "How

Glue Sticks," *Nature*, *180*, 262 (1957).

1433 Kelvin, Lord, *Baltimore Lectures on Molecular Dynamcis and the Wave Theory of Light*, C. J. Clay and Sons, London (1904).

1434 Reynolds, O., "Experiments Showing Dilatancy, A Property of Granular Material, Possibly Connected with Gravitation" in Ref. 156, pp. 217–227.

1435 Reynolds, O., "On the Dilatancy of Media Composed of Rigid Particles in Contact. With Experimental Illustrations" in Ref. 156, pp. 203–216.

1436 Schaefer, H. J., "Radiation Exposure in Air Travel," *Science*, *173*, 780 (1971).

1437 O'Brien, K., and J. E. McLaughlin, "The Radiation Dose to Man from Galactic Cosmic Rays," *Health Phys.*, *22*, 225 (1972).

1438 Cornelius A. T., T. F. Budinger, and J. T. Lyman, "Radiation-Induced Light Flashes Observed by Human Subjects in Fast Neutron, X-Ray and Positive Pion Beams," *Nature*, *230*, 596 (1971).

1439 Budinger, T. F., H. Bichsel, and C. A. Tobias, "Visual Phenomena Noted by Human Subjects in Exposure to Neutrons of Energies Less than 25 Million Electron Volts," *Science*, *172*, 868 (1971).

1440 McNulty, P. J., "Light Flashes Produced in the Human Eye by Extremely Relativistic Muons," *Nature*, *234*, 110 (1971).

1441 Fremlin, J. H., "Cosmic Ray Flashes," *New Scientist*, *47*, 42 (1970).

1442 Charman, W. N., J. A. Dennis, G. G. Fazio, and J. V. Jelley, "Visual Sensations Produced by Single Fast Particle," *Nature*, *230*, 522 (1971).

1443 D'Arcy, F. J., and N. A. Porter, "Detection of Cosmic Ray μ-Mesons by the Human Eye," *Nature*, *196*, 1013 (1962).

1444 Fazio, G. G., J. V. Jelley, and W. N. Charman, "Generation of Cherenkov Light Flashes by Cosmic Radiation within the Eyes of the Apollo Astronauts," *Nature*, *228*, 260 (1970).

1445 McAulay, I. R., "Cosmic Ray Flashes in the Eye," *Nature*, *232*, 421 (1971).

1446 Charman, W. N., and C. M. Rowlands, "Visual Sensations Produced by Cosmic Ray Muons," *Nature*, *232*, 574 (1971).

1447 Wang, T. J., "Visual Response of the Human Eye to X Radiation," *Am. J. Phys.*, *35*, 779 (1967).

1448 Young, P. S., and K. Fukui, "Predicting Light Flashes due to Alpha-Particle Flux on SST Planes," *Nature*, *241*, 112 (1973).

1449 McNulty, P. J., V. P. Pease, L. S. Pinsky, V. P. Bond, W. Schimmerling, and K. G. Vosburgh, "Visual Sensations Induced by Relativistic Nitrogen Nuclei," *Science*, *178*, 160 (1972).

1450 Budinger, T. F., J. T. Lyman, and C. A. Tobias, "Visual Perception of Accelerated Nitrogen Nuclei Interacting with the Human Retina," *Nature*, *239*, 209 (1972).

1451 Burch, W. M., "Cerenkov Light from ^{32}P as an Aid to Diagnosis of Eye Tumours," *Nature*, *234*, 358 (1971).

1452 Johnson, B. B., and T. Cairns, "Art Conservation: Culture under Analysis. Part 1," *Analytical Chem.*, *44*, 24A (1972).

1453 Werner, A. E., "Scientific Techniques in Art and Archaeology," *Nature*, *186*, 674 (1960).

1454 Gettens, R. J., "Science in the Art Museum," *Sci. Amer.*, *187*, 22 (Jul. 1952).

1455 Bohren, C. F., and R. L. Beschta, "Comment on 'Wave Propagation in Snow'," *Am. J. Phys.*, *42*, 69 (1974).

1456 LeMone, M. A., "The Structure and Dynamics of Horizontal Roll Vortices in the Planetary Boundary Layer," *J. Atmos. Sci.*, *30*, 1077 (1973).

1457 Murrow, R. B., *Saturday Review*, *50*, 51 (3 June 1967).

1458 Zahl, P. A., "Nature's Night Lights," *Nat. Geographic Mag.*, *140*, 45 (July 1971).

1459 Alyea, H. N., "Lithium Spectrum" in "Chemical Principles Exemplified," R. C. Plumb, ed., *J. Chem. Ed.*, *48*, 389 (1971).

1460 Herbert, F., *Dune*, Ace Books, New York (1965).

1461 Saylor, C. P., "Case of the Flowing Roof," *Chemistry*, *44*, 19 (Dec. 1971).

1462 Palmer, F., "What about Friction?", *Am. J. Phys.*, *17*, 181, 327, 336 (1949).

1463 Brewington, G. P., "Comments on Several Friction Phenomena," *Am. J. Phys.*, *19*, 357 (1951).

1464 Rabinowicz, E., "Direction of the Friction Force," *Nature*, *179*, 1073 (1957).

1465 Rabinowicz, E., "Resource Letter F-1 on Friction," *Am. J. Phys.*, *31*, 897 (1963).

1466 Guy, A. G., *Introduction to Materials Science*, McGraw-Hill, New York (1972), Chapter 10.

1467 Gilman, J. J., "Fracture in Solids," *Sci. Amer.*, *202*, 94 (Feb. 1960).

1468 Corten, H. T., and F. R. Park, "Fracture," *Intern. Sci. Tech.* *15* (Mar. 1963), p. 24.

1469 Holloway, D. G., "The Fracture of Glass," *Phys. Ed.*, *3*, 317 (1968).

1470 Field, J. E., "Brittle Fracture: Its Study and Application," *Contemporary Phys.*, *12*, 1 (1971).

1471 Field, J. E., "Fracture of Solids," *Phys. Teacher*, *2*, 215 (1964).

1472 Plumb, R. C., "Auto Windows—Strong but Self-Destructing," *J. Chem. Ed.*, *50*, 131 (1973).

1473 Gilman, J. J., "Fracture in Solids," *Sci. Amer.*, *202*, 94 (Feb. 1960).

1474 Holloway, D. G., "The Strength of Glass," *Phys. Bul.*, *23*, 654 (1972).

1475 Plumb, R. C., "Durable Chrome Plating" in "Chemical Principles Exemplified," *J. Chem. Ed.*, *49*, 626 (1972).

1476 Rayleigh, Lord, "Polish" in Ref. 129, Vol. 5, pp. 392–403.

1477 Rabinowicz, E., "Polishing," *Sci. Amer.*, *218*, 91 (June 1968).

1478 Sharpe, L. H., H. Schonhorn, and C. J. Lynch, "Adhesives," *Intern. Sci. Tech.*, *28* (Apr. 1964), p. 26.

1479 Baier, R. E., E. G. Shafrin, and W. A. Zisman, "Adhesion: Mechanisms That Assist or Impede It," *Science*, *162*, 1360 (1968).

1480 de Bruyne, N. A., "The Physics of Adhesion," *J. Sci. Instruments*, *24*, 29 (1947).

1481 McCutchen, C. W., "Ghost Wakes Caused by Aerial Vortices," *Weather*, *27*, 33 (1972).

1482 Taylor, G. I., "The Motion of a Sphere in a Rotating Liquid," *Proc. Roy. Soc. Lond.*, *A102*, 180 (1922); also in Ref. 255, Vol. IV, pp. 24–33.

1483 Takahara, H., "Sounding Mechanism of Singing Sand," *J. Acoust. Soc. Am.*, *53*, 634 (1973).

1484 Knott, C. G., *Life and Scientific Work of Peter Guthrie Tate*, Cambridge at the Univ. Press (1911), pp. 58–60, 329–345.

1485 Rothschild, M., Y. Schlein, K. Parker, and S. Sternberg, "Jump of the Oriental Rat Flea Xenopsylla cheopis (Roths.)," *Nature*, *239*, 45 (1972).

1486 Sharp, C. M., and M. J. F. Bowyer, *Mosquito*, Faber and Faber, London (1971), pp. 74–78.

1487 Welch, S., "What Makes It Turn?", *Phys. Teacher*, *11*, 303 (1973).

1488 Thompson, O. E., "Coriolis Deflection of a Ballastic Projectile," *Am. J. Phys.*, *40*, 1477 (1972).

1489 Anastassiades, A. J., "Infrasonic Resonances Observed in Small Passenger Cars Travelling on Motorways," *J. Sound Vibration*, *29*, 257 (1973).

1490 Evans, M. J., and W. Tempest, "Some Effects of Infrasonic Noise in Transportation," *J. Sound Vibration*, *22*, 19 (1972).

1491 Tempest, W., and M. E. Bryan, "Low Frequency Sound Measurement in Vehicles," *Appl. Acoustics*, *5*, 133 (1972).

1492 Johnson, C. C., and A. W. Guy, "Nonionizing Electromagnetic Wave Effects in Biological Materials and Systems," *Proc. IEEE*, *60*, 692 (1972).

1493 Griffin, D. R., "Comments on Animal Sonar Symposium," *J. Acoust. Soc. Am.*, *54*, 137 (1973).

1494 Novick, A., "Echolocation in Bats: A Zoologist's View," *J. Acoust. Soc. Am.*, *54*, 139 (1973).

1495 Grinnell, A. D., "Neural Processing Mechanisms in Echolocating Bats, Corrleated with Differences in Emitted Sounds," *J. Acoust. Soc. Am.*, *54*, 147 (1973).

1496 Simmons, J. A., "The Resolution of Target Range by Echolocating Bats," *J. Acoust. Soc. Am.*, *54*, 157 (1973).

1497 Suga, N., and P. Schlegel, "Coding and Processing in the Auditory Systems of FM-Signal-Producing Bats," *J. Acoust. Soc. Am.*, *54*, 174 (1973).

1498 Lunde, B. K., "Clouds Associated with an Interstate Highway," *Weatherwise*, *26*, 122 (1973); also see A. Rango, J. L. Foster, V. V. Salomonson, "Comments on 'Clouds Associated with an Interstate Highway'," *Weatherwise*, *26*, 222 (1973).

1499 Patitsas, A. J., "Rainbows, Glories, and the Scaler Field Approach," *Canadian J. Phys.*, *50*, 3172 (1972).

1500 Rossmann, F. O., "Banded Reflections from the Sea," *Weather*, *15*, 409 (1960).

1501 Swing, R. E., and D. P. Rooney, "General Transfer Function for the Pinhole Camera," *J. Opt. Soc. Amer.*, *58*, 629 (1968).

1502 Sayanagi, K., "Pinhole Imagery," *J. Opt. Soc. Am.*, *57*, 1091 (1967).

1503 Hardy, A. C., and F. H. Perrin, *The Principles of Optics*, McGraw-Hill, New York (1932), pp. 124–126.

1504 "Sun Pillar and Arc of Contact," *Weather*, *15*, 406 (1960).

1505 Weaver, K. F., "Journey to Mars," *Nat. Geographic Mag.*, *143*, 230 (1973); see pp. 258–259.

1506 Bates, D. R., "Auroral Audibility," *Nature*, *244*, 217 (1973).

1507 Buneman, O., "Excitation

of Field Aligned Sound Waves by Electron Streams," *Phys. Rev. Letters*, *10*, 285 (1963).

1508 Beals, C. S., "Audibility of the Aurora and Its Appearance at Low Atmospheric Levels," *Quart. J. Roy. Met. Soc.*, *59*, 71 (1933).

1509 Campbell, W. H., and J. M. Young, "Auroral-Zone Observations of Infrasonic Pressure Waves Related to Ionospheric Disturbances and Geomagnetic Activity," *J. Geophys. Res.*, *68*, 5909 (1963).

1510 Sverdrup, H. E., "Audibility of the Aurora Polaris," *Nature*, *128*, 457 (1931).

1511 Chapman, S., "The Audibility and Lowermost Altitude of the Aurora Polaris," *Nature*, *127*, 341 (1931).

1512 Thorp, J. R., "Sunbathing, Sunburn, and Suntan," *Weather*, *15*, 221 (1960).

1513 "Anticrepuscular Rays," *Weather*, *22*, 18 (1967).

1514 Tricker, R. A. R., "A Note on the Lowitz and Associated Arc," *Weather*, *25*, 503 (1970).

1515 Singleton, F., M. J. Kerley, and D. J. Smith, "Recent Observations from Aircraft of Some Rare Halo Phenomena," *Weather*, *15*, 98 (1960).

1516 Jacobs, S. F., and A. B. Stewart, "Chromatic Aberration in the Eye," *Am. J. Phys.*, *20*, 247 (1952).

1517 Piersa, H., "Fixing of Dust Figures in a Kundt's Tube," *J. Acoust. Soc. Am.*, *37*, 533 (1965).

1518 Sherwin, K., "Man-Powered Flight as a Sporting Activity," *Am. Inst. Aeronautics Astronautics Student J.*, *10*, 27 (Apr. 1972).

1519 Sherwin, K., "Man-Powered Flight: A New World's Records," *Am. Inst. Aeronautics Astronautics Student J.*, *10*, 23 (Oct. 1972).

1520 Foulkes, R. A., "Dowsing Experiments," *Nature*, *229*, 163 (1971).

1521 Smith, D. G., "More about Dowsing," *Nature*, *233*, 501 (1971).

1522 Merrylees, K. W., "Dowsing Experiments Criticized," *Nature*, *233*, 502 (1971).

1523 Vogt, E. Z., and R. Hyman, *Water-Witching USA*, Univ. Chicago Press, Chicago (1959).

1524 Carpenter, R. K., "A Favorite Experiment," *Phys. Teacher*, *11*, 428 (1973).

1525 Swinson, D. B., "Skiing and Angular Momentum: A Proposed Experiment," *Phys. Teacher*, *11*, 415 (1973).

1526 Moore, M., "Blue Sky and Red Sunsets," *Phys. Teacher*, *11*, 436 (1973).

1527 Kamra, A. K., "Measurements of the Electrical Properties of Dust Storms," *J. Geophys. Res.*, *77*, 5856 (1972).

1528 Shankland, R. S., "Acoustics of Greek Theatres," *Phys. Today*, *26*, 30 (Oct. 1973).

1529 Waller, M. D., *Chladni Figures*, G. Bell and Sons, Ltd., London (1961).

1530 Trees, H. J., "Some Distinctive Contours Worn on Alumina-Silica Refractory Faces by Different Molten Glasses: Surface Tension and the Mechanism of Refractory Attack," *J. Soc. Glass Tech.*, *38*, 89T (1954).

1531 Rothschild, M., Y. Schlein, K. Parker, C. Neville, and S. Sternberg, "The Flying Leap of the Flea," *Sci. Amer.*, *229*, 92 (Nov. 1973).

1932 Silverman, S. M., and T. F. Tuan, "Auroral Audibility," *Advances in Geophysics*, Vol. 16, H. E. Landsberg and J. Van Mieghem, eds., Academic Press (1973), pp. 156-266.

1533 Blackwood, O. H., W. C. Kelly, and R. M. Bell, *Central Physics*, Fourth Ed., Wiley, New York (1973), p. 258.

1534 Brown, R., "What Levels of Infrasound Are Safe?," *New Scientist*, *60*, 414 (1973).

1535 Hanlon, J., "Can Some People Hear the Jet Stream?," *New Scientist*, *60*, 415 (1973).

1536 Gable, J., "Ultrasonic Anti-theft Devices—A New Hazard?," *New Scientist*, *60*, 416 (1973).

1537 Chigier, N. A., "Vortexes in Aircraft Wakes," *Sci. Amer.*, *230*, 76 (Mar. 1974).

1538 Golden, J. H., "Some Statistical Aspects of Waterspout Formation," *Weatherwise*, *26*, 108 (1973).

1539 Barcilon, A., "Dust Devil Formation," *Geophysical Fluid Dynamics*, *4*, 147 (1972).

1540 Sinclair, P. C., "The Lower Structure of Dust Devils," *J. Atmos. Sci.*, *30*, 1599 (1973).

1541 Mansfield, R. J. W., "Latency Functions in Human Vision," *Vision Res.*, *13*, 2219 (1973).

1542 Harker, G. S., "Assessment of Binocular Vision Utilizing Pulfrich and Ventian Blind Effects," *Am. J. Optom. & Archives Am. Acad. Optom.*, *50*, 435 (1973).

1543 Rogers, B. J., M. J. Steinbach, and H. Ono, "Eye Movements and the Pulfrich Phenomenon," *Vision Res.*, *14*, 181 (1974).

1544 McElroy, M. B., S. C. Wofsy, J. E. Penner, and J. C.

McConnell, "Atmospheric Ozone; Possible Impact of Stratospheric Aviation," *J. Atmos. Sci., 31,* 287 (1974).

1545 Foley, H. M., and M. A. Ruderman, "Stratospheric NO Production from Past Nuclear Explosions," *J. Geophys. Res., 78,* 4441 (1973).

1546 Levenspiel, O, and N. de Nevers, "The Osmotic Pump," *Science, 183,* 157 (1974).

1547 Collyer, A. A., "Time Dependent Fluids," *Phys. Educ., 9,* 38 (1974).

1548 Ette, A. I. I., and E. U. Utah, "Measurement of Point-Discharge Current Density in the Atmosphere," *J. Atmos. Terrestrial Phys., 35,* 785 (1973).

1549 Ette, A. I. I., and E. U. Utah, "Studies of Point-Discharge Characteristics in the Atmosphere," *J. Atmos. Terrestrial Phys., 35,* 1799 (1973).

1550 Gannon, R., "New Scientific Research May Take the Sting Out of Lightning," *Popular Science, 204,* 76 (Jan. 1974).

1551 Pinkston, E. R., and L. A. Crum, "Lecture Demonstrations in Acoustics," *J. Acoust. Soc. Am., 55,* 2 (1974).

1552 Schelleng, J. C., "The Physics of the Bowed String," *Sci. Amer., 230,* 87 (Jan. 1974).

1553 Elder, S. A., "On the Mechanism of Sound Production in Organ Pipes," *J. Acoust. Soc. Am., 54,* 1554 (1973).

1554 Konishi, M., "How the Owl Tracks Its Prey," *Am. Scientist, 61,* 414 (1973).

1555 Steven, S. S., and E. B. Newman, "The Localization of Pure Tones," *Proc. Nat. Acad. Sci., 20,* 593 (1934).

1556 Greenslade, Jr., T. B., "Suspension Bridges," *Phys. Teacher, 12,* 7 (1974).

1557 Gribakin, F. G., "Perception of Polarised Light in Insects by Filter Mechanism," *Nature, 246,* 357 (1973).

1558 "Leaking Electricity," *Time, 102,* 87 (19 Nov. 1973).

1559 Boas, Jr., R. P. "Cantilevered Books," *Am. J. Phys., 41,* 715 (1973).

1560 Mallette, V., "Comment on 'Speckle Patterns in Unfiltered Sunlight'," *Am. J. Phys., 41,* 844 (1973).

1561 Stanford, Jr., A. L., "On Shadow Bands Accompanying Total Solar Eclipses," *Am. J. Phys., 41,* 731 (1973).

1562 "Tacoma Narrows Bridge Collapse," Film Loop 80-2181/1, Holt, Rinehart and Winston, Inc., Media Dept., 383 Madison Avenue, New York, NY 10017.

1563 Photos, advertising for Old Town Canoe Company, Old Town, Maine 04468.

1564 Musgrove, P., "Many Happy Returns," *New Scientist, 61,* 186 (1974).

1565 Crawford, F. S., "Coille Effect: A Manifestation of the Reversibility of Light Rays," *Am. J. Phys., 41,* 1370 (1973).

1566 Land, E. H., "The Retinex," *Amer. Scientist, 52,* 247 (1964).

1567 Land, E. H., and J.J. Cann, "Lightness and Retinex Theory," *J. Opt. Soc. Am., 61,* 1 (1971).

1568 Wahlin, L., "A Possible Origin of Atmospheric Electricity," *Foundations of Phys., 3,* 459 (1973).

1569 Mehra, J., "Quantum Mechanics and the Explanation of Life," *Amer. Scientist, 61,* 722 (1973).

1570 Warner, C. P., "Nature's Alert Eyes," *Nat. Geographic Mag., 115,* 568 (April 1959).

1571 Barrett, R. P., "The Dynamic Response of a Suspension Bridge," film loop, Amer. Assoc. Phys. Teachers, Drawer AW, Stony Brook, NY 11790 (1972).

1572 Dobelle, W. H., M. G. Mladejovsky, and J. P. Girvin, "Artificial Vision for the Blind: Electrical Stimulation of Visual Cortex Offers Hope for a Functional Prosthesis," *Science, 183,* 440 (1974).

1573 "Seeing by Phosphene," *Sci. Amer., 230,* 45 (1974).

1574 "Archimedes' Weapon," *Time, 102,* 60 (26 Nov. 1973).

1575 "Re-enacting History—With Mirrors," *Newsweek, 82,* 64 (26 Nov. 1973).

1576 Mielenz, K. D., "Eureka!", *Appl. Optics, 13,* A14 (Feb. 1974).

1577 Phillips, E. A., "Arthur C. Clarke's Burning Glass," *Appl. Optics, 13,* A16 (Feb. 1974).

1578 Deirmendjian, D., "Archimedes's Burning Glass," *Appl. Optics, 13,* 452 (Feb. 1974).

1579 Claus, A. C., "On Archimedes' Burning Glass," *Appl. Optics, 12,* A14 (Oct. 1973).

1580 Stavroudis, O. N., "Comments on: On Archimedes' Burning Glass," *Appl. Optics, 12,* A16 (Oct. 1973).

1581 Setright, L. J. K. "The Gastropod," in *Motor Racing. The International Way Number 2,* ed. by N. Brittan, A. S. Barnes & Co., New York (1971), pp. 39-44.

1582 Swezey, K. M., *After-Dinner Science,* Revised Ed., McGraw-Hill, New York (1961); pp. 118-119.

1583 Levi, L., "Blackbody Temperature for Threshold Visibility,"

Appl. Optics, *13*, 221 (1974).
1584 Wellington, W. G., "Bumblebee Ocelli and Navigation at Dusk," *Science*, *183*, 550 (1974).
1585 Lloyd, J. E., "Model for the Mating Protocol of Synchronously Flashing Fireflies," *Nature*, *245*, 268 (1973).
1586 Floyd, Bill, "The Fantastic Floydmar Camera—The Pinnacle of Pinholery," *Petersen's Photographic Mag.*, *1*, 42 (Sep. 1972).
1587 Harrison, E. R., "Why the Sky is Dark at Night," *Phys. Today*, *27*, 30 (Feb. 1974).
1588 Crawford, F. S., "Singing Corrugated Pipes," *Am. J. Phys.* *42*, 278 (1974).
1589 Tsantes, E., "Notes on the Tides," *Am. J. Phys*, *42*, 330 (1974).
1590 Krider, E. P., "An Unusual Photograph of an Air Lightning Discharge," *Weather*, *29*, 24 (1974).
1591 American Cinematographer *55*, 406–455 (April 1974), entire issue.
1592 Symmes, D. L., "3-D: Cinema's Slowest Revolution" in Ref. 1591, pp. 406 ff.
1593 Wales, K., "The Video West, Inc. Three Dimensional Photographic System" in Ref. 1591, pp. 410 ff.
1594 Williams, A., "A 3-D Primer" in Ref. 1591, pp. 412 ff.
1595 "The Stereovision 3-D System" in Ref. 1591, p. 414.
1596 "An Alphabetical Listing of Three-Dimensional Motion Pictures" in Ref. 1591, p. 415.
1597 Williams, A. D., "3-D Motion Picture Techniques" in Ref. 1591, pp. 420 ff.
1598 Symmes, D. L., "3-D Cine Systems" in Ref. 1591, pp. 421 ff.
1599 "Calculations for Stereo Cimenatography" in Ref. 1591, pp. 424–425.
1600 Hoch, W. C., "Challenges of Stereoscopic Motion Picture Photography" in Ref. 1591, pp. 426 ff.
1601 "The Spacevision 3-D System" in Ref. 1591, pp. 432 ff.
1602 Vlahos, P., "The Role of 3-D in Motion Pictures" in Ref. 1591, pp. 435 ff.
1603 Layer, H. A., "Stereo Kinematics: The Merging of Time and Space in the Cinema" in Ref. 1591, pp. 438–441.
1604 Dunn, L. G., and D. W. Weed, "The Third Dimension in Dynavision" in Ref. 1591, p. 450.
1605 Biroc, J., "Hollywood Launches 3-D Production" in Ref. 1591, pp. 452 ff.
1606 Weed, D., and K. H. Oakley, "Cine-Ortho: 3-D Movies for Eye Training" in Ref. 1591, pp. 454 ff.
1607 Holm, W. R., "Holographic Motion Pictures for Theatre and Television" in Ref. 1591, pp. 454 ff.
1608 Stong, C. L., ed., "The Amateur Scientist," *Sci. Amer.*, *230*, 116 (April 1974).
1609 Hickman, K., Jer Ru Maa, A. Davidhazy, and O. Mady, "Floating Drops and Liquid Boules—A Further Look," *Industrial & Engineering Chemistry*, *59*, 18 (Oct. 1967).
1610 Mattsson, J. O., "Experiments on Horizontal Halos in Divergent Light," *Weather*, *29*, 148 (1974).
1611 Corliss, William R., *Strange Phenomena. A Sourcebook of Unusual Natural Phenomena*, P. O. Box 107, Glen Arm, Maryland, 21057, USA.
1612 Higbie, J., "The Motorcycle as a Gyroscope," *Am. J. Phys. 42*, 701 (1974).
1613 Carlsoo, S., "A Kinetic Analysis of the Golf Swing," *J. Sports Medicine and Physical Fitness*, *7*, 76 (June 1967).
1614 "Observations of the Green Flash," *Sky and Telescope*, *48*, 61 (1974).
1615 Simms, D. L., "More on That Burning Glass of Archimedes," *Appl. Optics*, *13*, A14 (May 1974).
1616 Denton, R. A., "The Last Word," *Appl. Optics*, *13*, A16 (May 1974).
1617 Few, A. A., "Thunder Signatures," *EOS*, *Trans. Am. Geophys. Union*, *55*, 508 (1974).
1618 Guza, R. T., and R. E. Davis, "Excitation of Edge Waves by Waves Incident on a Beach," *J. Geophys. Res.*, *79*, 1285 (1974).
1619 Levin Z., and A. Ziv, "The Electrification of Thunderclouds and the Rain Gush," *J. Geophys. Res.*, *79*, 2699 (1974).
1620 Collyer, A. A., "Viscoelastic Fluids," *Phys. Educ.*, *9*, 313 (1974).
1621 Shute, C. C. D., "Haidinger's Brushes and Predominant Orientation of Collagen in Corneal Stroma," *Nature*, *250*, 163 (1974).
1622 Robertson, G. W., "Unusual Halo Phenomenon at Swift Current," *Weather*, *29*, 113 (1974).
1623 Teer, T. L., and A. A. Few, "Horizontal Lightning," *J. Geophys. Res.*, *79*, 3436 (1974).
1624 Buck, J., and J. F. Case, "Control of Flashing in Fireflies. I. The Latern as a Neuroeffector Organ," *Bio. Bull.*, *121*, 234 (1961).
1625 Mason, C. W., "Structural Colors in Insects, II," *J. Phys. Chem.*, *31*, 321 (1927).
1626 Bryant, H. C., and N. Jar-

mie, "The Glory," *Sci. Amer.*, *231*, 60 (July 1974).

1627 Whitaker, R. J., "Physics of the Rainbow," *Phys. Teacher*, *12*, 283 (1974).

1628 Thompson, A. H., "Water Bows: White Bows and Red Bows," *Weather*, *29*, 178 (1974).

1629 Liou, K., and J. E. Hansen, "Intensity and Polarization for Single Scattering by Polydisperse Spheres: A Comparison of Ray Optics and Mie Theory." *J. Atmos. Sci.*, *28*, 995 (1971).

1630 Roesch, S., "Der Regenbogen in Wissenschaft und Kunst," *Appl. Optics*, *7*, 233 (1968).

1631 Harsch, J., and J. D. Walker, "Double Rainbow and Dark Band in Searchlight Beam," *Am. J. Phys.*, *43*, 453 (1975).

1632 Walker, J. D., "Karate Strikes," *Am. J. Phys.*, *43*, 845 (1975).

Index

Short Answers

There are several very real dangers in writing answers, even short ones, to *The Flying Circus*. First of all, my references and my physics may be wrong. This danger may especially be true with those questions dealing with topics still under research, such as ball lightning, whose nature is sometimes tackled in every other issue of some of the journals. All I can say about my answers here is that they are the best I can do in a very small amount of space and with the currently available literature. Please remember that these short answers are but the tip of the iceberg; there is much physics below each. And you should not take them as the last word; instead, you should take them as starting points and then update them as new papers and articles appear.

The second danger in writing answers is more serious and makes me hesitate in even trying. You may flip to the answer section of this book so quickly that you miss the excitement of the question. Unless you savor the questions, even to the point of some frustration, you will miss the point of this book. The potential for learning how to examine the world in which you live lies much more in the first part of this book than in the second. So, please spend as much time as you can worrying over your own answers before you turn to this section or search the library for references.

1.1 The squeal heard in these several situations results from "stick and slip." For example, the incorrectly held chalk first sticks on the chalkboard, but when the writer bends the chalk sufficiently, it suddenly slips and then vibrates, periodically striking the chalkboard and producing the squeal we hear. As the vibrations decrease, the friction between the chalk and the board increases until the chalk sticks once again.

1.2 The finger excites the longitudinal oscillations, that is, the vibrations along the perimeter of the rim. The rim also oscillates transversely, that is, perpendicular to the rim. This second type of oscillation causes the fluid motion since it is a motion into and out of the liquid. The antinodes of the transverse oscillations and hence of the fluid motion are 45° from the antinodes of the longitudinal oscillations. Since the finger's position must be a place of maximum longitudinal motion and thus a longitudinal antinode, the fluid motion must have an antinode 45° behind the finger.

1.3 Imagine that one membrane is oscillating and the other is not. The moving membrane begins to excite the other by pushing on the air between them. As the second membrane begins to oscillate, however, the air thereafter hinders the oscillations of the first membrane, eventually stopping it. By the time the air has maximized the oscillations of the second membrane and stopped the first, the situation is reversed, and the air then transfers the oscillation back to the first membrane.

1.4 To press the bass into the records at the same intensity level as the higher frequencies would require an oscillation of the needle that would carry it into the adjacent groove.

1.5 and 1.6 In both of these cases the sound apparently results from the oscillation of the sand when parts of it are forced to move under a shearing stress. In the footstep the sand beneath the foot is forced downward; in the sand dune a small avalanche causes some sand to slide over another part. Although the noise production is not understood, it is apparently confined to sand having mostly spherical grains of uniform size.

1.7 The drawing of the bowstring on the edge of the plate sets the plate into vibrations. The pattern of the vibration, that is, where there are the maximum and minimum amplitudes, depends on the shape of the plate and where it is forced to have no vibrations because it is held in place. During the bowing, the sand, initially in the areas of maximum vibration (called antinodes), is thrown to the areas of no vibration (called nodes), eventually collecting in order to indicate the vibrational pattern. The fine dust would do the same except that it is carried by the air currents set in motion by the vibration. Since these currents blow across the plate from the nodes to the antinodes and then upward, the dust is carried to the antinodes and deposited.

1.8 More of the higher frequency harmonics are excited when the banjo is plucked with the fingernail or with a sharp pick than when the string is pulled with a finger. These higher frequencies give a twangy quality to the banjo's music.

1.9 The voice vibrates the can, which then excites waves on the string. These waves excite the second can in a reverse way to be heard. The can does not respond to the lower frequencies in your talk, and thus those frequencies will be missing at the other end, making your words thin.

1.10 The bow alternately sticks to the strings and then slips, leaving the strings oscillating during the intermediate time of sliding.

1.11 The frequency of a vibrating string depends on the string's density, length, and tension. If a string is tightened, the first two remain constant, and the increased tension raises the frequency of vibration. If a rubber band is stretched, however, all three of the quantities change so that the frequency is essentially unchanged.

1.12 The first sound comes when the bottom of the pan is heated and small bubbles form, each with a click and collectively with a hiss. With further heating, the bubbles detach from the bottom, rise into the cooler water, and then collapse, creating a louder noise. This noise continues until the water is sufficiently hot for the bubbles to reach the surface to break. Then the water is in full boil, and the noise of the bubbles reaching the surface is a softer, splashing sound.

1.13 Part of the sound of the brook is due to the bubbles formed in the running water. The creation of a bubble has little sound, but the volume oscillations and the collapse of a bubble produce much more sound.

1.14 If the ground is very cold (−10°F or lower), the ice beneath your feet cannot melt due to your weight and will snap instead. (Also see FC 3.46–3.49*.)

1.15 The small spaces in the snow's surface absorb the sound just as acoustic tile does in most modern offices. As the snow becomes more packed, this sound absorption is reduced.

1.16 Any periodic motion can produce sound waves. The periodic jerking as the individual threads snap when you rip cloth will set up sound waves for you to hear the ripping.

1.17 The popping is due to the bursting of tiny gas bubbles in the fluid lubricating the finger joints when the fingers are pulled and the fluid pressure is thereby reduced. Several minutes are needed before the gas is reabsorbed and ready for an encore.

1.18 The sound is caused by the release of air from the cereal grains as those grains become soggy in the milk and eventually burst.

1.19 The cracking is due to the thermal stresses in the ice as the ice warms. The "frying" sound, however, is due to the bursting of tiny air bubbles trapped in the ice as the surface reaches those bubbles. Ice free of such bubbles will melt with only the cracking.

1.20 The fact that the speed of sound is greater in ground than in air is not important since horses travel much slower than sound. The main advantage in listening to the ground is that there are less objects in the sound's path to scatter and attenuate the sound.

1.21 The resonant frequencies of your mouth (as any other resonating volume of gas) depend directly on the speed of sound in the gas in it. The speed of sound is greater in helium than in air, and therefore your voice will have a higher pitch.

1.22 Air trapped on the powder is released as the powder dissolves. Since the speed of sound is lower in air than in water, the speed of sound in the air-water mixture is lower than in pure water. During that period while the air escapes the container, the resonant frequencies of the water, which depend directly on the speed of sound, will also be lower. Hence, you hear a lower tone until the air escapes.

1.23 Because resonant frequencies in a wind instrument depend directly on the speed of sound, those frequencies increase as the player warms the instrument with his or her breath and thereby increases the speed of sound in it. Warming of the string instruments by friction will expand the string, thus decreasing the tension on the strings, which decreases the resonant frequencies of the string vibrations.

1.24 Within a couple of meters from the ground, the sound directly from the airplane can interfere with the sound reflected from the ground to enhance certain frequencies. The height at which constructive interference can occur (where one sound wave will reinforce another) depends on the wavelength of the sound. The nearer the ground, the shorter the wavelength will be for constructive interference. So, near the ground you will hear more of the short wavelengths, and thus high

* These citations refer to the answers in this book.

frequencies. As you raise your ear more, longer wavelengths and thus lower frequencies will be heard.

1.25 The sound that is reinforced in its travel through the culvert must reflect from the walls at a certain angle, which depends on the wavelength of the sound. The longer wavelengths that are reinforced and eventually heard must reflect at a greater angle than the shorter wavelengths and therefore travel less along the culvert's length between each reflection and take longer to reach the end. A listener at the end will thus first hear the shorter wavelengths (higher frequencies) and then progressively longer wavelengths (lower frequencies).

1.26 To have a distinct echo, the reflected sound must come no closer than 50 msec to the direct sound. Eliminating reflections from the walls will drastically reduce the acoustic richness of the room. The walls are designed to diffuse the sound throughout the room. Some of the scatterers on the walls should be small to scatter the short wavelengths (high frequencies) whereas other features should be larger to scatter the longer wavelengths (lower frequencies). The scattering should be diffuse enough to eliminate dead spots of destructive interference in the room.

1.27 Sound rays emitted from one focus in an elliptical room will cross at the other focus, thus making a conversation at the first focus audible at the second.

1.28 Sound travels faster in warmer air than cooler. If the air temperature decreases upward, the top portion of an initially horizontally traveling sound wave will propagate slower than the lower part, and the wave path will bend upward. With such a normal temperature distribution, the sound will not be able to travel very far along the ground before this refraction has bent it away from the ground. On a cold day the air temperature may increase upward, especially just over a body of water, thus refracting the sound downward instead of upward. Hence the sound will be kept near the ground longer.

1.29 The speed of sound in air increases as the air temperature increases. Thus, as a sound wave reaches the increasingly warmer air in the stratosphere, the higher portion of the wave will travel faster than the lower portion, causing the wave path to bend over and eventually turn downward. Sound will be heard where it reaches the ground. In the meantime, some direct sound from the source will travel horizontally until it is scattered or absorbed by the objects on the ground. Between the outer region of this horizontal propagation and the region where sound is returned from the stratosphere, there is a region in which sound from the source is not heard. If the sound returned from the stratosphere is reflected by the ground sufficiently well that it can return to the stratosphere, it can be bent over and returned to the ground once again to give yet another region of sound, the outer white area in Figure 1.29.

1.30 The scattering of sound by objects small with respect to the wavelength of the sound will depend inversely on the fourth power of the wavelength. The shorter wavelengths (higher frequencies) will therefore be scattered more than the longer wavelengths (lower frequencies). An echo of a yell will thus be higher pitched because the higher frequencies are more strongly returned.

1.31 In continuously reflecting from the walls of the dome, the sound waves are reinforced in a narrow belt around the perimeter of the wall. If the listener stands inside this belt, he or she can hear the whisper. Further from the wall, however, the reinforcement decreases and the whisper becomes inaudible. A whisper works better because it has more of the high frequency sounds than does normal talking, and the audible belt is wider with higher frequencies.

1.32 Suppose you are facing a picket fence. The sound reflected from a particular picket on your left, say, will return to you slightly sooner than the sound reflected from the next picket more to your left, because that next picket is a little further away from you. The sound returned first comes from the nearer pickets, and then progressively later sound returns from progressively more distant pickets, producing a musical tone to the echo. The frequency of the tone is the inverse of the time interval between reflections from adjacent pickets.

1.33 Sound from outside the funnel may not be able to reach the inside, because it is refracted so strongly by the high velocity winds around the funnel. The interior of the funnel may also appear to be silent, because hearing is difficult if one is suddenly thrust into a low pressure region. (Hearing is difficult in flying until the ears have a chance to adjust to the pressure changes in ascending and descending.)

1.34 There is no one answer to this question since, depending on the bridge, either effect may dominate or both may be equally present.

1.35 Sound goes further along the ground downwind not because there is less attenuation, but because the sound waves are refracted downward in that direction, whereas they are refracted upward in the upwind direction. Because of the obstacles on the ground, the wind velocity normally increases with height. An initially horizontally traveling wave moving downwind will have its upper portion moving at a greater speed than its lower portion, causing the wave to refract downward. By similar reasoning, a sound wave traveling upwind will be refracted upward.

1.36 Brontides are most likely the anomolous sound propagation of FC 1.29, with the unseen and distant sources being anything from explosions to thunder.

1.37 Sound waves are diffracted by the aperture between the stacks and thus sent into the area behind the stacks. The angular width of the diffraction pattern (the angle over which the sound will be spread by the aperture) is greater for longer wavelengths. Thus, if the birds had a lower frequency squeal, the sound would spread more.

1.38 For reasons similar to those in FC 1.28 and 1.29, the sound waves from a lightning stroke will be refracted upward by the warmer air near the ground. Beyond a range of about 15 miles, the sound is refracted so much that it is traveling upward and then, of course, cannot be heard by a ground observer.

1.39 This problem involves a similar refraction of sound waves by a change in temperature with height. Normally the water temperature decreases downward with depth. Hence, a sound wave issued horizontally will be bent over and sent downward because its upper portion, being in warmer water, will travel faster than its lower portion, being in cooler water. The refraction of the sound may be so severe that all the sound is bent downward and away from the submarine.

1.40 Sound is diffracted by the aperture just as in FC 1.37. Although the door may be almost closed, the sound coming through the remaining opening spreads through the room.

1.41 Sound from the guitar player's speaker would be picked up by his or her guitar, reamplified, and then emitted by the speaker a short time later. The frequency of the ringing is the inverse of the time needed to emit a sound once the guitar pickup is activated.

1.42 The angular width of the diffraction pattern (i.e., the angle over which the sound is spread when it propagates through the opening) is greater the narrower the opening is. Hence, the narrower width provides more spread and should be oriented horizontally as shown in the figure.

1.43 Barring reflection of the sound from nearby objects, you hear your friend when your friend is turned away because the sound diffracts around his or her head. The angular width of the diffraction pattern is larger for longer wavelengths (smaller frequencies). Since whispering is composed largely of high frequencies, it is heard poorly from behind because the whispering is diffracted less.

1.44 The effective length of a tube is increased by about one third of the tube's diameter for each open end. This lengthening will increase the wavelengths of the tube's harmonics (thus decreasing the resonant frequencies) and can be noticeable in wider tubes.

1.45 The low frequency sound can oscillate your chest, for example, because the variations in air pressure are sufficiently slow that your chest can follow them. Internal hemorrhage can result from organs being forced to rub against each other during their oscillations. Lower intensity levels may just cause dizziness and nausea. Common car sickness may partially result from the infrasound produced by the car.

1.46 With increased flow rates through the restricted portions of the pipes, turbulence can occur, leading to cavitation (formation of bubbles). Oscillations of the air bubbles can be amplified by the pipe and the walls, ceilings, and floors to which the pipes are connected.

1.47 The vibrations of the rod create a standing sound wave in the tube. Powder at the antinodes of air motion in the tube is gradually shaken toward the nodes where it collects in the larger piles shown in the figure. If the air flow is sufficiently rapid, vortices are created. The smaller ripples of dust are formed where two adjacent vortices rise or descend together.

1.48 From the noise of the pouring water, those frequencies that will resonantly excite the air column in the bottle will be picked out, enhanced, and heard. Of these the loudest frequency will be the lowest one, but the value of that frequency depends on the volume of the air column. The greater the volume, the lower the frequency. Thus, as water is poured from the bottle, the resonant frequency that is heard decreases.

1.49 Noises from the environment, including the slight whispers of a breeze passing the shell, will excite the shell's air volume at its resonant frequencies. The coming and going of these resonant frequencies gives the listener the illusion of hearing the ocean waves come and go.

1.50 The length and tension of the vocal cords determines the pitch of the voice. As the air pressure in the trachea increases, the cords are suddenly forced apart and then returned to their normal position. Continued oscillations of the cords produce the air pressure variations that excite the resonant harmonics of the mouth and nasal cavities. Usually a man's voice has lower pitch than a woman's because a man usually has thicker and longer vocal cords which will vibrate at lower frequencies. A boy's voice "breaks" during the rapid growth of the larynx, and the vocal cords change from being short and thin to those of a man. During whispering the vocal cords are removed from the larynx by relaxing them. The frequency of the sound will then depend on the oscillations created by other obstacles in the air stream and on the resonant frequencies of the mouth and nasal cavities.

1.51 Were you to sing in a large open area, you would hear your voice only as it is being produced. In the shower stall each sound will reflect many times from the close walls, prolonging the sensation when it is heard, and thus adding brilliance (continuation of the high frequency sounds) and fullness (continuation of the low frequency sounds) to your singing.

1.52 A glass will oscillate at certain resonant frequencies. Should a singer sing for several seconds at one of these frequencies, the oscillations of the glass can build up to the level where the glass cracks.

1.53 The wind can howl by whistling through wires and bare tree limbs (FC 1.55) or by creating edge tones (FC 1.56) on roof corners or other sharp obstacles.

1.54 Air flows from the held end to the circling end, because the rapid motion of the circling end reduces the air pressure there, whereas the air pressure at the held end is atmospheric pressure. As the air flows through the tube and over the corrugations, it begins to oscillate. The frequency of these oscillations is determined by the spacing of the corrugations and the flow speed of the air. From the small range of oscillations produced with a given twirling speed, the tube picks out its own resonant frequency to enhance, and that is the tone you hear from the tube. A faster twirling speed moves the range of air oscillations into higher frequencies, and a higher frequency harmonic of the tube is then exicted and heard.

1.55 As the wind passes a wire or bare tree limb, the air can become unstable and shed vortices from the obstacle. On a telephone wire, for example, vortices will be released alternately on the top and bottom of the wire. These vortices create the pressure variations that eventually reach our ears to be heard. If the wind is strong enough, the pressure variations on the two sides of the wire can cause the wire to vibrate, but that vibration is not necessary to the production of the sound. Because the pressure variations are due to the vortices produced on the two sides, the wire is forced to vibrate perpendicular to the air flow.

1.56 In the edge-tone setup, vortices are shed from the edge when the air stream strikes it. The reaction force of the edge then creates the sound we hear. Some of the sound returns to the source of the air stream, causing instability in the stream, which creates more vortices downstream. When the vortices reach the edge, more sound is produced, and the whole procedure is repeated. In the hole-tone production, the sound that is returned to the source of the air stream changes the speed of the air stream and causes vortex rings to form (such as in cigar smoke rings). When these rings strike the hole, more sound is produced and, again, the procedure is repeated.

The common tea kettle whistles

by hole-tone production. The cap on the kettle has two small holes spaced with a small cavity. When air from inside the kettle passes through the first hole, that hole acts as the air stream source for the second hole. At first the air stream is too slow for the necessary instability of the air at the second hole to cause sound. As the water approaches boiling, however, the air moves faster so that the vortices at the second hole are strong enough for the sound to be heard.

1.57 A coke bottle, a flute, and a recorder are different whistles than the ones in FC 1.56 because of the addition of a resonant cavity adjacent to the edge or hole where the instability is produced. From the range of frequencies in the sound made at the edge or hole, the cavity selects its resonant frequency to enhance, and that is the frequency heard.

1.58 The air stream blown into the police whistle creates an edge tone from which the adjacent cavity selects its resonant frequency. The ball inside the cavity periodically blocks the air holes and causes the whistling to warble.

1.59 Normal whistling through your lips appears to be a hole tone (FC 1.56) with an adjacent resonating cavity (the mouth), but the details of the air flow do not seem to be worked out.

1.60 The horn and the air contained in it provided resistance against which the diaphragm could work and which would transform high velocity motion over a small area to low velocity motion over a larger area. A long narrow tube would store the energy in standing waves and select only its resonant frequencies. A diaphragm opened to the room without a horn would be able to move so freely that relatively little of its oscillation energy would be transferred to air motion.

1.61 The sound is apparently due to pressure variations in the unstable vortices emerging from the central stem.

1.62 There are two main reasons for the difference in size. A large paper cone cannot respond quickly to a high frequency sound, breaking up into assorted waves over its surface instead. Thus, a smaller cone is used for the higher frequency range. Second, the speaker should spread its sound over a wide angle to fill the room. The angle of the diffraction pattern will depend on the relative size of the wavelength and the speaker cone. Short wavelengths (high frequencies) on a large cone will have a small diffraction pattern and thus will be beamed into the room. Therefore, the short wavelengths must come from a small speaker to be spread well into the room.

1.63 Normally the sound directly from the mouth diffracts to spread nearly uniformly in all directions. A cheerleading horn with a large front end will have considerably less diffraction because the open end is larger than the wavelengths in the cheerleader's yell. Thus, the sound in the direction of the horn will be louder than if no horn is used.

1.64 The bass notes will be produced in the ear even if none are issued from the speaker. If two different frequencies are incident on the ear, its nonlinear response produces vibrations having frequencies equal to the sum or difference of the incident frequencies, or the sum or difference of some integral multiple of those frequencies. The most prominent of these extra notes is the difference frequency, which gives the listener a bass note.

1.65 The frequency heard by a listener depends on the relative speeds of the listener and the source. Such a shift in the detected frequency from the frequency issued according to the source is called the Doppler shift. As a race car approaches a stationary listener, the car's whine is Doppler shifted up in frequency; as the car recedes, its whine is Doppler shifted down in frequency.

1.66 Exactly how a bat extracts information from its signal is current research and not well understood. Some bats emit a short constant frequency (CF) signal whose return indicates the target's presence and whose return frequency can indicate the target's speed (*see* the Doppler shift in FC 1.65). Other bats emit a frequency modulated (FM) signal. The frequency response of the target is analyzed to indicate the target's shape, size, surface texture, and range. Because the FM signal is swept over a range of frequencies, the bat cannot analyze it for a Doppler shift for the target speed. Thus, some bats emit a combination FM-CF signal to obtain all the possible information about the target.

1.67 The fluctuations are not quite loud enough (i.e., the pressure changes on the ear drum not quite large enough) to be heard. Even if the Brownian motion were more intense, it probably still would not be heard because the brain will ignore any continuous, constant noise signal.

1.68 Suppose you are talking to someone in a social room. You hear that person's voice both directly and in a diffuse way after it has reflected from the room. If there are other such conversations occurring in the room, there is a general diffuse level of talk with which your acquaintance must compete. The power of that diffuse background depends on the volume of the room, the acoustic absorption of the walls and the other objects in the room, some characteristic mean free path of sound between the walls, and the number of other conversations. At a critical number of conversations, the diffuse background begins to drown the direct level of talk from your friend. Any additional number of conversations means your friend will raise his or her voice, but all the other people will do the same and the party will become hopeless.

1.69 Because the rockets exceeded the speed of sound, the sound from their explosions could reach a listener before the sound of the flight through the air.

1.70 Such a live conversation in a noisy party provided at least one additional piece of information that the tape does not: directionality. You can distinguish a particular conversation from a very noisy background if you can distinguish the direction of the conversation through your normal binaural hearing.

1.71 Much of the sound you hear when you talk, especially the low frequency components, comes through bone conduction. Others hear you without those low frequency tones that, to you, add fullness to your voice. Listening to a tape of yourself on a good quality tape player reproduces what others normally hear of your speech.

1.72. You can determine the direction of a sound source by three means: comparison of the intensity, phase, or arrival time of the signals at your two ears. Intensity differences are useful only with short wavelength sound because long wavelengths diffract around the head to give about equal intensities at the ears. However, the long wavelength sounds will have different phases at the ears, the phase difference depending on how much from straight forward the sound source is angled. At intermediate wavelengths, corresponding to about 4000 Hz, neither technique works very well, and determination of the source direction is more difficult.

1.73 If an airplane exceeds the speed of sound at the height at which it is flying, the air it compresses will form a shock wave. Behind the airplane is a cone whose outer boundary is the shock wave. As the cone sweeps the ground, an observer experiences first a pressure rise and decline and then another rise until normal pressure is resumed. The second shock wave is made by the plane's tail. Sometimes the two increases in pressure are indistinguishable; other times they will come as two distinct booms. A shock wave may never reach the ground if it is sufficiently refracted by the warmer air it encounters during its descent. (For similar refractions of sound waves by variation in the air temperature, see FC 1.28 and 1.29.)

1.74 Current research is being done on thunder production and characteristics. The tremendous heating of the discharge column in the lightning stroke rapidly expands the air, creating a cylindrical shock wave that is the basic noise source for the thunder. Near the stroke, one may distinguish a hissing, probably due to corona discharge (FC 6.46), and a click, probably somehow due to an upward moving discharge in the stroke (FC 6.32). Continuation of the thunder (rolls and rumbles) may result from echoes of the initial sound from the environment.

1.75 The attenuation of sound by the atmosphere (due to viscosity and thermal conduction) is too severe for a sound made at a height of 80 km or more to reach the ground. If the sound is due to the collision of ice crystals in one's breath, the temperature should be at least as low as −40°C.

1.76 Shock waves from the artillery cause the visible bands because the refraction of the air is altered on their passage or because there is momentarily increased condensation in a cloud or fog bank through which the waves pass (similar to FC 3.27).

1.77 The whip crack may be due to the slap of the tip against itself or from the shock wave as the tip exceeds the speed of sound.

2.1 To simplify the problem, let us assume you are wearing a rainhat and need not worry about the rain on your head. If the rain is toward your front or directly overhead, then you should run as fast as possible to shelter. If, however, the rain is on your back, then you should run with a speed equal to the horizontal velocity of the rain, that is, along with the rain.

2.2 Experience surely helps greatly in the outfielder's ability to extrapolate a trajectory. The reference suggests that the elevation angle may also be of help. By running to keep the rate of increase in the ball's elevation angle constant, the player will arrive at the proper place at the proper time. The player may have this procedure so ingrained in his or her playing that he or she may not even be aware of the change in the elevation angle.

2.3 Upon approaching an intersection whose light has just turned yellow, you can stop at a maximum negative acceleration, race through at some maximum positive acceleration, or maintain your same speed. For example, let us consider the following parameters: your car is moving at 20 mph (30 ft/s) when the light turns yellow, the intersection is 30 ft wide, the duration of the yellow light is 2 s, and the maximum acceleration is either $+10 \mathrm{m/s^2}$ (for increasing your speed) or $-10 \mathrm{ft/s^2}$ (for stopping). Assuming ideal conditions (e.g., that your engine instantly responds to a push on the acceleration pedal), we can calculate the distances to the intersection needed for your three options. In order to race through successfully, you would have to be closer than 50 ft when the yellow appears. To stop successfully, you have to be further away than 45 ft. Between 45 ft and 50 ft, you have either option.

2.4 Since the ball is over the plate for only about 0.01 s, the timing should be to about ± 0.01 s. The error along the vertical must be less than 1 mm. "The 1962 world championship was finally determined by an otherwise perfect swing of a bat which came to the collision 1 mm too high to effect the transfer of title (4)."

2.5 If the wall is such that you cannot just drive around it, and assuming ideal conditions of brakes, road conditions, and so on, and ignoring any considerations of how people in the car might get hurt for impacts on certain sides of the car, then calculations indicate that you should steer directly for the wall and attempt to stop as quickly as possible. Twice as much force would be needed to turn the car in a circular arc in attempting to avoid the wall as would be needed to stop the car in a straight stop.

2.6 In driving, the greatest speed will be given to the clubhead for the greatest torque you apply to the club. For a given torque, however, the clubhead speed will depend on how the torque is applied. According to one study (5), the more one hinders the uncocking of the wrists, by using a negative torque on the wrists during part of the swing, the greater will be the clubhead speed. The proper negative torque and this hindrance may be part of the "timing" sought by golfers.

2.7 The beans contain a small worm that jumps upward.

2.8 A pole valuter must maximize the kinetic energy in order to maximize the height to which he or she travels, but a high jumper obtains most of the height from the last spring rather than from the kinetic energy in the approach. A broad jumper may bicycle his or her legs to correct an initial forward tilt.

2.9 Ruth could have hit home runs off stationary balls. So, unless he was fooled by the ball and swung too soon, a slow ball merely added to the chances of a home run.

2.10 The follow-through will do comparatively little damage since it is primarily pushing on the opponent. The karate strike is focused a couple of centimeters into the opponent's body, so that initial contact is made when the striking hand is at its maximum speed and thus so that the impact force is maximum.

2.11 If the point is to deform the struck object, such as in forging, then an inelastic collision is desired. A greater fraction of the hammer's energy will be lost in each collision the lighter the hammer's mass. So, in forging you should use a light hammer. In pile driving you want to transfer kinetic energy to the pile and avoid any loss of energy to deformation. So, use a heavy hammer there.

2.12 The softer the ball, the longer it will be in contact with the bat, and the more work the batter can do on the ball during the contact. Thus, follow-through should definitely be used on a softball.

2.13 The optimum bat size would be one that requires the least energy from the batter to give a particular speed to the ball. However, batters typically choose heavier bats for home runs or lighter bats for more ease in wielding the bat, giving little thought as to the energy imparted to the ball. With several simplifying assumptions, one source (4) finds that the optimum bat mass is about 3.4 times the mass of the ball. More exact calculations would raise this result but not to the ratio of 7 common in baseball or 5 common in softball.

2.14 The frictional force from the floor is the external force.

2.15 When the beetle is on its back, a peg prevents the jump muscle from swinging the front half of the body upward to initiate jumping. Tension is slowly built up in the muscle until the peg finally slips, and then the rapid rotation of the front half of the body hurls the beetle upward. Before the beetle can jump again, the tension will have to be rebuilt.

2.16 During the falling of the sand, each hourglass weighs the same in spite of the fact that some of the sand is in the air. The additional force on one side is due to the impact force of the sand. What will the balance say when the sand first begins to fall or just after the last of the sand has hit?

2.17 Each hole has a different cross-sectional area. The same cylinder's weight spread over different areas will maintain different pressures. For example, the largest area hole has the least weight per unit area, thus requiring the least pressure to raise it. Therefore, it will regulate the lowest pressure in the pan.

2.18 The large ball collides with the small ball after rebounding from the floor, transferring its momentum and kinetic energy to the small ball. The small ball is then able to reach a greater height than from which it started. The maximum gain in velocity for the small ball from the collision is three times what it has upon reaching the floor. Thus, the maximum height to which it can return is nine times the height from which it was dropped. The closer the small ball's mass is to the larger ball's mass, the less this return height will be.

2.19 The coefficient of sliding friction is less than the coefficient of static friction. Thus, there is greater friction to stop the car if the tires turn smoothly on the road rather than slide over the road. On dry, smooth asphalt the friction coefficients for rolling may be as much as .8 whereas for sliding .6 or less. If sliding begins, the asphalt and tire melt and the car then skims along on a thin layer of liquid. All other things being equal, the sliding would require about 20% more distance than the rolling stop. Thus, you will have the quickest stop if you apply the brakes just slightly less than what will lock them.

2.20 The frictional force on the tires does not depend on the surface area in contact with the pavement, and so a wide slick is as effective as a narrow one. If the tires are spun over the surface as is done in drag racing, then the wide tire has an advantage in that it has a wider surface to heat and is less likely to melt. (Melting greatly reduces the coefficient of friction. See FC 2.19.)

2.21 The first part of the run is limited by the traction with the pavement. Greater traction reduces the time spent in this part but will not affect the final speed more than a few percent. That final speed will be determined by the power limitation of the dragster in the second part of the run.

2.22 The finger that is first moved slides beneath the stick with a kinetic coefficient of friction. The stick does not slide over the other finger because the static coefficient of friction there is larger. The magnitude of friction on either finger depends not only on the friction coefficient, but also on the weight of the stick on the finger. As the moving finger is brought toward the center, more and more of the stick's weight is on that finger. Eventually the friction on the finger is greater than on the other finger in spite of the difference in friction coefficients. Then the first finger stops and the other finger begins to slide. Such an exchange of motion can occur several times before both fingers are at the center.

2.23 Sudden braking in a turn throws the car forward, decreasing the weight on the rear wheels, which then are more likely to slide outward. Conversely, accelerating rocks the car backward to increase the rear wheel traction.

2.24 The initial speed must be slow or the maximum static friction with the road will be exceeded and the tires will slip. Hence, a small amount of torque is first needed. What gear is best depends on the driver's ability and the smoothness of the clutch. If the driver always spins the wheels in first, the torque can be cut in half by going to second gear.

2.25 If left untied, throw a shoe or some other article opposite the direction you want to go. If the ice is ideally frictionless, then the total linear momentum of the system must remain zero and you will therefore slide off.

2.26 The angular momentum of the motorcycle wheels is significant and much larger than that of the bicycle wheels. To turn the motorcycle you lean it. The torque of the front wheel's weight, calculated with respect to the point of contact with the road, causes the wheel to precess, thereby turning the motorcycle. (Similar precession is common in tops. See FC 2.69) A bicycle cannot depend on precession because the angular momenta of its wheels are much less. To turn a bicycle you have to lean it and also turn the handlebars. If you want to turn left, for example, do you first turn the handlebars to the left or right?

2.27 The kinetic energy of the center of mass of the cue ball (the translational kinetic energy) is tranferred, but the cue ball retains its rotational kinetic energy. Therefore, the cue ball continues to rotate just after the collision but slips and does not move across the table. Eventually the rotation slows because of the friction from the table and the ball begins to roll. If it was originally struck high, it will then follow the ball with which it had collided.

2.28 Consider a superball thrown at an angle to the floor with some spin. In addition to the spin, the center of the ball has a velocity component downward and one parallel to the floor. Collision with the floor merely reverses the vertical component so that is upward afterward. The spin and the velocity component parallel to the floor change in a more complicated way. Consider the point of the ball making contact with the floor. The sum of its velocity around the ball's center plus the center's velocity component parallel to the floor gives the total velocity of this particular point parallel to the floor. The collision reverses that total parallel component of the point, as well as changes both the spin and the center's parallel component. The recoil direction of the ball can then be found by determining the new full velocity vector of the center. For example, suppose a ball is thrown to the floor at 45° and with no spin. After the collision it will move at 23.2° with respect to the perpendicular to the floor and with a spin in which the front rotates downward. Multiple collisions with appropriate initial spins account for the several tricks shown in the figures.

2.29 and 2.31 A stable bicycle is one whose forkpoint (the point of intersection of a projection along the front steering axis and a horizontal line through the wheel center) falls as the wheel turns into a lean when the bike is tilted. Of the three other designs drawn in Figure 2.29b, the third is unstable while the second is too stable, being too unresponsive to the rider's changes of direction. Gyroscopic effects have little to do with riding stability, although if the bike is pushed off riderless, then the gyroscopic effect from the wheels will help stabilize the bike for a while.

2.30 The oscillatory motion of the point of contact between the Hula-Hoop and the person keeps the hoop moving. That point of contact leads the hoop in the trip around the person. The initial hoop speed should be greater than the eventual driving speed.

2.31 See 2.29.

2.32 The forces in a spinning lasso have apparently not been analyzed, and you might try to investigate them either theoretically or experimentally. The short length of rope from the hand to the circular section both lifts and drags the circular section around in a rotational motion. Once a fast rotation is created, there is some gyroscopic stability due to the angular momentum.

2.33 The rotation about the axes of maximum and minimum moments of inertia are stable against small deviations. Rotation about the axis of the intermediate moment of inertia is not, and any small perturbation sends the book wobbling.

2.34 Friction from the stick prevents the ring from just falling. Part of the stabilization of the spinning rings comes from the perpendicular force from the stick as the ring spins around the stick. The spinning rate increases as the ring falls because some of the potential energy is converted into rotational kinetic energy. Since nothing has been written on this toy, why don't you try some experiments or develop some models to predict the increase in rotational speed?

2.35 When inverted in water, the kayaker extends the paddle and pushes toward the bottom to provide a torque that rotates himself or herself and the kayak toward the surface. To maintain sufficient angular speed to right himself or herself, the kayaker will attempt to keep his or her body as close to the axis of rotation as possible so as to reduce the moment of inertia around the rotational axis.

2.36 The car can probably produce a certain maximum angular acceleration of the wheel. The larger the tire's diameter, the greater the distance covered in each revolution of the wheel, and thus the greater the linear acceleration of the car. If the car is power limited, then adding greater diameter tires will decrease the angular acceleration, and the car will have the same linear acceleration.

2.37 The best procedure depends on several factors: the speed of the spinning as compared to the linear speed of the car's center of mass, which wheels have traction, and whether or not stopping the spinning is more or less important than stopping the linear motion. For example, consider that your car has its rear spinning to your right, that your speed down the road is negligible, and that your front wheels still have traction. To stop the spinning, you should turn your front wheels into the spin (i.e., to the right) and modestly accelerate the car. The torque from the tires about the center of mass of the car will decrease the angular momentum of the spinning. As the spinning decreases, ease the front wheels back to the left so that you are oriented properly on the road once again.

2.38 A tire statically balanced with a single weight will not be dynamically balanced when spinning. On the other hand, the tire can be dynamically balanced and thus have no wobble, but if only a single weight is used, the tire will still suffer vibrations. The usual tire balancing is a compromise between the two types of balancing. If two weights are used, both types of balancing can be obtained.

2.39 The torque you apply by pulling on a sheet of paper must overcome the torque due to the friction between the dispenser and the cardboard tube of the toilet paper roll. If the required force on the sheet is too much, then you will tear the paper each time you attempt to rotate the roll. The fatter the roll, the smaller the force you will need for a given torque but the greater the torque from the friction because of the increased mass. The critical radius for most dispensers is about 2 cm.

2.40 A stone that skips on sand usually has its trailing edge strike first, causing a torque that produces both the short hop and a rotation to bring the leading edge down for contact. After the front end strikes, the stone takes a long jump. The short hops appear to be missing from the skipping over water. Again the trailing edge hits first, but now the stone planes along the water, tilting back as a crest develops in front of it, and then finally takes a long jump. You might try high-speed photography to analyze the forces and torques more carefully.

2.41 An inside wheel is not rigidly connected to an outside wheel. Instead, there is a differential between them that, with a set of four bevel gears, allows the outside wheel to turn faster than the inside wheel.

2.42 A center-mounted engine will have less moment of inertia about the car's center and will therefore require less torque to turn.

2.43 For a thin rope or wire, the walker must constantly oscillate left and right, first falling one way, then moving the support point beneath the center of mass, overshooting, falling the other way, then moving the support point that way, and so on. Using a pole makes the balancing easier. By shifting the pole left or right, the walker can position the center of mass of himself or herself and the pole over the wire. With the center of mass over the support point, the walker is balanced.

2.44 The ball will orbit around the bottle but will not hit it unless the ball is swung directly at the bottle. There is no torque on the ball perpendicular to the plane of its orbit, and therefore the angular momentum vector perpendicular to the plane must remain constant. With the bottle beneath the string's point of attachment, this conservation of momentum means that the ball must orbit *around* the bottle except for the direct collision swing. You can hit the bottle in a sneaky way, however. Twist the string before you release the ball so that the ball spins while in flight and encounters a force such as the one used in throwing a curve (see FC 4.39).

2.45 The net angular momentum of the cat is constant throughout the free fall because there are no external torques on it. By extending or retracting its legs, the cat can make the front half of its body have a different moment of inertia about its body axis than the rear half. For example, if it extends its front legs and retracts its hind legs and then rotates the rear half, the front half will rotate in the opposite direction but not as far. So there is a net rotation in the direction that the rear half rotated. The cat then extends the rear legs and retracts its front legs and repeats the process to gain a further net rotation in that direction. By then the net rotation is sufficient that the cat can grab at the floor and finally right itself completely.

2.46 The Austrian turn involves rotations similar to those the cat uses in righting itself. If there is no net torque on the skier, then rotation in one sense by the upper half of the body must be accompanied by a rotation in the opposite sense by the lower half in order to conserve angular momentum. Turning can also be caused by shifting one's weight forward or backward. Consider the first for skiing diagonally across a slope. With the center of mass forward, the friction on the rear of the ski will have larger lever arms from the center of mass as compared to the friction on the front of the ski. Thus, there is a net torque on the ski that will rotate the ski and the skier.

2.47 When the yoyo spins at the end of the string, the kinetic friction between the string and the bottom of the yoyo in contact with the string is not very large. Suddenly moving the string upward increases the contact force between the yoyo and the string, with a sudden increase in the friction that may stop the sliding. The static friction is more than the kinetic friction and thus, instead of the sliding then resuming, the yoyo wraps itself up the string.

2.48 I have seen no studies on the physics of judo, and you might experiment with this question and with other aspects of the sport. The slapping will increase the contact area of the body with the floor at the moment of impact, thereby decreasing the impact force per unit area, a decrease that is especially desirable on the rib cage. The slap may also rotate the trunk of the body away from the impact and further protect it.

2.49 The spinning bullet will act as a gyroscope and attempt to maintain its spin orientation throughout the flight. Therefore during most of the parabolic path, the wind blows not along the bullet's body axis but at an angle to that axis. The resulting torque causes a precession of the spin in the same way a top is made to precess. With the bullet slightly broadside to the left or right, it is deflected from its intended path.

2.50 The stack will not fall if the following rule is met: the center of mass of all the books above any particular book must lie on a vertical axis that cuts through that particular book. This must be true for each book in the stack. You might try determining, either theoretically or experimentally, the maximum shift possible for a given number of identical books. Or, vice versa, how many identical books are needed for a given overhang. For an overhang of 1 book length, you need at least 5 books. For 3 book lengths you need 227 books, and for 10 book lengths you need 1.5×10^{44} books.

2.51 The top part of the chimney would have less angular acceleration than the bottom part were it not for the rigid coupling in the chimney. Consequently, stress develops along the chimney's length during the fall. The maximum stress during the first part of the fall is approximately halfway along the length, and it is there that the break will most likely occur. Should the break occur during the last part of the fall, it will be approximately a third of the way up and due to shearing.

2.52 Projectiles are deflected by the Coriolis force from what should be a straight sighting. That force is a fictitious one an observer on the rotating earth will invoke to explain the path of a projectile as the earth and the observer turn underneath the projectile. Gunners will sight their guns to take this apparent deflection into account, but the extent of the correction will depend on the latitude and will be in the opposite senses in the two hemispheres. The British gun sights were set for Britain's northern latitude, not for a southern 50°.

2.53 The Coriolis force (FC 2.52) also causes a small deflection of a river: to the right in the northern hemisphere and to the left in the southern hemisphere. This deflection supposedly increases the erosion on those particular sides.

2.54 The force that increases the spin is the Coriolis force (FC 2.52).

2.55 The right-handed boomerang is thrown in a vertical plane so that it spins about a horizontal axis. Since it is an air-foil, there is a sideways "lift" on it, the lift being larger on the top half than on the bottom half, because the top half is turning in the same direction that the boomerang is traveling, whereas the bottom half is turning in the opposite direction. Therefore, there is a torque attempting to tilt the boomerang, but instead of tilting, the boomerang veers to the left and remains vertical. Sufficient veering causes the boomerang to turn full circle in its flight.

2.56 You can pump the swing by raising your center of mass (e.g., raising your legs) each time you go through the lowest point of the swing. Your work will add energy to the swing and thereby increase the amplitude. Starting the swing from rest is harder to explain. By leaning backward and momentarily falling, you gain kinetic energy and give angular momentum to the swing, you and the swing acting as a double pendulum. Upon reaching arm's length, you stop your fall, and swing with the swing as a single compound pendulum until you have the opportunity to fall backwards again.

2.57 People fear that the periodic pounding of the bridge might match a resonant frequency of oscillation of the bridge. Although each pounding of the feet would add only a little energy to the oscillation, if there is resonance between the pounding and the oscillation, the energy will be stored and built up, perhaps leading to such an extreme oscillation amplitude that the bridge collapses. (Also see FC 4.84.)

2.58 The incense swing is pumped as is explained in the answer to FC 2.56.

2.59 Imagine an initial single bump in the road that sets the front end of passing cars oscillating. When the front end descends during the oscillation, it may force the tires to dig in just then. If many cars do this at approximately the same place, a new bump will develop.

2.60 The ship's rolling is 90° out of phase with the ocean waves striking the ship. The water tank's oscillations are at the same resonant frequency as the ship but will lag the ship's rolling by another 90°. (Why?) The tank's oscillations are therefore a net 180° out of phase with the external waves and will oppose the rolling of the ship.

2.61 The pendulum will not topple if the vertical acceleration due to the oscillation is greater than the acceleration of gravity. If there is no friction in the pendulum system, the pendulum will then oscillate to and fro as long as the end is still forced to oscillate vertically. If there is significant friction, then the pendulum will assume a stationary vertical position.

2.62 You have to choose the masses and the length such that the frequency of the strictly spring oscillation matches the frequency of the strictly pendulum oscillation. Then, if the system is started in one of the types of oscillations, the twisting of the spring will feed energy into the other type of oscillation until there is a complete transfer of energy. The transfer will then reverse to feed energy the other way.

2.63 A double compound pendulum, in which both pendula are hinged together, and where one has a smaller mass and length than the other, will swing together. If this is a bell and its clapper, then the bell will never ring. One solution, and the one apparently used at Cologne Cathedral, is to greatly lengthen the clapper.

2.64 The oscillations of the balance wheel are near the resonant frequency of the swinging of the pocket watch. If the watch case oscillates with a frequency somewhat greater than the balance wheel, then the case and balance wheel swing in opposite phase, and the watch gains time. The opposite result occurs for a watch case having a lower frequency.

2.65 The dominant frequency may result from an acoustic standing wave set up in the falling water column, much as a standing wave can be produced in a tube with one open end and one closed end. The factor of one fourth results from the speed of sound in water being one fourth that in air.

2.66 Vibrational standing waves are produced on the bat when the ball strikes the places of antinodes. This result is undesirable because the vibration "stings" the batter, wastes energy that could have been given to the ball, and may break the bat.

2.67 When the arrow is released, it receives a lateral impulse from the string and the bow's stock. The resulting vibrations cause the arrow to snake around the bow's stock without touching it, but since there is no interference from rubbing with the stock, the oscillations are around the direction of flight, and the arrow flies true to the aim.

2.68 The horizontal and vertical vibrations of the notched stick are not the same in frequency or amplitude because of the difference in shape vertically and horizontally and because of the pressure from a finger or thumb. The resulting vibrational motion of the stick, and thus of the pin, is elliptical. Depending on the finger pressure and on which side the stick is rubbed, the elliptical oscillations will be either clockwise or counterclockwise, and therefore so will the pin oscillations. The friction between the pin and the propeller sets the propeller into a corresponding motion.

2.69 I have no general rules for the behavior of tops, although some of the equations of motion are worked out in the advanced mechanics books. An asymmetrical top will certainly not be stable and its behavior will be erratic. A top will precess (i.e., its spin axis will rotate around the vertical) because of the torque on it due to its weight. Superimposed on the precession is a wobbling called nutation. If the asymmetric top has its spin axis initially vertical, it will remain there as long as its spinning speed is above a certain value. Once friction decreases the spinning speed below that critical value, the top will begin to wobble.

2.70 The spinning diabolo is in essence a top or gyroscope. To orient it properly, you must pull on the cord in the correct directions. For example, if it is spinning counterclockwise with its axis outward from your body and then its far end dips, which way should you pull on the cord? You should slightly loosen and lower the left-hand cord while pulling the right-hand cord upward and toward you. In this way the torque from your pulling will change the angular momentum of the diabolo to make it horizontal again.

2.71 The **raw** egg is not stable because it is asymmetric, and it therefore cannot rise like the tippy tops in FC 2.73. If the shell is briefly touched during rotation, the fluid inside the raw egg will still be rotating and will restart the rotation of the shell when you remove your finger.

2.72 The bottom surface of the stones is not exactly ellipsoidal, being somewhat slanted toward one side. When the stone is displaced from its equilibrium position by a tap on one end, the perpendicular force from the table provides a torque that causes the celt to rotate. A certain celt will rotate in a certain direction depending on which way its bottom surface is slanted.

2.73 If the top is spun on a rough surface, the friction on the point in contact produces a torque that precesses the top to an inversion.

2.74 The mass distribution of the moon is not spherically symmetric. The earth's gravitational field therefore produces a torque on the moon that causes synchronous rotation of the moon about its axis. With such a forced synchronous rotation, the moon will always present us with the same face.

2.75 An orbit must be around the center of the earth, because the gravitational force on the satellite is directed to that point. There is no way to put a satellite at the Moscow latitude so that it remains at that latitude because the center of its orbit would then not be at the center of the earth.

2.76 Less energy is required with the figure 8 path. To reach the moon, the ship must at least reach the line beyond which the moon's gravitational attraction dominates the earth's. To reach that line with the least energy, the ship should stay as close as possible to the line through the centers of the earth and moon.

2.77 The moon does primarily orbit the sun, just as the earth does, the pull from the earth causing a perturbation on that orbit.

2.78 A plumb line can be deviated from the vertical by perhaps several tens of arc seconds due to adjacent mountains and also due to adjacent absences of mass such as with a lake.

2.79 The atmospheric friction reduces the total energy of the satellite, but only half of the decrease in potential energy goes into heating. The other half is transformed to kinetic energy, and, as a result, the satellite has a greater speed in spite of the air drag. The orbit axis is reduced, of course, so this procedure continues only until the satellite eventually burns up.

3.1 Assuming the air is trapped in the bra, the volume of the air will depend inversely on the pressure in the plane, which would be less than that at ground level and therefore would result in a larger bra. Had the cabin been suddenly opened to the atmosphere, the quick reduction in cabin pressure probably would have exploded the bra.

3.2 For the same oven temperature more water will evaporate from the cake at the higher altitude than the lower because of the reduced barometric pressure at the higher altitude. Thus, you will need more water in the recipe. Also because of the reduced barometric pressure the gas inside the cake (consider an angel food cake) will cause the cake to rise more, perhaps increasing the volume beyond the strength of the cell walls and causing the cake to fall. To avoid losing the cake in this way, less sugar can be added to decrease the production of the internal gas. However, rather than decrease the sugar and therefore the sweetness of the cake, recipes increase the flour to obtain the same results. An angel food cake will not brown as well at high altitudes for the same oven temperature because of the lower boiling temperature of water (FC 3.62). To obtain the same browning, high altitude recipes increase the oven temperature. None of these factors should reduce the tenderness of a cake. More flour will increase tensile strength and thus the toughness, but the greater expansion and the lower internal temperature (which will retard the coagulation of the protein) will decrease the tensile strength about as much.

3.3 The cottage does not measure the barometric pressure but is sensitive to gradual changes in the humidity that may accompany pressure changes. The movement of the figures is caused by a twisted piece of catgut whose length changes with the humidity.

3.4 Although the general water level in a well is governed by the local rainfall or snowmelt, changes in the barometric pressure can vary the water level by several inches. When the barometric pressure drops during a storm, the well level will rise. The resulting increased water flow through the ground may pick up enough sediment to make the water unfit to drink.

3.5 The smaller balloon has a smaller radius of curvature and therefore the elastic forces tangent to the surface on any small surface area have a greater net component toward the center of the balloon than on the larger balloon. With a greater inward force, there will be a greater internal pressure. Hence, the smaller balloon has greater internal pressure. This result also explains why a balloon is initially more difficult to blow up but becomes easier as the balloon expands: the components of the elastic forces toward the center become progressively less.

3.6 With the greater atmospheric pressure at the bottom of the tunnel, much of the carbon dioxide remained in solution. When the dignitaries returned to the surface, the gas then came out of solution. They had to return underground to reduce the release of the gas to a tolerable rate, thus indicating to what depths drinking can drive a person.

3.7 If you do not release air continuously during the ascent, you will very likely rupture your lungs because of the volume expansion of the air in the lungs as the external pressure on them decreases. An ascent from a mere 15 ft after inhaling air from an air tank can be fatal.

The partial pressure from the CO_2 in the lungs does not increase linearly with time as you ascend because you are continuously exhaling part of it. The depth of maximum partial pressure of the CO_2 has been determined from the following: from the maximum depth (where air was inhaled from a tank or submarine), expressed in feet, subtract 33 ft and then divide the result by 2.

3.8 The air currents are primarily due to changes in the barometric pressure. If the cave has more than one entrance, air may circulate between the two entrances because of the temperature difference between the cave's interior and the outside.

3.9 The tissues of the body will not saturate or desaturate at the same rate. Consider, for example, a dive with an essentially instantaneous descent, followed by a 30-min stay at bottom and then a programmed ascent. Those tissues that absorbed the nitrogen quickly will desaturate quickly as the diver begins the ascent. Those tissues that absorbed slowly will not initially have as much pressure difference between the absorbed nitrogen in the tissues and the nitrogen in the blood and will therefore desaturate much slower. Hence, the initial stages of the ascent are quick to desaturate the fast tissues, and the final stages must be slow to desaturate the slow tissues.

3.10 As the hot water heats the faucet valve, the valve's metal expands and shuts off the water flow.

3.11 Ice will first form on the inside wall of the pipe and gradually grow radially until there is a solid plug blocking the pipe. Until that situation is reached, the expansion of the freezing water merely pushes water back into the water main. But once the plug is in place, further expansion of the freezing water between the plug and the valve will significantly increase the water pressure unless the valve is open. If the valve is closed, the pipe will eventually burst at its weakest point. Hot water pipes will be more likely to burst because the initially higher temperatures decrease the ability of the freezing nuclei to initiate the freezing, thus lowering the freezing temperature. The water in the hot water pipe then supercools, that is, cools below 0°C, until freezing is suddenly and very quickly initiated. The resulting rapid expansion of the ice plug traps more water between itself and the (closed) valve, making bursting more likely.

3.12 The constriction is sufficiently narrow for the mercury to pass it only under pressure, either from thermal expansion or from the "centrifugal" force when it is swung in an arc. When cooling, the mercury thread breaks at the constriction because the intermolecular forces in the mercury are not strong enough to pull the upper column of mercury back through the constriction. If you stick the thermometer into hot water, the glass surrounding the bulb of mercury will expand before the mercury itself.

3.13 The rubber molecules are stretched chains that, when agitated by thermal motion when the rubber is heated, will pull more on their ends and thereby decrease the length of the rubber. When you stretch the rubber and those molecular chains, you do work, part of which goes into heat. If the rubber is then allowed to contract, part of the work done by the elastic force decreases the internal energy of the rubber and thus its temperature.

3.14 A watch would run at different rates when at different temperatures were the balance wheel not properly compensated for a temperature change. Suppose the watch is warmed. The metallic wheel would then expand, have a greater moment of inertia about the center, and therefore oscillate less quickly and slow the watch. However, the expansion is compensated such that the oscillation is kept approximately constant. The rim of the wheel is in two or three pieces, each having one free end and one end mounted to a spoke of the wheel. The rim pieces are bimetallic. If the temperature increases, the free end of each rim piece curls inward because its metal strips have different thermal expansions, the strip on the outside expanding more and thus bending the rim piece inward. This bending inward offsets the expansion of the spokes. Although the wheel changes shape because of the heating, its moment of inertia is about the same and so are its oscillations.

3.15 Under ideal conditions the U tube is initially in equilibrium. If there is a disturbance from the outside, the equilibrium will be broken, and the oscillations will begin. Suppose the disturbance displaces a small amount of water from the left to the right. The left-hand vertical section then has more cold water than the right-hand vertical section and the denser cold water on the left will push into the lighter warm water on the right. Eventually the difference in water levels on the two sides will balance out this difference in buoyancy due to the temperatures, and the flow will stop. The heating and cooling on the tube then brings the water back to the original temperature equilibrium, the buoyancy force is therefore reduced, and water flows from right to left because of the difference in water heights. This procedure continues periodically. By experimenting you could determine how the oscillation depends on the size of the U tube. You will also notice that the reservoir must be larger than some critical value or the argument above will not be valid.

3.16 When you pump and compress the gas, the compression is essentially adiabatic (no heat transfer with the outside) and the internal energy of the air increases, thus increasing its temperature. The hot air heats the valve. The station's compressed air was hot when first compressed but since then has cooled to room temperature and will not heat the valve.

3.17 The prevailing winds in the United States are from the west. As the moist winds from the Pacific are forced upward by mountains (for example the Rocky Mountains), the air adiabatically cools because of the reduced atmospheric pressure and cannot retain as much moisture. The western sides of the mountains therefore receive the released moisture. The eastern sides will receive comparatively much less.

3.18 As the winds descend from the mountains and into the greater atmospheric pressure, the moving air is adiabatically compressed and thus heated (FC 3.16). If the descent is rapid, little of this heat will be exchanged with the local air, and therefore the wind will be warmer than the local air. The wind will be relatively dry because of the answer in FC 3.17. How dry, warm winds affect humans is not understood, but the positive and negative ions in such winds may be the source of the irrational behavior (FC 6.14).

3.19 The pressurized gas in the bottle rapidly and adiabatically expands when the bottle is opened, doing work in expanding against the atmospheric pressure (FC 3.16). The energy for that work comes from the internal energy of the gas, thereby reducing its temperature and causing some of the water vapor in it to condense out as a fog.

3.20 The reference suggests that the air current over the top of the car reduces the air pressure in the passenger area, implying that the air there had expanded and therefore cooled slightly. This effect would be similar to the cooling of the air in the rapidly flowing air stream above an airplane wing, an effect sometimes made apparent by the fog formed above the wing.

3.21 The winds from the Pacific have dumped their moisture on the west side of the Rockies lying between the Pacific and Death Valley (see FC 3.17) and are adiabatically heated as they descend the eastern slopes (see FC 3.18). The hot, dry wind entering the valley reduces it to a desert. The lack of shade and the greater reflection from sand than would be obtained from a vegetation-covered terrain also increases the air temperature just above the ground.

3.22 The air ascending a mountain expands and therefore cools as it moves into less atmospheric pressure.

3.23 Rising columns of warm, moist air cool as they expand in their ascent to lower atmospheric pressures. The lower temperature eventually condenses out some of the water to form the clouds and also to warm the rising air somewhat by the released latent heat in the condensation. The clouds, therefore, are not held together at all, but are constantly being formed. Sit sometime and watch them form, change, and decay.

3.24 The atomic fireball very rapidly heats the air, which then quickly rises, pulling ground-level air, dirt, debris, and water moisture up in its tail to form the stem of the mushroom. As the hot air rises and cools by expansion, it eventually reaches the temperature of the local air and thereafter spreads horizontally to form the top of the mushroom.

3.25 The cause of the holes in the clouds is not well understood. One of several explanations is that a natural or artificial accumulation of freezing nuclei in an altocumulus layer quickly brings about freezing. The resulting thin cirrus filaments further act as nucleation agents, spreading laterally to widen the hole, while ice crystals fall from the central cirrus and thus remove water from that area.

3.26 The air forced upward over a mountain expands and cools and then condenses out some of its water moisture. If the condensation occurs just at the mountain top, a lenticular cloud will result. If the wind is strong, the condensation will occur on the leeward side of the mountain in the turbulent wake. In both cases, the clouds may appear to be permanent but are, in fact, constantly being created and destroyed.

The air forced over the mountain may oscillate on the leeward side to produce clouds on each upturn in the oscillation. These lee waves have a wavelength dependent on the wind speed and also on the change in air density with height. If there is little change in air density with height, the atmosphere is relatively stable, and the lee wave oscillations will be slow, giving long wavelengths and thus large distances between the clouds. Large changes in air density will cause more rapid oscillations of the moving air, resulting in short wavelengths and thus short distances between the clouds. The faster the wind speed, the greater the distance will be between each upturn of the lee waves; therefore, a stronger wind will spread the clouds.

3.27 The rapid heating of the air by the blast generates a shock wave, with high pressure preceding low pressure. During the low pressure the air expands and cools, and some of its water vapor condenses. After passage of the shock wave, the air pressure returns to normal, and the cloud disappears. Thus, the cloud is relatively narrow and expands radially from the blast area.

3.28 Much of the visible light incident on the clouds from the sun passes through the clouds and is absorbed by the earth. As the ground warms, its thermal emission (long wavelength light) increases. When the clouds absorb this radiation, the temperatue difference between the base and top of the clouds becomes sufficient to cause turbulence, which then destroys the clouds.

3.29 The mamma are formed when a dense cloud layer overlays and then descends into a dry layer as a downward moving thermal.

3.30 Fogs are classified into several types according to how they are formed. Radiation fog results when the moist air cools by radiating its heat into space and condensing out its excess water vapor as the humidity increases. Advection fog forms when warm, moist air flows over a cold ground or body of water. The humidity does not have to reach 100% for a fog to form because in the modern atmosphere there are many condensation nuclei that will attract the water and initiate the fog at humidities as low as 60%. Near the ocean, these nuclei may be salt particles. Near cities they are more likely to be the particulate matter released by industrial smokestacks. The open-coal burning of London supplied a great many condensation nuclei. Once the burning diminished, the condensation and the resulting fog became less likely. Sometimes a thermal inversion (layer of warm air overlaying a layer of cooler air) will trap industrial pollutants near the ground. During such an inversion in December 1952 in London, the pollution turned the fog that was present black and within a few days reduced visibility to just a few inches. Approximately 4000 people died from this smog.

3.31 When your warm breath strikes the cold glass, it cannot hold as much water vapor and droplets form. These drops will initiate on condensation nuclei either on the glass or in the adjacent air. A fresh hot piece of toast will "exhale" onto a cool plate in the same manner.

3.32 A vortex is shed by each wing so that there is downflow in the center behind the fuselage and upflow on the outside behind the wings on each side. Condensation may come directly from the water vapor in the engine exhausts or from the cooling of the air during the vortex motion. Since most planes have two prominent wings, there will initially be two trails. The downflow in the center between the two vortex centers decreases in momentum as it travels downward because of the buoyancy force on it upward. To reduce its momentum, this air decreases in volume, which pulls the two vortices closer together, eventually making the two trails indistinguishable. The speed of the descending air increases in spite of the decrease in overall downward momentum. This downward increase in speed will magnify irregularities in the trails: the parts of the trails descending first will descend even more quickly because of the increase in speed, making those parts appear to have burst downward. Soon mixing within the vortices reaches the cores, and the descent stops. Then an underview of the trails may show burst popcorn areas connected by loops where two trails are still distinguishable.

3.33 The salt apparently acts as nuclei for the bubble formation.

3.34 The air heated by the fire is lighter than the cooler air in the room and will be pushed into the chimney by the cooler air. Once the circulation is initiated, it will continue even if the chimney is not exactly over the fire. The taller the chimney is, the better the draft into the chimney will be, because a taller chimney will hold more warm, light air. Consider a parcel of air at the entrance to the chimney, just over the fire. It is being pushed into the entrance by the cooler room air, but it is also being pushed downward by the air above it. If the air above it is warm, however, that downward push will be much less than the upward push. The more overhead hot air there is, that is, the higher the chimney is, the easier it will be for the room air to push each parcel of hot air into the chimney. Some chimneys will puff if the draft is low and if cool outside air periodically pours into the chimney top.

3.35 The hot smoke and gases are more likely to rise in the cooler evening than during the warmer day.

3.36 Initially the hot gases have a laminar flow because their ascent is relatively slow. During the ascent, however, the net upward force on them (resulting from the buoyancy on the hot gases in the cooler surrounding air) accelerates them sufficiently that the flow begins to break up into eddies. Typically the acceleration requires about 2 cm to obtain the necessary speed for turbulence.

3.37 The general characteristics

of a stack plume depend largely on the change in atmospheric temperature with height near the level of the chimney emission. If the temperature increases quickly with height (a situation that is called an inversion), the hot gases emitted by the chimney will not be able to rise and thus will stream horizontally in the wind as shown in the first of Figure 3.37*a*. If the temperature decreases from the ground to the chimney height and then increases for greater heights, the gases will not be able to rise but can mix downward as in the second figure. If the temperature decreases moderately with height from the ground up, the plume will be like the third figure. If there is a very marked decrease in temperature with height, the plume will attempt to rise but can also be brought back to the ground in a loop by thermal eddies.

A plume can be split into two parts if the breeze is light enough not to distort the vortex pair that is created as the hot gases leave the chimney top. The gas flow in the center of the chimney exit is strongly upward, but the gas on the outside of an emerging puff is downward. In a light breeze, this pattern splits down the middle to form the two plumes as shown in Figure 3.37*b*.

3.38 Ice crystals will grow preferentially in one plane, called the basal plane, and grow much slower along an axis, called the c axis, perpendicular to that plane. The actual orientations of the ice crystals in the ice over a lake will vary from place to place and will result in different melting rates over the lake. An area with mostly horizontal c axes will melt such that the vertical ice crystals become isolated and look like candles jutting upward. Lake water seeps upward between these candles by capillary action to give that area a dark appearance. Areas with vertical c axes will have larger horizontal crystals that melt internally to form a honeycomb of hollow tubes. These areas will be brighter. The darker areas absorb more sunlight than the brighter areas, heat faster, and therefore will be weaker and more dangerous than the brighter areas.

3.39 The melting point of water is 0°C, but the freezing point will likely be lower, depending on the purity of the water. Very pure water can be supercooled (i.e., cooled below the melting point without freezing) to about −40°C. Impurities will initiate freezing at higher temperatures (but not higher than the melting point), the exact value depending on the type and concentration of the impurity. As the water is cooled, random fluctuations in the free energy of the system (due to its microscopic thermal motion) produce small ice islands. If the water is still relatively far from the "freezing temperature," these islands quickly disappear. At "freezing temperature," the islands grow to a certain critical radius beyond which they can continue growing because the continued freezing will lower the free energy of the whole system. This critical growth will occur at −40°C for pure water but at a higher (subzero) temperature for impure water, because the impurities reduce the necessary critical radius of the ice islands.

3.40 The critical feature is the increased evaporation from the initially warmer water. If equal masses of warmer and cooler water are set outside in freezing weather and in open-topped containers, the evaporation from the warmer water will reduce the mass remaining in that container. With less mass to cool, the water in that container can overtake the cooling of the initially cooler water and reach the freezing point sooner. The actual cooling rate can depend somewhat on the composition of the containers, the circulation above the containers, and the circulation in the water. Although Bacon commented on the effect and although the result is well known in Canada, people in warmer countries find it mysterious. The physics journals rediscovered it recently only after a high school student in Tanzania convinced his skeptical teacher of the result.

3.41 Thunderstorms occur most frequently in the midafternoon to early evening and over land masses. A graph of the worldwide thunderstorm activity will be dominated by the thunderstorms over Africa and Europe, making the maximum activity at approximately 7P.M. London time. A graph of the earth's electric field (see FC 6.33) will have the same shape, with the maximum at about the same time, because the thunderstorm activity recharges the earth, bringing negative charge to the ground and positive charge to the upper atmosphere by lightning and point discharges (see FC 6.32 and 6.46).

3.42 The moisture on your skin can freeze to bond your skin to the metallic object. The freezing is more likely on metal than wood, say, because the thermal conductivity of metal is high, and your finger tip will be rapidly cooled. (See FC 3.78.)

3.43 Normally the water drains from ice as soon as some of it melts. With the wet paper around the ice, the external heat will have to be conducted through the water layer, thereby slowing the supply of heat to the ice.

3.44 Water has the greatest density when it is about 4°C. Thus, when water begins to freeze on a pond, the lighter ice floats, the water just above freezing rises, and the water slightly warmer (near 4°C) sinks. The surface will therefore be colder and will freeze first. The surface will also cool faster, because it radiates its heat into the atmosphere and because the air circulation over the surface removes heat. The ground at the bottom of the pond is warmer and will supply heat to the bottom water. The bubbles of relatively warm air will supply heat to prevent, delay, or decrease the freezing.

3.45 If the snow is near the melting point, friction from the skis initially melts a thin layer of snow on which they can glide. The continuous shearing of the water layer by the moving skis provides the continuous heat needed to maintain a water layer. The type of material used in the skis, whether they are metal or ebonite, does not directly matter as far as the initial melting goes. However, if the skis conduct heat well, as metal skis would, the heat will be lost too quickly to maintain the water layer. Ebonite skis (and the wooden skis used years ago) conduct poorly enough to maintain the layer. If the snow is well below the melting point, there will be no water layer, and the skis will have to be waxed to reduce the friction.

3.46 Like the skis in FC 3.45, the ice skates glide over a thin water layer, but unlike the skis the water layer is due to pressure melting. The weight of a skater supported over the two thin blades puts a pressure on the ice of 7000 lb/sq in. or more. To know the pressure accurately, you would have to calculate the weight distribution over the actual points of contact with the ice, not the total bottom surface area of a blade. Other materials are unlike water and ice in that they do not melt by increased pressure as ice does and thus could not be used for skating. Since skiing does not depend on pressure melting, presumably you could ski on other materials such as dry ice.

3.47 A sudden warming can melt some of the snow to provide enough water to lubricate the sliding of the remaining snow. A sudden cooling can be equally dangerous. At sunset, for example, the cooling can freeze liquid water already present, and the resulting 11% expansion by the water can trigger an avalanche.

3.48 When you grasp and firm a snowball, you pressure melt at least the surface, which then refreezes to bond the snowball together.

3.49 Snow tires are designed to bite into snow to increase the traction. Studded snow tires depend on melting the ice and snow beneath each stud by the pressure from the car's weight. With that weight distributed over a smaller contact area, the stud's contact area, there is increased pressure on the ice and more melting. If the snow and ice are below 0°F, the increased pressure is insufficient to cause melting. Sand is useful if pressure melting will embed it in the snow and ice but, again, that will not happen at very cold temperatures.

3.50 and 3.51 When salt is added to the water, more heat must be removed from the mixture to reach freezing, and thus the freezing temperature is lowered. Not only must the water molecules be slowed sufficiently for ice crystallization to begin but also for them to overcome the adhesion to the salt molecules. Salt will also raise the boiling temperature of water. Because the water molecules are attracted to the salt, they will have to be moving even faster than normal in order to escape into the vapor state. Similar lowering of the freezing point and raising of the boiling point of water is behind the use of antifreeze in car radiators.

3.52 The evaporation of the water requires heat, which is removed from your body if you stand wet and nude in a breeze. To remove a water molecule from a layer of water, in other words, to evaporate some water, energy must be supplied to that molecule to overcome the attraction to the other water molecules in the water layer. In the meantime, some of the water molecules already in the vapor state will randomly run into the water layer and so return to the liquid state and give up energy. If both the liquid and vapor are in a closed system under equilibrium, then as much energy is removed by evaporation as is returned by condensation. In a breeze, however, the water vapor is constantly being blown away and there will then be a net energy loss from the water layer. If the layer is on your skin, the net loss in energy will come from your skin, and you will feel cool. Methyl alcohol evaporates faster than water and will cool the skin quicker. The porous canvas water bag is cooled by the evaporation on the bag's surface, especially if a wind is constantly blowing across the bag.

3.53 The fuel evaporation (which is enhanced by the increase in the air speed when the air is forced through the central aperature past the fuel jet) takes energy for its phase change from the air. As the air cools, it can saturate in humidity and begin to condense excess moisture. If the outdoor humidity is between 65% and 100% and the outdoor temperature between 25° and 50°F, then the condensing water can freeze on the throttle plate.

3.54 The brine (salt water solution) in the ice blocks is in cells that will migrate downward because of gravity and also in the direction of the higher temperature on the block because of progressive melting and refreezing. Usually the latter is also downward since the ice block is either floating in the ocean (the ocean, which is at freezing temperature, is warmer than the air above) or on the ground (the ground will also be warmer than the air). For both effects, the brine drains from the block, which becomes potable after about a year and has nearly no salt after several years.

To understand the drainage because of the temperature difference from top to bottom on the block, consider a vertical cell of brine in the block. The salinity of the solution will be enough that its temperature matches the average temperature of the adjacent ice. At the lower, warmer end, this salinity will be too much (see FC 3.50) and there will be melting to reduce it. At the upper, cooler end, there is too little salinity and there will be freezing to increase it. The cell as a whole then moves downward, eventually reaching the bottom of the block.

3.55 A relatively large amount of heat is required to vaporize the water in the pan. If the pan is open, this heat of vaporization is constantly lost as the water vapor rises in the air currents. The pan top will trap the vapor and thus keep that heat in the pan.

3.56 Other than the one reference describing this effect, apparently nothing has been published on it. Why not try experimenting with your own oven? Will the increased moisture content of the air in the oven mean that the air heats any faster? Is less heat needed to increase the air's temperature by, say, one degree? Will the circulation of the air change?

3.57 Often people will prevent the freezing of their car radiators by similarly placing a large tub of water near the radiator in the garage. As the room temperature cools and approaches freezing, the water will act as a heat reservoir for the room. When ice begins to form in the tub, a relatively large amount of energy (latent heat) is released, which will then aid in preventing further cooling of the room.

3.58 The outside air that is cooled during the night will flow down into the icehouse during the early morning as the outside begins to warm. This air flowing into the icehouse will saturate in humidity because of the colder temperature there, condensing out some of its moisture. The phase change from vapor to liquid releases a large amount of heat. There is less heating of the icehouse if direct sunlight is allowed to warm the air in it to prevent the inflow of fresh air and the consequent condensation.

3.59 The lower, wider end of the device absorbs heat from the oven and heats the water in the lower end of the hollow interior. Eventually the water is converted to steam, which requires a relatively large amount of heat for the phase change to vapor state. The hot steam rises through the tube to the upper end that is in the cooler interior of the meat. There the water vapor condenses, releasing the (latent) heat it had abosrbed upon changing to steam. The liquid water then runs down the tube to begin the cycle again. The net heat transfer to the inside of the meat is 100 to 1000 times faster than the conduction through a solid rod of the same metal because of the large amount of heat involved in the phase changes from liquid to vapor and then back again.

3.60 If a mountain were higher than a critical height of about 30 km, the pressure at the base of the mountain due to its weight would be great enough to liquefy the base, causing the mountain to sink below the critical height. Hence, mountains must be less than about

30 km high. Because the gravitation on the Martian surface is less than that on the Earth's surface, Martian mountains have a greater critical height.

3.61 If the demonstration comes soon after the first loud sounds come from the cauldron, the water is not at the boiling temperature (see FC 1.12) and, althought hot, it should not be dangerous. The water breaks into droplets when the performer flings it into the air. Those drops should cool somewhat by the time they reach the skin. If the performer is sweating throughout the performance, as I certainly would be, the sweat will protect him from the droplets.

3.62 There is evaporation from a body of water even before you begin to heat it. Molecules in the liquid state having sufficiently high energy escape the liquid. Some of those vapor molecules eventually reenter the liquid, but there is a net loss. As the water is heated, the number of molecules in the vapor state just above the liquid surface increases until finally the pressure of this water vapor reaches its maximum value, called the saturated vapor pressure. The corresponding temperature is called the boiling temperature. The value of the maximum water vapor pressure depends on the atmospheric pressure. In the reduced atmospheric pressure on a mountain top, water boils at a lower temperature because saturation is reached at a lower value. Whether or not the surface is rolling with boiling bubbles depends also on the nucleation of water vapor bubbles in the body of the water. If the water is especially clean and left carefully undisturbed, the water can be heated to above the normal boiling temperature for the existing atmospheric pressure. A small perturbation, perhaps a dust particle, can then cause an explosive eruption of boiling.

3.63 Splashing and capillary action puts relatively thin layers of water on the edges of the puddles or lakes. As that water evaporates, the dissolved salt is left behind on the edges.

3.64 The tube connecting the bird's head and base extends into the base. There is a liquid in the base that is deep enough to submerge the lower end of the tube. Above the liquid, both in the base and in the rest of the tube and the head, is the vapor from that liquid. Those two pockets of vapor are therefore not connected. As water evaporates from the felt on the head, the head and the vapor inside the head cool (see FC 3.52), lowering the pressure of the vapor in the head. Then there is greater pressure in the vapor pocket in the base than in the head and liquid is pushed slowly up the tube toward the head. Eventually this makes the bird so unstable that it tilts forward and dunks its head in the water glass. Just when the bird is horizontal, those two pockets of vapor are connected and equalize in pressure. With equal pressures, there is nothing to force the liquid up the tube and the instability is removed. During the jostling of the dunk, the bird rights itself, and then the whole cycle begins again.

In one study by the RAND corporation (1457), large birds were considered for use in transporting water between irrigation canals in the Middle East.

3.65 As a water drop approaches a very hot skillet, its bottom portion instantly vaporizes to provide a vapor layer between the skillet and the remaining water drop. The drop is then heated by radiation through the vapor layer, convection currents in the layer, and conduction by the layer, but these three processes will take up to 1 or 2 min to bring the drop to boiling temperature. Thus, the vapor layer protects the drop for that long, allowing it to dance and skim over the skillet surface.

3.66 Superheated water (liquid water hotter than its boiling temperature) seeps upward into a geyser's cavity and main column from heated rock as much as a thousand meters below the surface. Once the water is in the cavity, small vapor bubbles form and then grow as they ascend. The water through which the bubbles pass flashes to steam. The resulting pressure forces some of the remaining water to erupt into the air. This procedure then repeats itself, sometimes, as in the case of Old Faithful, with a definite periodicity.

3.67 In the older-style coffee makers, water was placed in a bottom container, and then the coffee holder was snugly fitted over the top with a rubber gasket. When the water was heated, the air and water vapor above the water expanded and pushed water up the central tube to pour out onto the coffee. After about 5 min, the heat was turned off. As the air in the bottom container cooled and contracted, its lowered pressure would allow the outside atmospheric pressure to push the water through the grounds and back into the lower container. In many of the modern percolators, there are several differences. The lower end of the tube is conical and fits over the bottom of the pot. The water trapped in that cone is quickly heated and forced up the tube by the air bubbles that are released. Were the tube wider, the bubbles could not do this. Once the water is spilled onto the coffee, gravity draws it through the grounds and back to the water below. Each time water spurts up the tube, the tube and cone are momentarily lifted, allowing fresh (and cooler) water to flow under and into the cone.

3.68 Steam rises through the pipe and into the radiator where it condenses and then flows back down the pipe. The phase change, rather than a temperature change in the water, releases heat to be given off by the radiator.

3.69 When I do this demonstration with the lead, I wet my hands first. When my fingers enter the molten lead, some of that water immediately vaporizes to form (at least momentarily) a protective sheath around my fingers, similar to the protective vapor layer in FC 3.65. Normal moisture on the skin (especially if I am scared and sweating) works almost as well. Fire walking probably involves protection by the moisture on the feet and may depend as much on sweating between each footfall as having callouses on the feet. Although having wet feet helps, I find I can walk on hot coals with no special preparation of my feet.

3.70 Water hammering occurs when water collects somewhere in the pipes. When the steam flows over the water and suddenly cools, the pressure is very suddenly reduced. The water is quickly drawn into the area of lower pressure, striking the pipes with a loud thump. To get rid of the water hammering, you should drain the collected water.

3.71 The dull side will radiate and absorb heat better than the shiny side (see FC 3.75). Therefore, having the dull side out might help cook a potato faster and then cool it quicker at the table. Since apparently nothing is published on this effect, why not check it?

3.72 The metal evaporates from the filament to darken the bulb. Convection currents in the small amount of gas in the bulb will carry those metal molecules upward to darken the top of the bulb.

3.73 When viewed by a dark-adapted observer against a perfectly dark background, an incandescent blackbody radiator becomes visible at about 650 to 800 K, depending on the angle the object subtends in the observer's field of view.

3.74 You might cool yourself momentarily by opening the refrigerator door, but by then the cooling system comes on to attempt to cool the interior again. More heat is released by the motor than is absorbed by the released cool air, so the room will become even hotter. You can be sneaky, however. You can unplug the refrigerator as soon as you open the door. Of course, then your beer will not stay cool, and you will have to drink it all.

3.75 A black surface will absorb the radiated heat faster than a shiny surface. Thus, the black pie pan will heat the pie faster. Glass absorbs most of the thermal (infrared) radiation incident on it and will therefore heat the pie faster than the shiny pie pan.

3.76 Archimedes' feat was reconstructed in 1973 by a Greek engineer who had 70 flat mirrors (each about 5 ft by 3 ft) held by soldiers who focused the sun's image on a rowboat about 160 ft off shore. Once the soldiers properly aimed their mirrors, the rowboat begun to burn within a few seconds, eventually being engulfed in flames. Arthur C. Clarke independently used the idea in one of his science fiction short stories ("A Slight Case of Sunstroke"). The home-town spectators at a soccer match were each given a shiny souvenir program. When one of the referees called an unpopular decision in favor of the visiting team, the home-town spectators burned the referee to a crisp by directing the sunlight on him with their programs.

3.77 The candle converts some of the water in the boiler to steam, which then pushes the water column back through the tube to emerge behind the boat in a jet. Upon leaving the boiler, some of the steam condenses in the cooler tube and contracts, thereby pulling water back into the tube. However, the key feature is that the water entering the tube comes from a hemisphere of directions, not from a single direction. There is a net propulsion forward because of the asymmetry in the jet emission rearward and the inflow from all directions in the rear hemisphere.

3.78 How cold an object feels depends not only on its temperature but also its thermal conductivity. The quicker a relatively cool object conducts heat away from your finger when you touch it, the colder it will feel.

3.79 Clothes are needed in hot climates to protect the wearer from the direct sunlight. Dark clothes will absorb more visible and infrared light than white clothes, and thus white clothes should be worn in hot climates. If water is plentiful, then the clothes should be porous so that sweating and evaporation can cool the skin. However, if water is scarce, then the clothes must be less porous to protect against rapid dehydration of the wearer. In Herbert's science fiction classic, *Dune* (1460), water is so scarce that the desert people wear suits to seal in all the precious body moisture.

3.80 The more massive, thicker iron pots and pans should have a more uniform temperature across their bottoms than the modern thin steel ones, which will have hot spots directly over the burner. Sticking is often produced by the hot spots.

3.81 The Northern Hemisphere winters are cold not because the earth is further from the sun (for it is, in fact, closer) but because the tilt of the earth's axis shortens the days and lowers the sun in the sky. Both reduce the amount of heat deposited on the surface during the day. However, the change in temperature will lag behind changes in these factors by about a month because the ground and atmosphere take time to cool.

3.82 The side of the astronaut facing the sun is absorbing thermal radiation, which heats the astronaut. He is also radiating thermal radiation over his entire surface area. So, the sunny side of his suit will be warm, the shadow side cool. (Actually, the suits have air conditioning units.) A thermometer left in space will warm until it is absorbing as much radiation as it is emitting. At the earth's distance from the sun, the thermometer should read approximately room temperature, depending on how much of its area is facing the sun. You experience the same thing when you face a fire.

3.83 The so-called "greenhouse effect" is commonly misunderstood. Greenhouses are not hot because of any radiation trapping by the glass, but because the cooling by air circulation is diminished or eliminated. In fact, the glass may reduce the radiation flux into the greenhouse rather than increase the radiation inside by trapping. There is trapping of radiation by the earth's atmosphere, however, because it is more transparent to the short wavelength solar radiation than the long wavelength radiation. Part of the transmitted short wavelength radiation is absorbed by the earth's surface, heating it, after which long-wavelength radiation is emitted by the surface. Because the atmosphere will not transmit that second radiation well, some of it will be trapped in the atmosphere.

3.84 If there is little or no wind, most of your heat loss is by thermal radiation. Any object with a temperature above absolute zero will radiate heat, the hotter it is, the more it radiates. It will also absorb heat from the surroundings, the amount available depending on the temperature of the surroundings. Since your body is almost always warmer than your environment, there is a net radiation loss. Outside on a cold day, or facing toward a window with the cold on the outside, the absorbed radiation is less because of the reduced radiation from the environment. Consequently, your net radiation loss is more, and you feel cold. An astronaut space walking without a suit in empty space and far from the sun should feel a profound coldness since there would be no environment to radiate to him.

People can adapt to continuously cold conditions by adjusting their diet and the rate at which blood flows to their skin. Eskimos have a higher protein diet than do most people living at lower latitudes in order to maintain a higher basal metabolism to counter the cold. To protect against rapid heat loss through the skin, the capillaries carrying blood to the skin contract. If the temperature in the limbs drops too much, the person will shiver so that the increased activity will warm the limbs.

Besides radiation, a person can lose heat by conduction (e.g., through the feet to a cold ground) and by convection, including evaporation losses discussed in FC 3.52. A fur coat will help keep a person warm because the air pockets in the fur are very poor heat conductors. With increased wind speeds, however, the air pockets become less effective. The best way to wear a fur coat to minimize heat losses, especially on a windy day, is to wear it inside out so that the wind will not destroy the pockets of insulating stagnant air in the fur.

3.85 The metal pipe wall conducts heat well and would constantly be losing a significant amount of heat to the convection currents flowing over the pipes. By placing a layer of asbestos around the pipe, the rate at which heat is conducted to the surface to be lost to convection is decreased, because the asbestos will not conduct as well as the metal wall.

3.86 If the thundercloud is several miles away, the wind you experience will blow toward it because of the updraft in the front part of the cloud. When the thundercloud is closer, the wind will blow away from it because of the downdraft of cold air dragged down by the rain.

3.87 The warmth from the finger suddenly decreases the density of the fluid next to where the dish is touched. The nearby denser fluid then displaces the warmed fluid, which spreads out over the surface until it cools, whereupon it sinks. The circulation waves are made apparent by the aluminum powder.

3.88 In the early evening the tree is a reservoir of warm air that will rise from the tree in a convection plume. The insects are attracted to the warmer air and perhaps also to the condensation that may occur in the plume as the air cools during the rise.

3.89 The sunlight heats the stone on the bottom, and the lighter warm water rises in a convection plume to the surface. The shrimp apparently like the warm water (and possibly any organic compounds it may carry) but do not like sunlight. Hence they ride the convection current upward but steer away from the sunlight and, on reaching the sunlit surface, descend again.

3.90 Increased heat flow by the blood to the skin surface and increased sweating carry off most of the additional heat. But these can lead to several minor and also serious disorders. The increased blood flow to the skin may decrease the flow to the brain, causing faintness, especially when a person suddenly stands. Nausea, cramps, and circulatory failure can result from the salt depletion caused by the increased sweating. If about 2% of the body's water weight is sweated away, the person becomes very thirsty. If the losses are about 7%, the circulation can fail, with the person quickly dying. Overheating in the body results in the same symptoms, leading to collapse and possibly death.

3.91 If you want your coffee as hot as possible when class begins, add the cream just before class because it will cool the coffee. The dissolving of the sugar will also cool the coffee since energy will be used in the dissolving. Stirring will cool the coffee because it brings warmer fluid to the surface and to the walls quicker than the normal convection currents would. A metal spoon will absorb heat and will also conduct the heat out of the coffee to be released to the convection currents in the air or to be radiated to the room. Since black objects radiate heat more than white objects, white coffee would cool more slowly. A similar consideration can be made for the cup being black or white. The relative importance of these factors has apparently not been determined for coffee. Why not experiment?

3.92 Dye molecules must diffuse from the negative through about 250 microns to the positive, arriving in the proper amount of processing time (usually a minute in today's Polaroid pictures). The rate at which the molecules cross a given distance will depend on how fast they are going which in turn depends on the temperature. A colder environment slows these molecules and thus the diffusion to the positive.

3.93 Cities will be warmer than the countryside and even their suburbs because of several factors. There is less evaporation in the city, which would cool the city by the loss of the latent heat of vaporization (see FC 3.52). The paving and building materials store more heat than does soil. There is usually less wind in cities because of the taller and more complicated structures. Less important factors are the relatively fast snow removal in the winter and the heat generation by the machines (including cars) in the cities.

3.94 The total kinetic energy of the room's air molecules (what is called the translational kinetic energy) is proportional to the product of the number of molecules in the room and the temperature. With the assumption that the air is an ideal gas, it is also proportional to the product of the pressure of the air and the volume of the room. When you heat the air in the room, the volume of the room certainly does not change. Less obvious is that neither does the pressure because the room is not sealed and will always have leaks to the outside. The room's pressure then must be the outside atmospheric pressure. As you heat the room, both pressure and volume remain constant, and thus so does the total energy of the room's air molecules. But this is possible only because as the temperature increases, some of the air molecules are forced outside.

3.95 The fruit grower will set up pots at the end of the day after the ground has warmed from the sunshine. The clouds produced by the smudge pots absorb the thermal radiation emitted by the ground and reradiate it to the ground. The heat is therefore trapped between the ground and the clouds, and the orchard does not cool as quickly as it would were the ground's heat radiated to the atmosphere with no return. Natural clouds can do the same thing.

3.96 The snow layer is not a good conductor of heat and will help maintain the ground's heat by insulating it (and the crops) from the colder air above the snow.

3.97 The fires result from the visible and infrared radiation emitted by the nuclear-blast fireball. During about the first second after the burst, the fireball is so hot (initially about 500,000 K) that most of the electromagnetic radiation from it is in the ultraviolet. But ultraviolet light is readily absorbed by air and therefore does not leave the immediate area of the burst. As the fireball expands and cools, more of the electromagnetic radiation will be in the visible and infrared because the shift in the emitted radiation will be toward longer wavelengths. Both of these are transmitted by the air. Their intensity about 2 or 3 s after the burst will be sufficient to set materials such as wood on fire. Virtually anything shading a person from the direct light from the burst can diminish the burning of the flesh. There were examples in Hiroshima and Nagasaki where uncovered skin was badly burned whereas adjacent skin shaded by the person's clothes suffered essentially no burning.

3.98 An impurity can nucleate the crystal growth, serving as an initial point of attraction for the molecules.

3.99 The hexagonal structure of snowflakes is determined by the hexagonal bonding of the water molecules that make up a snowflake. Once the initial crystal is formed, water vapor molecules will diffuse to and collect on the corners of the crystal to begin the growth outward to form the branches. Why one particular pattern is produced instead of another will depend on the falling speed, the temperature, and the availability of the vapor molecules, but not in a known manner. Since symmetry is so common in snowflakes, the conditions for adding molecules and branching must be the same across the width of the flake.

3.100 Milk rises between two nearby Cheerios by capillary action, and the surface tension of the milk has a horizontal component that will pull the Cheerios closer.

3.101 The soil must be broken up in order to retain its moisture. A packed ground has many small openings that will act as capillary tubes. As the water climbs to the surface, it is lost to evaporation. Cultivated ground has much larger openings and thus less capillary action.

3.102 If the liquid surface curves upward on the vertical sides of a container, then the adhesive force between container molecules and liquid molecules is stronger than between liquid molecules. The opposite is true if the surface curves downward on the vertical walls. Similar consideration of the molecular forces explains whether a drop will spread on a particular surface or will remain a drop.

3.103 The atmospheric pressure does not push the sap up the trees as was thought last century. The transport is believed to be due to negative pressure. As the water molecules evaporate from the leaves, and other molecules move onto the leaf surface to take their place, a strong intermolecular force pulls the column of sap upward all the way from the roots.

3.104 See FC 3.106.

3.105 The rocks rise because of the freeze-thaw cycles that occur in winter. When the ground freezes, the freeze line descends through the soil, drawing water vapor upward to the line by diffusion. When the freeze line reaches a rock, it descends faster through the rock than through the adjacent soil, because of the higher thermal conductivity of the rock. Therefore, the bottom of the rock freezes sooner than the nearby soil at the same level. The earlier freezing means more water vapor will be drawn there to freeze, and thus the rock will be pushed upward by the expansion of the ice beneath it more than the adjacent soil will be pushed upward. When the ground thaws, loose soil next to the rock fills in beneath it to keep it in its new position. Many freeze-thaw cycles eventually brings the rock to the surface.

3.106 The expansion of the freezing water initially just beneath the pavement can account for only 11% of the observed "frost heave." The expansion is greatly increased by the freezing and expansion of water migrating through the pores in the soil to the freeze area beneath the pavement. If the top of the soil is free, the ice crystallization will push the ice upward to form ice columns projecting 1 or 2 in. from the ground (FC 3.104).

3.107 Capillary action draws water a certain distance upward into the wall. The dissolved salts that are deposited at the top of this capillary column as the water evaporates will draw water further up the wall by osmotic pressure. The short circuit grounds the positive area in this region of higher salt concentration to eliminate the osmotic effect.

3.108 Soap bubbles are held together by surface tension. Because the fluid drains to the bottom of the bubble, the top will thin quicker and be more likely to burst. The air pressure inside the bubble is greater than the ambient air pressure because part of the surface tension is directed toward the center of the bubble, thus pushing the surface inward.

3.109 Inverted bubbles, sometimes called antibubbles, have not been analyzed in much detail. Surface tension pulls the inner water into a sphere and helps prevent the liquid at the two surfaces from flowing across the air layer.

3.110 If the flame is too large, the capillary transport of the fuel to the wick's top will be insufficient to maintain the flame. Once the flame diminishes there is less evaporation from the wick, and the transport provides more fuel than the flame consumes. So the flame increases. With a wick of about 2.5 mm, oscillations occur for wick lengths between about 1.5 mm and 5.0 mm, the shorter lengths giving the higher oscillation frequencies because the shorter transport distance results in a more rapid response to the flame changes.

3.111 Because of the great increase in the ratio of surface area to volume of the individual particles as compared to the original lump of material, a flame or spark can quickly bring the individual particles to their combustion temperature. The readily available air then results in rapid, explosive combustion.

3.112 The screen quickly conducts away the flame's heat so that the flame cannot extend outside the screen. Flammable gases may still leak inside to the flame, but the volume then ignited by the flame is insufficient to explode.

3.113 When mud dries, it contracts, setting up stresses over the surface and for some depth below the surface. These stresses rupture the soil. On the average, the intersection of two rupture lines should be at right angles because the second of the two cracks formed should be perpendicular to the greatest tension in the first crack.

3.114 Similar to the mud cracks (FC 3.113), ice cracks form in the contraction of frozen ground during a sudden cooling.

3.115 The exact cause of the stone nets is not known, although there are several theories. For example, frost heave of the ground (FC 3.106) may roll an initially uniform distribution of stones outward to form a circle. Or, if there originally was a bare area in the midst of stones, that area could have absorbed more water than the adjacent area and then pushed the stones radially outward when it froze and expanded.

3.116 Although this problem is popular among physics students, it overlooks an important fact: a biological system (you, for example) is not an isolated system under thermodynamic equilibrium since it must constantly have energy inputs to maintain itself. The energy flow through the system can organize the system and thus reduce its entropy, but the overall entropy change of the world will increase. The detailed mathematical analysis of the entropy reduction in a biological system is found in current research.

4.1 The pressure on the boy's finger depended only on the density of the sea water and how far below the sea's surface the hole was. The overall size of the ocean did not matter.

4.2 The further you are below the water surface, the more pressure there is on your lungs. Around a depth of 3 ft the water pressure is large enough to prevent your inhaling through a tube to the surface.

4.3 In order to standardize blood pressure readings, they are all made level with the heart. If they were made, say, at the ankle level, then the readings would depend on the heights of the people, and the results would be more difficult to interpret.

4.4 On the inside of the gate is fresh water from the lakes that feed the Canal, while on the outside is the salt water of the ocean. When the pressures are equalized on the two sides of the gate, and the gate is opened, the fresh water will still be higher than the salt water because of the greater density of the salt water. The higher fresh water flows seaward as the levels equalize, giving the boat a free ride.

4.5 Part of the difference in ocean levels at the two Canal ends is due to the difference in salinity of the two oceans. The Pacific is saltier and therefore its water denser. Using identical reasoning as in FC 4.4, the Pacific end of the Canal should be lower than the lighter Atlantic end.

4.6 The answer is simple, perhaps mischievous. The buoyancy on the hourglass must certainly be the same in the two cases because its volume did not change. But with the hourglass inverted, it tilts against the side of the tube, and friction holds it in place.

4.7 The stone in the boat displaces a volume of water whose weight equals the weight of the stone. Since the stone is denser than water, it displaces **more** than its own volume of water. When it lies on the pool's bottom, it can displace only its own volume of water. Hence, less water is displaced when the stone is thrown into the pool, and the pool's level must drop.

In the case of the sinking boat, the water level remains the same until the boat is fully submerged, and then it drops.

4.8 As the first water pours in and fills the first loop, some pours over into the second loop. Soon that loop fills, and there is an air cavity left at the top of the first loop. The water flow then stops until the water column beneath the funnel increases. If that column is sufficiently high, then the procedure is repeated for the next loop. With a limited water column, and several loops, eventually the column is unable to eliminate the air cavities and there is no further water flow.

4.9 The water can be removed until there is a centimeter or less left between the ship and the dry dock walls. The hydrostatic pressure from the ship is independent of the extent of water below or to the sides of the ship. Of course, if the water layer becomes very thin, then the water will climb the walls by capillary action.

4.10 A submarine can descend by taking on water to increase its mass. Blowing the water back out with compressed air decreases the mass for ascension. In order for the submarine to be stable while submerged, the density of the seawater must increase with depth at that level. If the submarine then moves upward slightly, a net downward force returns it to the previous depth. If it moves downward, the net force is upward. The density depends inversely on the water temperature and directly on the salinity, both of which decrease with depth. Between 25 and 200 m in depth, a submarine can find several layers in which the temperature decreases rapidly enough with depth to offset the decrease in salinity and thus to provide stability.

4.11 If the ratio of the bar's density to the fluid's density is close to 1 or 0, the bar floats in stable equilibrium as in the first figure. If the ratio is some intermediate value, then the bar floats in stable equilibrium with its sides at 45° to the fluid's surface. In each case, the orientation of stable equilibrium is determined by the position in which the potential energy of the system is least.

4.12 Fish use their swim bladders to give themselves neutral buoyancy so that they do not float to the top or drop to the bottom of the sea. Suppose a fish swims downward. The increased water pressure would compress its gas cavity. Then, because of the decrease in the fish's volume, it would have less buoyancy and would have to swim in order to avoid falling further. Instead, the fish secretes gas into its swim bladder to keep its volume approximately constant. So, in spite of the increased pressure, it maintains its same volume and thus its same buoyancy force. If it ascends, the fish reabsorbs some of the gas, again to keep the same buoyancy force.

4.13 Two forces hold the cardboard in place: atmospheric pressure and surface tension. Once the glass is inverted, the water column descends slightly, leaving the air trapped in the glass at a lower pressure than the air outside the glass. The pressure difference between top and bottom of the water column provides a force to hold the water up against its own weight. Additional force is provided by the surface tension between the water and cardboard and between the water and glass.

4.14 Gas produced in the victims' bodies buoys them up.

4.15 Because the arrangement is unstable, any small perturbation of the water surface, any small wave, will grow quickly in amplitude. A bubble develops to rise to the tube's top, allowing water to fall along the sides of the tube. The bubble's upward velocity, and therefore the speed at which the glass empties, depends on the square root of the gravitational acceleration (9.8 m/s^2) and the radius of the upper portion of the bubble.

4.16 Both the temperature and the salinity of the water decrease with depth. As the cold, fresher water from the bottom rises, it warms from the surrounding water and then is lighter than the saltier top water. Thus, the flow will continue. In fact, even without the tube it would continue since the rising water would exchange heat much faster than salt with the surrounding water.

4.17 The instability forming the fingers is the same as in FC 4.16. The initial motion comes from small perturbations, small waves, on the surface between the two layers of water. The dyed water loses heat to the undyed water as it descends. That descending water will then be denser than the undyed water and will continue its descent. Undyed water initially pushed upward by a small wave will warm and then be lighter than the surrounding dyed water. Its protrusion upward will thus continue.

4.18 The interface between the salt water and the fresh water undergoes the same instability as in FC 4.15 and 4.17

4.19 The volume flow rate (so many cubic meters of fluid passing through a cross section of the stream each second) must remain constant throughout the stream in order to conserve mass. Since the water speeds up because it is falling, the same volume flow rate requires less cross-sectional area the further down the stream you consider.

4.20 The ball is held in place against gravity by a pressure difference due to the passage of the air jet: the pressure beneath the ball is greater than that above it. The ball deflects most of the jet over its top. The pressure in that deflected air is reduced, creating a pressure difference between top and bottom of the ball. (See FC 4.25 for a similar reduction in pressure.) As a result, there is lift on the ball. Also as a result, the air stream is deflected downward on leaving the ball. Since the ball is likely to be turning, the Magnus effect that deflects a spinning baseball (FC 4.39) can contribute lift. One possible wrong answer is to attribute the lift to a reduced pressure in the free air jet just because the air is moving. Such a conclusion is a misuse of the Bernoulli principle. The kinetic energy in the air jet comes from mechanical work in the machine, not from a reduction of pressure in the air. The pressure in the free jet is, in fact, just atmospheric pressure.

4.21 The ball is suspended by the pressure of the air stream on its bottom side and gets its stability according to FC 4.20. The air stream blown into the toy entrains air inside the tube, causing an air flow from the higher opening to the lower one. As the ball passes the higher opening, it is merely sucked into the tube by the air flow.

4.22 The water impact supports the ball and also supplies its stability. Most of the time the ball is off center, and the impact forces it to spin in a certain direction. Part of the water that adheres to the ball's surface is carried around half a revolution, say, and then thrown off. As the water leaves, it pushes backward on the ball (in other words, there is a reaction force on the ball), thereby holding it in the stream. Even if the ball leaves the stream, some water is still thrown off in the next half revolution, and the ball returns to the stream as a result.

4.23 Apparently nothing more than a description has been published on this demonstration. Why not try experimenting with it? What is the pressure just above and just below the egg? Does turbulence matter? Suppose an egg that would float in static water were in a narrow, horizontal water jet. Would the egg move upstream in the jet?

4.24 The boundary layer of the stream next to the spoon develops a narrow eddy having reduced pressure. With atmospheric pressure on the side opposite the spoon and this reduced pressure adjacent to the spoon, the stream is held against the spoon. (This phenomenon is called the Coanda effect.) Turning the bottom corner is aided by the "teapot effect" of FC 4.118.

4.25 The passage of the air jet reduces the air pressure at the mouth of the tube. The water surface outside the tube is at atmospheric pressure. Thus, the pressure difference forces water up the tube. The real question is why there is reduced pressure due to the air jet. One wrong answer is to attribute a low pressure to the free jet. However, as is discussed in the answer to FC 4.20, the free jet has atmospheric pressure. So, the jet must suffer a pressure reduction because of its deflection by the tube.

Two factors should be considered. Some of the air may be forced up and over the top of the tube. The air adjacent to the tube in this deflected stream would move faster and be reduced in pressure. If the flow is turbulent, as is likely, then the stream develops eddies above the tube, which also reduces the pressure there. Either way, the pressure at the tube's top is reduced.

4.26 A high speed train produces a high pressure pulse immediately in front of itself and a low pressure area in its wake. When trains are passing, the low pressure area between them can suck windows outward.

4.27 As the air is forced up and over these structures, the pressures at the tops of the structures are reduced (see FC 4.25). Air can then be pulled from the ventilator shaft or the prairie dog tunnel.

4.28 The acceleration of the air up and over the front of the car is so great that the insects rupture from the force.

4.29 Imagine the flag perfectly smooth and fully spread in a strong wind. A small perturbation develops that, on one side of the flag, forces the air outward slightly in order to move over the ripple. That air stream must speed up as it crosses the ripple. The faster air has less air pressure, and thus the ripple grows because of the difference in air pressure on its sides: the reduced pressure of the air crossing the ripple and the normal pressure on the other side of the flag. The ripple also moves down the length of the flag in the direction of the wind, so the flag eventually flaps.

4.30 The wing was tilted downward so that it forced the car downward and therefore increased the traction of the tires on the road. With greater traction, the car could take a turn faster. The aerodynamic force from the wing was just like on an airplane (see FC 4.31), but downward instead of upward.

The car with fans in the rear also received a downward force to increase traction. The air that was forced beneath the car had to speed up because it was being made to enter a restricted opening. With greater speed, the air had less air pressure according to the Bernoulli principle. Hence, there was greater air pressure above the car than below it, and the car was pushed onto the road more. The weight of the car was effectively increased by about 50%.

4.31 The air passing the top of the wing moves faster than the air passing below the wing. The pressure above the wing is less than the pressure below the wing. As a result, a net upward force is on the wing.

Whether or not the Bernoulli principle applies to the calculation of this lift is not always clear in the references. The principle is a statement of the conservation of energy (here, pressure and kinetic energy) along a stream line in the air flow. Since the air flow around a wing is affected by adhesion to the wing and viscosity, both of which do work on the air, the principle should not be applicable. However, it can still be used if the adhesion and viscosity are accounted for by superimposing a circulation of air (forward underneath the wing, rearward on top) on the irrotational flow of the air passing the wing. In the work of Kutta and Joukowsky on lift, such a superposition of a circulation is made. Above the wing, the circulation speed adds to the irrotational speed past the wing to give a greater speed. Below the wing the circulation opposes the irrotational flow, and the air speed is reduced. By Bernoulli's principle, the pressure above the wing is less than that below the wing, thus there is lift. The application of the Bernoulli principle in obtaining lift on the wing is therefore somewhat subtle.

The actual lift on a wing is calculated in the Kutta-Joukowsky theory by determining the momentum change in the air stream as it is deflected by the superimposed circulation. According to Newton's law, the force necessary to deflect the air stream downward is equal to the lift on the wing. Some references erroneously describe lift on the wing but then show an air stream that leaves the wing moving in exactly its original direction.

4.32 As the pilot attempts to pull out of a dive, the weight of the plane effectively increases because of the centripetal acceleration in the turn upward. The lift on the wings, previously inadequate, will now have to be even larger because of this effective increase in weight. To gain greater lift, the plane's air speed will have to be greater than normal.

4.33 The passing wind produces a "horizontal lift" on the sail toward the convex side. (See FC 4.31.) This force is most efficient and gives the greatest boat speed if the boat is sailing 90° to the wind.

4.34 Normally the frisbee sails through the air with its front edge upward, thereby gaining lift as a wing does (FC 4.31). In addition, the frisbee's orientation is somewhat stabilized by its rotation, just as a gyroscope is stabilized by its rotation.

4.35 There are two types of attempts at man-powered flight: where planes are powered by men and where people (wisely or unwisely) leap from tall structures flapping their arms and attached wings. The latter is unlikely to be successful for more than 10 or 20 ft, with the landing unlikely to be soon forgotten. In contrast, building lightweight aircraft in which one or two people paddle for power to lift and propel the craft seems promising. The first such flight occurred in 1961, lasting for about 50 yd. Numerous plane designs have appeared since then. Wing spans, for example, have ranged from 60 to 120 ft. The primary concern in these designs is to reduce the power necessary for lift. Presently, even a good athlete cannot power a plane beyond about 100 yd. However, sporting planes of 50 ft wing span and appropriate wing shape should be feasible for man-powered flight if the plane also takes advantage of thermals and wind currents. The craft would then be powered by one or two people only until it is sufficiently high that it can act partially as a sailplane (see FC 4.98).

4.36 With bottom spin the golf ball gains lift in the same way that a spinning baseball is deflected sideways. (See FC 4.39.)

4.37 The passing wind pushed the cylinders sideways in the same way that a spinning baseball is deflected in FC 4.39. Appropriate orientation of the ship would result in the ship moving forward through the water.

4.38 Some of the wind incident on the building is forced through the opening, having to speed up to do so.

4.39 A curve is thrown by spinning the baseball about a vertical axis. The passing air then exerts a horizontal force (called the Magnus effect) that deflects the ball. The force on the ball is due to the unequal pressures on the ball: the side turning into the passing air has greater pressure than the side turning with the passing air.

The application of the Bernoulli principle is as difficult here as in explaining the lift on airplane wing in FC 4.31. Again, the principle should not apply because the air streams passing the spinning ball experience both adhesion to the ball and viscosity. The side turning into the passing air decreases the stream's kinetic energy and thus its speed. The side turning with the passing air increases the air's kinetic energy and thus its speed. The Bernoulli principle can be applied, however, if the effects of the adhesion and viscosity are accounted for by superimposing on the irrotational flow of air pass the ball a circulation of air that turns in the same sense as the ball's spin. On one side the irrotational speed adds to the circulation flow's speed, giving a greater speed. On the other side, the two speeds oppose each other, giving a lesser speed. Since the Bernoulli principle now works (because we no longer have to include external forces doing work on the passing air once we superimpose the circulation flow), the pressure on the first side must be less than the pressure on the second side. The pressure difference deflects the ball.

The actual deflection force (the horizontal lift) can be calculated with the Kutta-Joukowsky theory as with the airplane wing (FC 4.31). Again, some books err in their description of the ball's deflection by giving the ball a deflecting force without giving the air stream a deflection.

4.40 A reverse effect can be produced for a slowly spinning, slowly moving, smooth ball. Under certain conditions, the side of the ball spinning with the direction of the passing air may remain laminar (smoothly flowing) whereas on the other side there may be turbulent mixing. The pressure in the turbulence would be less than the pressure on the other side, causing the ball to deflect in the opposite way as in FC 4.39. Upon leaving the ball, the air stream would, of course, be deflected in the opposite sense also.

4.41 The vertical forces initiate small amplitude waves. As the air passes over these waves, it is forced upward slightly by the peaks and then flows downward into the troughs. At the peaks the air speed is greater and thus the pressure is less. The opposite would be true in the troughs if the flow were ideal. Such an ideal situation, with low pressure on the peaks and high pressure in the troughs, would not transfer energy from the wind to the waves and thus the wave would not grow. In the nonideal flow the air circulates in a reverse direction in the bottom of the trough and thereby shifts the high pressure point closer to the preceding peak. The pressure variations are therefore no longer in phase with the water wave and a net amount of energy is transferred from the air to the water. The water waves then grow.

4.42 The monster waves are the chance meeting of many ocean waves in phase. They are not giant waves that traverse the ocean. Instead, they quickly disappear as the composite waves go off in their own directions and leave with their slightly different speeds.

4.43 Wind speeds above about 5m/s produce water surface turbulence that then produces air bubbles. Rafts of these bubbles are called whitecaps. The group velocity of ocean waves is about half the phase velocity. This result means that individual waves form at the rear of a group of waves, move foreward at about twice the speed of the group as a whole, and then disappear at the front of the group. The greatest amplitude occurs in the center of the group. So, each wave in turn moves through the maximum amplitude position. If that amplitude is more than a certain critical value, then breaking and subsequent foaming occurs. But the foaming will take place only when an individual wave happens to move through the center of the group. Thus, the whitecaps will appear periodically downwind of each other.

4.44 A slowly moving boat produces bow waves of relatively small wavelength. Several of these waves will be along the length of the ship at any given moment. As the boat goes faster, the wavelength of the bow waves increases until eventually the wavelength is equal to the boat's length. Then the bow and stern waves reinforce each other, and the ship is essentially trapped between two crests, one at its bow and the other at its stern. For faster speeds the resistance from the waves increases considerably, requiring much more power from the boat. The hydroplane avoids this problem by lifting the hull from the water. Supports extending into the water act as airfoils do on airplanes: the deflected water currents over the moving supports give them lift. (See FC 4.31.) As far as the lift is concerned, these boats are just airplanes flying through water

4.45 There are two types of water waves: capillary waves, which are governed primarily by surface tension, and gravity waves, which are controlled mainly by gravity. Longer wavelength water waves are of the second type; shorter wavelengths are of the first type. Neither of these can propagate with speeds below 0.23 m/s. If the beetle skims slower than that speed, no wave pattern is produced. For faster skimming, the beetle creates both types of waves. The capillary waves have group velocities greater than the wavespeed, and they are therefore in front of the beetle. The gravity waves have group velocities less than the wavespeed and thus are behind the beetle. Only the beetle's capillary waves are prominent, but the gravity waves are visible with close inspection.

4.46 Were the ship to generate waves of a single wavelength, then the angle of its wake could be found just as the angle θ is found for the shock wave cone left by a supersonic aircraft: $\sin \theta = c/v$ where c is the speed of the sound wave and v is the speed of the aircraft. In contrast, the ship generates waves of a large range of wavelengths that travel at different speeds. From any particular position of the ship, these waves are sent outward in all directions, the longer wavelength waves traveling faster than the shorter wavelength waves. However, these waves destructively interfere except on a circle that expands forward in the ship's direction of motion. The ship's progression leaves a trail of these expanding circles of constructive interference, the ones further back larger than the more recent ones. These circles develop the V-shaped area in the figure such that the angle of this V is independent of the ship's speed.

Consider a particular point on the center axis directed from the ship backward through the wake. The distance from that point to the ship is always three times the distance from the point to the edge of the wake (the outer limit of the spreading circles) along a perpendicular to the central axis. As a result, the sine of the angle of the V, and thus the angle itself, must always be the same. Inside the V the expanding circles of constructive interference produce the particular pattern of crests shown in the figure.

4.47 Apparently there is no published elementary explanation for the edge waves. The recent publications suggest that they may be caused primarily by the nonpropagating oscillations near the oscillator rather than the waves propagated through the basin.

4.48 The wave speed depends on the depth of the water, the shallower the water is, the slower the waves move. If a wavefront approaches the shore at some angle, the inshore portion of the wavefront slows before the offshore portion. As the wave progressively slows, the wavefront is swung around until it is close to being parallel to the shore line (or at least the line of shallow water).

4.49 The front end of the board is tilted upward (as the rider assumes the characteristic surfing stance by leaning backward slightly), and water is forced beneath the passing board. If the skimming is quick enough, then the board passes before the water beneath it can be squeezed out. For example, if the water is 1 in. deep, then the water waves will move at about 0.5 m/s. Thus, skimming faster than that speed constantly supplies fresh water beneath the board to avoid stalling. The force supporting the rider is not buoyancy, instead, it is the impact force from the water.

4.50 To ride the waves, the surfer must move with the wave speed. Normally, in deep water, the wave speed is greater than the speed of the water particles in the wave. If the wave is nearly breaking, the water has almost the same speed as the wave, and the surfer needs only a little more in order to keep up with the wave. The additional speed comes from the continuous falling downhill of the surfer on the side of the wave. Thus, in order to surf, one needs a beach with waves that are either breaking or almost breaking. The water speed is greatest at the crest of the wave. Therefore, the speed of the rear of the board through the water should be less than the speed of the front of the board, creating an unstable situation. The shorter the board, the less this difference in speeds is a problem.

4.51 The bow of the moving ship creates a high pressure area in front of the ship. The porpoises ride between that high pressure region and the normal water pressure further ahead of the bow.

4.52 The tides are not due to the moon or sun pulling the water radially outward from the earth. Instead the bulges are caused by the horizontal components of the gravitational forces from the moon or sun collecting the water in the bulges. Since the horizontal components are less below the moon's or sun's position in the sky and on the opposite side of the earth, the bulges collect there. Were the moon to revolve about the earth always directly above the earth's equator, there would be no diurnal (once-a-day) tides. However, with the moon's orbit off from the earth's equator, some low latitude areas can have a dominant diurnal tide.

4.53 The tidal generating force depends on the inverse **cube** of the distance to the sun or moon. As a result, the moon's effect dominates in spite of the fact that the sun has a greater gravitational pull.

4.54 To conserve the total angular momentum of the earth-moon system, the separation between the two increases in order to compensate for the earth's loss of spin.

4.55 The wind, barometric pressure variations, and seismic events oscillate these bodies of water. From the range of oscillation frequencies in the disturbance, a body of water picks out its resonant frequency. Standing waves then develop in the body of water, just as standing sound waves are produced in an organ pipe excited with a range of sound frequencies.

4.56 See FC 4.58.

4.57 The natural period of oscillation of the bay is about 13 hr, and so the semidiurnal tide forces resonant oscillations of the bay, much as sound waves can force an organ pipe to resonantly oscillate. As a result, the bay's oscillation energy has been enhanced, and the amplitude of oscillation increased.

4.58 The bore and the sink jump are both examples of an hydraulic jump, which is a surface water wave analagous to an atmospheric shock wave. Normal (sinusoidal) gravity waves can propagate upstream on a moving stream of water if the speed of the water is less than the speed of the waves. (See the answer to FC 4.45 for the distinction between gravity and capillary waves.) The ratio of the stream speed to the wave speed is called the Froude number. If the Froude number is less than 1, then the stream is "subcritical." If it is more than 1, the stream is "supercritical." The hydraulic jump is a wave that occurs where the water flow changes between being supercritical and subcritical. There is a change in height because the wave speed depends on the square root of the water depth. For example, in the sink hydraulic jump the depth is shallow inside the circle, the gravity wave speed is low, and the flow is supercritical. Outside the circle, the depth increases, hence the wave speed is more, and the flow is subcritical. In the case of the bore, the inflow of tidal water through a channel that narrows and rises makes the stream supercritical to any wave initiated by obstacles in the stream. The bore changes the flow from supercritical to subcritical by increasing the depth of the water and thus increasing the speed of the water waves.

4.59 Apparently nothing has been published on this demonstration. So, you might like to experiment with it yourself.

4.60 The cause of beach cusps is still current research. Although many theories have been postulated, none are generally accepted. The larger cusps appear to be due to rip current flows, which have fairly regular spacing along the beach. According to the theory, points (horns) of the giant cusps are between the rip currents where there is least transport of the bottom material parallel to the shore. The rounded portion (bays) of the giant cusps develop where the rip currents flow outward to sea and thus where there is maximum material transport. The cause of the smaller beach cusps is not known. One of the more recent theories describes the incident ocean waves creating standing waves oblique to the beach. The crests and troughs of these oblique waves shape the beach into the cusps.

4.61 The rotation of the earth causes an apparent force, the Coriolis force, to deviate the surface flow from the direction of the surface wind. This deviation is about 45° to the right in the Northern Hemisphere and 45° to the left in the Southern Hemisphere. If the flow is laminar (smooth), the deviation increases with depth. A plot of the velocity vectors with depth is called the Ekman spiral. To find the net transport through the extent of this spiral, one must integrate the flow over the depth. The result of the calculation indicates that the net transport is about 90° from the direction of the surface wind.

4.62 The change of the Coriolis force with latitude shifts the general circulation of the oceans to the west. Since the streamlines in the west are then more crowded, the current flow there is more intense.

4.63 When the tea is rotating around the center of the cup, the centripetal acceleration for such circular motion comes from the pressure difference between the tea nearer the wall and the tea nearer the central axis. This pressure difference also leads to an additional flow, called the secondary flow, that deposits the tea leaves in the center of the cup. Consider two horizontal surfaces through the tea, the top layer and the bottom layer. In both layers there is greater pressure at larger radii from the center. But in the bottom layer less pressure difference is needed to provide the centripetal acceleration because the friction from the cup's bottom prevents the tea from circling as fast as it does higher up. In both top and bottom layers there is a pressure difference, but the difference is greater at the top. If a small parcel of tea is initially at the outer top part of the top surface, not only does it circle the central axis, but it also descends along the wall to the bottom because of the pressure difference between the outer top and the outer bottom. To replace the fluid lost from the outer top, there is a flow of fluid from the central bottom upward along the central axis and then to the outer top. Thus, while the tea is circling, it is also flowing from outer top to outer bottom, to inner bottom, then to inner top, and finally to outer top again. Tea leaves lying on the bottom are captured by this secondary flow and deposited in the center of the cup where the fluid begins its ascent.

4.64 The secondary flow in the preceding problem is also responsible for the meandering of rivers. Perpendicular to the flow of the stream in an initially slight bend is a secondary flow circulating from outer top to outer bottom, then to inner bottom, up to inner top, and then finally back to outer top. This flow removes material from the outer stream bed wall and deposits it on the inner bed wall somewhat downstream. Although a young stream may start relatively straight, the slight turns in it are enhanced, and the stream begins to meander.

4.65 If the ball is to rise at its normal rate, it will have to push the water above it outward to the sides. But such motion of the water will be against the pressure difference that keeps the fluid in circular motion. (The centripetal acceleration of the water in this circular motion about the central axis is provided by the pressure difference between the water at larger and smaller radii: there is greater pressure on the outside.) If the sphere's upward speed is too small to provide this outward displacement of the water above it, then the sphere ascends with a column of water that rises at the same rate as the sphere. In other words, the sphere pushes and pulls upward a column of water its own diameter in size. The friction on this column and the greater mass that is moving both increase the time needed for the sphere to rise.

4.66 The dye displaces some of the water when it enters. Part of the water is pushed inward toward the central axis. But that particular parcel of water is now moving too quickly for its new radius, and thus it tends to press radially outward trying to regain its former position. The water that is pushed radially outward by the drop finds itself under too much pressure for the centripetal acceleration it has, and thus is pushed radially inward trying to regain its own old position. (The radial pressure difference is discussed in the preceding answer.) As a result, the dye is compressed radially. It mixes downward, but stays in a narrow sheet.

4.67 Comments on the direction of swirl in a draining bathtub are often as strong as in heated religious clashes: some people argue that all Northern Hemisphere tubs drain counterclockwise; others insist that only about half do. Shapiro (722) was the first to carefully test the swirl direction, although the arguments appear to have continued anyway. Unless extreme care is taken with a carefully designed tub, the rotation due to the Coriolis force cannot be seen. Normal bathtubs and sinks are by no means designed to show the Coriolis effect. Swirling in them could be in either sense, being due to such uncontrolled factors as the shape of the tub, the motion due to the pulling of the plug, the residual vorticity from the filling, the air currents above the water, and the shape and position of the drain. To show the relatively weak Coriolis force, you need a very symmetric tub with a central outlet that can be opened without swirling the water. Once the tub is filled, the water should sit for about one or two days for the vorticity from the filling to die out. There should be no air currents above the water or change in temperature in the room, both of which could produce motion that would swamp the motion due to the Coriolis effect. With these and other precautions taken, the proper rotation of the draining water can finally be seen.

4.68 The cause, nature, and behavior of tornadoes and waterspouts are poorly understood. Indeed, the distinction between the two vortices is not clear other than that the waterspout is over water, is weaker, travels faster, and lasts longer. True tornadoes, the type in the central plains of the United States, are highly destructive and accompany violent storms. The vertical motion appears to be upward through the funnel. (Dorothy was carried upward in "The Wizard of Oz".) The funnels are visible because of the water condensation in its low pressure or because of the dirt, debris, or spray it accumulates from the ground. They often occur in the spring when the cool, dry air from the north meets the warm, moist air from the Gulf of Mexico region. However, the mechanism that generates the vorticity is not known. Thermally induced rotation may be a cause. Existing rotational motion may converge to intensify the motion. Super thunderstorms may repeatedly produce electrical discharges that heat the air so severely as to generate the vorticity. The frequent occurrence of lightning in tornadoes (either stroke or ball—see FC 6.32 to 6.35) makes this latter proposal attractive.

4.69 The granular substance promotes the release of carbon dioxide gas by acting as nuclei for bubble formation. The bubbles form in the center of the flow, especially if the granular substance is dropped there, because the pressure in the water is less in the center than further out. (The pressure distribution is discussed in the answer to FC 4.63.) The released bubbles provide buoyancy to the central water, which then rises. Other water flows inward at the bottom, resulting in a concentration of angular momentum in the center. The rotational speed increases, and the swirl forms.

4.70 Because of the greater density of the cold milk, the milk stream descends into the hot coffee. Vortex tubes (vortex columns) in the rotating coffee become entrained in the milk and are stretched by the descent. As a result, the angular speed of the entrained vortices increases, perhaps enough to dimple the surface. If hot milk is poured into the coffee, it will not descend or at least will not descend as quickly. If the hot milk is less dense than the coffee, the entrained vortex tubes are shortened, and the rotational speed decreases.

4.71, 4.72, and 4.73 The cause and maintenance of dust devils are not well understood. Apparently superheated air initially lies in unstable equilibrium near the ground. Any small disturbance breaks this hot air out of the boundary layer so that it may rise. Once that break is made, the rising hot air will pull other hot air up through a chimneylike effort (FC 3.34). The rotational sense is entirely random and does not show the preferential rotation as do hurricanes. The whirlwinds developed above fires and over Lake Michigan are similar phenomena in that there is very unstable hot air beneath cooler air.

4.74 As a drop enters the water, its sides are retarded by the water and move slower than its center. The vortex forms as the faster moving center descends, and slower moving edges curl upward. The expansion of the ring as it approaches the bottom is similar to the expansion of smoke rings in FC 4.103.

4.75 Vortices are shed from both edges of the cardboard in the first arrangement. In the second, the air is swept along the length of the cardboard and then finally breaks into a vortex at the trailing edge.

4.76 The gas initially cools because it expands on entering the tube. Near the inlet a vortex is created that has greater speeds near the tube's axis and slower speeds closer to the tube's wall. As the flow spirals along the tube, the speed distribution over the width becomes more uniform as the inner air does work on the outer air due to viscous interaction. As a result, the outer region heats by the time it reaches the hot air exit. The core of the vortex flows toward the cold air exit, expanding as it passes the inlet and thus cooling. So, the increase in temperature in the outer layer of the swirl is due to viscous work in speeding up the outer layer. The decrease in temperature in the core is due to expansion as it flows in the opposite direction.

4.77 As a bird thrusts downward with its wings, it forces an updraft beyond the wing that then trails beyond the bird. The purpose of the V formation is to have another bird behind the first to take advantage of that trailing updraft. Thus, all but the central bird can save on energy by using the updraft left by the preceding bird.

4.78, 4.80, and 4.81 The behavior of all of these falling or rising objects is governed by the pattern and changes in the fluid flowing past them, but theoretical or even qualitative explanations of the results are not available. Instead, current research has attempted to correlate the types of behavior with the Reynolds number (which is related to the presence and degree of turbulence in the flow) or some other such parameter.

4.79 The trailing car is propelled forward by the vortex flow left by the leading car and encounters less air drag because the air flow has already been diverged by the leading car. The whiplash appears to occur when the trailing car begins to pass. Part of the air flowing past the leading car on that side is then forced to pass through the relatively narrow space between the cars, thus speeds up, and therefore is reduced in pressure. The trailing car then has greater pressure behind it than in front on the side closest to the leading car. The pressure difference accelerates the trailing car momentarily as it pulls out to pass. The leading car should experience a corresponding force rearward.

4.80, and 4.81 See FC 4.79.

4.82 The fish swim in a manner so that, like the birds in FC 4.77, they can take advantage of the wakes left by their leaders. Consider a fish inside the school. As it swims it leaves a trail of vortices that develop alternately on opposite sides of an axis extending directly behind the fish. The vortices turn such that on the axis the water flow is in the direction opposite the fish's motion. Were another fish to swim directly behind this particular fish, the trailing fish would have to expend more energy because it would be swimming against the flow of these vortices. However, if the trailing fish were to the side of the axis, it would be in the part of the vortex flow that moved forward. Imagine two leading fish with a trailing fish between the two body axes extending backwards from them. The trailing fish would be in the forward moving part of the vortices from both the leading fish, and thus would have to expend less energy than the leading fish in swimming. The purpose of the school is partly to decrease the energy expenditure of all but the leading fish by having the trailing fish take advantage of the vortex flows of the fish in front of them.

4.83 The wind breaks into vortices as it passes the building. On the windward side the wind is somewhat laminar (smoothly flowing), whereas on the opposite side the vortices make the wind gusty.

4.84 The large vertical plates on the bridge were ultimately responsible for the bridge oscillations. Such a broad face to the wind forced a large amount of air to divide and flow around the plate and then across the bridge. The pressure just above and just below the plate had decreased air pressure because of this rerouting and consequent speeding up of the passing air. Were the plate perfectly symmetric in the wind, the decrease in pressure on top and bottom would have been the same. However, the wind blew at fluctuating angles to the plate, and thus the pressures were different from moment to moment. This pressure difference flowed across the width of the bridge and was augmented by the turbulence shed by the windward plate. As a result of the pressure difference between top and bottom, the bridge began to oscillate. Similar oscillations are also developed in "galloping" telephone wires, where the pressure differences and vortex shedding also produce a whistling from the wires (FC 1.55).

4.85 Clear air turbulence (CAT) appears to be due to what is called Kelvin-Helmholtz instability. As a model of the instability, consider a dense and a light fluid in a basin with the lighter fluid on top and with the two fluid layers sliding over each other. If their relative speed is slow, any small disturbance in the interface is quickly eliminated. For greater speeds, however, a perturbation in the interface can result in an intrusion of one fluid in the other where the intruding fluid then develops a swirl. Similar vortex development can occur in the atmosphere where there is strong vertical wind shear (and thus relative motion of two layers) and large horizontal temperature gradients (and thus differences in densities of adjacent layers). The CAT is thought to be swirls developed at the interface.

4.86 Because there is reduced air pressure on the mountain tops, the air viscosity is less there, and thus the watch should run faster.

4.87 The mesh introduces turbulence and cavitation in the stream because it narrows the aperture through which the water passes. The sensation of softer water is probably due to the air bubbles that are formed.

4.88 To my knowledge there has been no systematic research on this question, although sports articles often refer to fast and slow pools. I could guess that the gutters designed to absorb surface waves would eliminate the reflected waves that may interfere with swimmers. Why not research this and other aspects yourself?

4.89 A stream passing over a spillway is similar to an air stream passing an edge (FC 1.56) in that oscillations are created in the stream. In the falling water the oscillations set up a standing wave with five-fourths wavelength from the spillway to the ground. Because of the resonant feeding of energy from the pressure variations induced by the edge to the oscillations of the falling stream, those oscillations can grow to a significant amplitude. This standing wave may be related to the earth's vibrations near a water fall (FC 2.65).

4.90 As the air passes the outer edge of the parachute, vortices are shed. Since the shedding alternates from one side to the other, and since they each have reduced air pressure, the chute experiences lower pressure first on one side, then on the other. This alternating of pressure begins to swing the parachute. If the oscillation frequency is close to the resonant pendulum frequency of the parachute and its load, the oscillation can be as large as 60°. The central hole allows some of the incident air to continue along the central axis of the parachute and break up the vortices on the top side. Stock cars can tolerate the oscillations even less than parachutists, so the parachutes on the cars have even more direct flow areas to further reduce the vortices.

4.91 The explanation of the boat drifting faster than the stream is still missing details about the water flow and momentum transport near the boat. However, the higher boat speed is partially justified by a simple analysis of the forces on the boat. Its weight is directly down, but the buoyancy is at an angle to the vertical, because the river is flowing downhill. Thus, there is a component of the boat's weight that is parallel to the stream's surface. This component is balanced by the drag from the water. An equivalent volume of water in the boat's place would experience drag also. But because of turbulent mixing in such a volume of water, it would meet greater resistance than the solid boat does at the same speed. As a result, the speed at which the drag cancels the parallel component of the weight is greater for the boat than for an equivalent volume of water.

4.92 A solid wall creates strong vortices that swirl the snow. A fence, on the other hand, creates milder vortices. If the air speed in these fence vortices is less than that needed to suspend the snow, then the snow is deposited on the leeward side of the fence.

4.93 In order for an obstacle to capture the snow, the snow has to be brought nearby. The wind begins to diverge tens or hundreds of meters in front of a large house, thus diverting the snow too early for it to be deposited at the house. A smaller obstacle, such as a pole, diverts the air much less, and the snow can get near.

4.94 The trailing edge is sharp so that the boundary layer on the top of the wing does not separate prematurely. Such a sepration would result in turbulent mixing at the rear of the wing, which would put the airplane in stall because it destroys the lift.

4.95 The aerodynamics of a skier are, of course, too complicated for an exact theoretical solution, so the choice of a particular stance without experimental data is largely just a guess.

4.96 The air drag on the ball comes from two factors: the pressure difference between front and back sides of the ball, and the friction between the air and the ball. With a smooth ball the boundary layer of air on the ball separates from the ball without entering the rear side much. After separation, the air develops vortices and leaves the rear in reduced pressure. Since there is higher pressure on the front side, the pressure difference retards the ball. A rougher surface delays the separation of the boundary layer. As a result, there is less pressure reduction on the rear side, less pressure difference between front and rear, and therefore less drag due to the pressure difference. The dimpled golf ball goes further.

4.97 There are two general aspects to a bird's ability to fly. Its wings act as airfoils (FC 4.31), and the bird can soar (FC 4.98). But when it flaps its wings to propel itself, the thrust comes not from pushing backward on the air, but from the feathers twirling in the air and acting as propellers. Perhaps a plucked bird could soar, but it could not propel itself.

4.98 Birds and sailplanes can soar by two techniques. They can fly into wind that is deviated upward by some obstacle such as hills and water waves. More practical for distance flying, however, is for them to fly into rising bubbles of hot air. Once lifted by such a bubble, they can then glide downward until they find yet another rising bubble. The bubbles are not tall columns of hot air. Instead, they are ring vortices that are developed as hot air in the boundary layer of air on the ground breaks loose from the ground. The circulation in the ring is upward in the center and downward on the outside (an upside down version of the vortices in FC 4.74). A bird can soar by circling around in the rising portion.

4.99 All kites act essentially as airfoils in that they force the air to diverge and there is less pressure on the top than on the bottom to give lift to the kite (see FC 4.31). The different bridling techniques distribute the stress from the handling string and also give stability to the kite. For example, the last three techniques shown in the figure will give a stabler flight than the first technique. The bridling can also be used to adjust the kite's angle of attach, that is, its angle with respect to the wind direction. In a light wind, the kite should be at more of an angle to the wind so as to diverge more air to get the proper lift. With stronger winds, the kite should be at less of an angle since less wind needs to be diverted. A kite tail has two purposes other than just being fun to watch. Its air drag stabilizes the kite, thus making the kite less subject to gusty winds. And, second, the drag helps adjust the kite to the angle of attack proper to the prevailing wind.

4.100 Cloud streets are due to longitudinal vortex rows, that is, rows of vortices whose axes of rotation are horizontal and in the direction of the wind. Where the circulation is upward between two adjacent rows, the air cools by expansion and condenses out some of its moisture to form a cloud (see FC 3.23). No cloud is formed where there is downward motion between two adjacent rows. The vortices are produced by a thermal circulation in which warmer air rises and cooler air descends, similar to the Bernard circulation cells of FC 4.101. The horizontal wind stretches these vortices so that they become horizontal vortex rows.

4.101 If the bottom fluid is sufficently hotter than the top fluid, the fluid is unstable and convection currents of rising hotter fluid and descending cooler fluid can develop into these patterns. For example, the hot fluid can rise in the interior of a hexagon while cool fluid descends on the boundary with other hexagons. For a given temperature difference and a given fluid, theory can determine those patterns (rolls or hexagons) that are steady-state solutions of the flow. Part of the visible appearance of the cells on the coffee is due to tiny drops suspended just above the areas of rising coffee. A charged comb disturbs these drops and partially destroys the cellular appearance.

4.102 The dune streets are due to the same type of horizontal vortex row formation in the air that is responsible for the cloud streets in FC 4.100. Where two adjacent rows have ascending air, sand collects in a dune street. Where two adjacent rows have descending air, there is no dune. Since the dominant winds in all the world's deserts are north or south, the streets run north and south.

Similar rows develop on the surface of the ocean because of similar vortex rows in a layer of water beneath the surface. Where two adjacent rows have descending water, material such as seaweed collects. There is a corresponding absence of material where two adjacent rows have ascending water. Although the dependence of these vortex rows on the direction and strength of the wind is well established, their actual production mechanism is not known.

4.103 To explain the expansion of a ring as it approaches a wall, imagine a mirror-image ring approaching from inside the wall at the same time. The parts of each ring's flow that are perpendicular to the wall cancel. The parts of their flow that are near the wall and parallel to it add. As a result, the ring expands parallel to the wall as it gets closer to the wall. Of course, there really is no second ring inside the wall, but the air flow caused by the presence of the wall is the same as if there were a mirror-image ring.

In spite of the descriptions of multiple passages of smoke rings in the literature, the trick may be impossible. Until 1972 those descriptions were common in some textbooks, but then Maxworthy (850) carefully investigated the effect with water rings. If the rings initially have almost equal speeds, then the rear one merely becomes entrained in the leading one to form a single vortex ring that does not separate. However, if the rear one initially has a much greater speed than the leading one, the composite ring becomes unstable and throws the former rear ring forward, leaving the former leader behind. The former rear ring then has the same or a somewhat greater speed, making a future encounter between the rings unlikely. If Maxworthy's descriptions are complete, then the multiple-passage descriptions are an example of where textbooks have continued to use an illustration that no one bothered verifying.

4.104 On an initially flat sand floor, the wind picks up and then drops sand grains, which then cause other grains to hop up. The result is to build deposits of sand that in turn modify the wind to further enhance the deposits. Ripples develop with a wavelength about equal to the average distance a sand grain hops when struck by other sand grains.

Sand waves on stream beds may be built in a similar way, or they may result from the flow of vortex rows (see FC 4.100 and 4.102). In the latter case, the sand ripples are oriented along the direction of motion of the stream's flow.

4.105 Contrary to much popular belief, the fluid is not pushed over the siphon by air pressure, as is disproven by the fact that siphons can operate in a vacuum. The force that pulls the fluid over the siphon is its own intermolecular force. When the siphon works, there is more fluid on the outlet side than on the inlet side, and the resulting imbalance of weight causes the fluid to flow up, over, and then down the siphon. As the fluid travels up the inlet side, its pressure is reduced the further up it goes. If the siphon is high enough, the fluid pressure is eventually reduced to the point where bubbles (of air or other gases) begin to form. Such bubble formation limits the height of the siphon because it breaks the intermolecular bonding between the fluid molecules and destroys the siphoning. Siphons work better at atmospheric pressure than in vacuum, because the pressure on the two ends of the siphon increase the fluid pressure at all points in the siphon. Thus, with atmospheric pressure outside the siphon, the height at which bubble formation occurs is increased.

4.106 The wind (which flows down and to the left in the diagram) picks up sand grains on the windward side of the dune and then dumps them as it spills onto the leeward side. Although slow, the net transport of the sand results in the dune formation moving downwind.

4.107 All modern toilets have a siphon between the toilet bowl and the plumbing to the sewer. As water is put into the bowl, the water level in it and the inlet side of the siphon rises. Eventually water spills over from the inlet side to the outlet side of the siphon and the siphoning begins. (You can flush a toilet by just pouring a bucket of water into it.) The siphoning flow and the general swirling of the water pouring into the bowl remove the waste. The extra hole at the bottom of many bowls is a water jet that entrains the fluid from the bowl and increases the speed and vigor of the siphoning.

4.108 As an oil drop is released from a car, air resistance stretches it out, inflates it like a chef's hat, and then bursts the center of that inflated shape. When the oil strikes the road, it is doughnut shaped.

4.109 These lines are small ridges pushed up by the viscous forces of the stream flowing beneath surface films (e.g., an oil film).

4.110 Nothing beyond a description of the clear band has been published. You might try experimenting with different fluids and solutions to understand this effect.

4.111 The oil forms a very thin film over the water. The surface tension of such a film is not constant, instead it changes as the film stretches and contracts. Waves passing over the film alternately stretch and contract the film and so produce an alternating tangential drag on the water below the film. This drag increases the energy loss of the wave to such an extent that the wave damps out quickly, leaving the film-covered area calm.

4.112 Oil films on the water surface collect and then damp out small waves to give streaks or patches of calm water. (See the preceding answer.) Apparently the oil comes from diatoms that have oil for assistance in flotation and for food. If the wind is strong, the oil patches arrange themselves in rows as is discussed in the answer to FC 4.102

4.113 Both the crown and the central breakup of the central jet are due to an amplification of unstable waves on the water. In the crown case the wave is around the rim.

4.114 The surface tension of the water holds the water in a thin layer and eventually pulls it back to the central support beneath the disc.

4.115 If the two water jets are exactly identical, then the vertical components of their momenta are canceled in the collision, and the pressure developed at the point of impact sends the thin water layer out horizontally. Breakup occurs when small holes develop and are then enlarged by the water's surface tension.

4.116 Surface tension holds the streams together.

4.117 The surface film on which the pepper grains reside contracts as the soap film develops and then spreads across the surface.

4.118 The turn of the stream around the edge of the can is stable because of the pressure difference across the width of the stream. An ideal incompressible fluid in a circular path has greater velocities at smaller radii. Thus, by the Bernoulli principle, there is less fluid pressure at smaller radii. Here the atmospheric pressure outside the stream is greater than the fluid pressure near the edge and therefore holds the stream to the edge. At some point on the side of the can the stream detaches because it is unstable to small perturbations.

4.119 Previous to Loewenthal's work the tears forming above strong alcoholic drinks were thought to be due to surface tension pulling the solution upward along the glass where then the alcohol would evaporate to leave just water. However, Lowenthal demonstrated that the water collecting at the top of the climbing film was condensation from the room air. Furthermore, the force responsible for the film's climbing was not surface tension pulling the film up but a pressure that developed in the fluid next to the glass surface.

4.120 There are several tire designs to decrease the probability of aquaplaning. The tread can channel the water at the rear of the contact area outward and eject it. Other, shorter channels can eject water to the sides. Finally, small holes in the tire can essentially blot up a water layer as they make contact with the road in the front part of the contact area. In each of these techniques the emphasis is on removing the water quickly to avoid aquaplaning.

4.121 The support for the drops is not fully understood but is thought to be an electrical repulsion between the water molecules in the drop and those in the main body of water. A water molecule has a positive side where the two hydrogen atoms are located and a negative side where the oxygen atom is. If the drop's bottom surface and the surface of the main body of water just beneath the drop present the same charged side to each other, then those two surfaces will electrically repulse each other. Another support mechanism is employed if the main body of water is superheated. In that case the evaporation of the bottom surface of the drop provides a continuous vapor layer to support the drop (just as is described in the answer to FC 3.65).

4.122 The soup flow reversal is an example of elastic recovery by a viscoelastic fluid. When the soup has almost come to rest because of friction from the sides of the pan, the top surface briefly continues to move after the lower soup has stopped. The surface layer is then pulled back by an elastic force between it and the lower soup, and the swirl is momentarily reversed. Oscillations around the equilibrium position would continue except that the soup is viscous enough to damp out the oscillations almost immediately.

4.123 Although this effect, called the Kaye effect, depends on the elastic nature of the fluid, its cause is not well understood. Collyer and Fisher suggest that the leap may be due to a rapid change in the viscosity of the stream as it strikes the heap on the main body of fluid. The fluids that display the Kaye effect are apparently shear-thinning ones, that is, their viscosity decreases when the fluid is sheared (FC 4.126). During its fall the stream is unsheared and has a relatively high viscosity. Upon striking the heap, however, the rapid changes in velocity will create large shearing in the fluid, thereby reducing its viscosity. Being elastic also, the stream then reflects from the heap.

4.124 As the viscoelastic fluid rotates, the shearing of its layers creates stresses that act around the circumference of the circular path of the fluid, tending to contract the fluid to the center of rotation. These stresses are not created in normal (Newtonian) fluids. Their result in this demonstration is to push the fluid to the center of rotation and up the rod.

4.125 The compression on the falling stream causes the stream to buckle. Since the stream cannot break under these conditions, the buckling makes the bottom of the stream circle around as more fluid falls than can be absorbed into the main body of the fluid.

4.126 A fundamental explanation of how the viscosity of a fluid is decreased when the fluid is under shearing stress is not currently available. Most suggestions involve a change in the molecular configuration because of the shearing. For example, the long molecules may be stretched along the flow lines created by the shearing. As a result, the viscosity is decreased. Once the shearing is removed, the molecules regain their previous orientations, and the viscosity increases.

4.127 The internal stresses in the viscoelastic fluid are relieved when the fluid emerges, thereby forcing the expansion at the tube's mouth. One model of this relief and consequent expansion considers the molecules as being stretched when forced through the tube. When they emerge they contract and consequently swell the fluid.

4.128 and 4.129 Both of these demonstrations are examples of elastic recovery by a fluid. The silicone putty is highly viscous, but the viscosity is lower for slow shearing rates. At high shearing rates it fractures.

4.130 The viscosity of quicksand increases with shearing, so trying to pull yourself out of the stuff quickly is impossible; the more you shear the quicksand, the more it will hold to you. Move slowly so as to keep the viscosity as low as possible. The eyes of trapped animals may bulge because of the large hydrostatic pressure on the lower part of the body due to the density of the quicksand. (A sand-water mixture is denser than just water.)

4.131 If the cylinder is rotated slowly, then the dye is pulled along a thin layer and spiraled inward with each turn. Provided the reversal is made before molecular diffusion (thermal motion of the molecules) can smear the dye, the spiral is unwound almost exactly by rotating the cylinder back an equal number of turns.

5.1 In order for there to be a clear image on your retina, the eye must refract the light rays. About two thirds of the refraction occurs at the surface of the eye. If water is on the eye, nearly all of that refraction is lost because the refractive index of the eye material is approximately the same as that of water. If you wear goggles, there is a layer of air in front of the eyes to give you normal refraction. The fish that sees in both water and air simultaneously has two retinas and an egg-shaped eye lens. In order to compensate for the reduced refraction for the submerged portion of the eye, the eye lens has more curvature for the rays coming from the underwater scenes.

5.2 The man would be invisible if his index of refraction matched the air's index, which is slightly greater than exactly 1, the index of refraction for vacuum. A greater index would result in some refraction of the rays coming from scenes behind the man, thus making his presence noticeable by the distortion of the images, especially when he walked. In order for the man to see, he has to absorb some of the incident light. Such absorption would have to be slight enough that the man does not appear as a shadowy figure. In short, his index of refraction would have to have a real part that is approximately equal to 1.0 and an imaginary part that is great enough that he would absorb enough light to see but not so much that the subtraction of the light would be noticeable.

5.3 Water rises along the side of the pencil by capillary action, curving the water surface next to the pencil. The curved surface refracts light into what would otherwise have been a shadow region of the pencil to give the white gap.

5.4 The coin's image is first visible on the water's top surface because the rays from the coin are reflected from the back surface of the container, directed to the top, and then are refracted out to be seen. If you put your wet hand on the back, you destroy that initial reflection there. A dry hand has much less effect because it has relatively few contact points with the glass. A wet hand fills the spaces between the contact points with water. Since the indices of water and glass are about the same, this filling of the spaces effectively increases the contact area of the hand with the container to about 100%. Much of the light rays from the coin that fall on this area are therefore absorbed and lost. As a result the image on the top surface disappears.

5.5 Light rays from the submerged object are refracted at the water-air surface, bending toward the surface as they emerge and, as a result, appear to originate from a place higher than the true position of the submerged object. For normal viewing (eyes on a horizontal line) the horizontal distance is not distorted. If you turn your head so that your eyes are along a vertical line, then the rays reaching one eye have been refracted at a different angle to the water surface than the rays reaching the other eye. As you mentally extrapolate the rays back to find the apparent position of the submerged object, you place it not only higher than its true depth but also closer.

5.6 Rays passing the first plane of glass will be partially reflected from the inside surface of the second plane. Normally this partial reflection is insignificant because most of the light continues through the second pane to be seen. However, if the external air pressure differs from the air pressure between the two panes, then the panes are not parallel, and part of the internally reflected light can produce a faint but noticeable ghost image. Consider a ray from an object outdoors entering the first pane initially horizontally. Most of that light goes through the second pane and enters the room. If this is the only light seen, it gives an undistorted image of the object. However, part of the light is reflected by the inside of the second pane, returned to the first pane, reflected again, returned to the second pane, and finally transmitted into the room to provide a second, fainter image. If the two panes are not parallel, this second image is displaced from the true image.

5.7 through 5.11 All of these problems (except for the story about the bird in FC 5.9) are examples of mirages and depend on the variations with height in the refractive index of the air near the earth's surface. That index depends primarily on the temperature of the air. Looming (which is an example of what is called a superior mirage) can occur when the air temperature increases with height. The observed light rays have originated from a distant object, say a mountain, at an upward angle to the horizon. They are then refracted enough by the increase in refractive index (due to the temperature increase with height) that they are bent over for you to see. An observer mentally extrapolates straight back along the observed rays and places the image above where the object really is. In other words, the image looms above the object and is an example of a "superior" mirage.

The oasis mirage is an "inferior" mirage in that the image is below the true position of the object. In that case the object is the sky. Light rays from the blue sky are refracted by the ground layer of air in which the temperature decreases with height. The rays are bent up to the observer, who then mentally extrapolates straight back along the rays to believe that there is a body of blue water on the ground somewhere ahead. The shimmering due to variations in the refraction by the hot air gives the illusion of flowing water. The pelican could not have seen such a mirage because the light rays from the sky could never be so refracted as to return at such a large angle to the ground.

The Fata Morgana is a more complicated mirage in that the temperature profile producing it does not change linearly with height. The temperature increases with height, but at some intermediate height the rate at which it increases is less. Such a temperature profile, but with a more noticeable drop-off at the intermediate height, can result in a three-image mirage.

5.12 Most one-way mirrors depend on one side (say the room in which a criminal is being questioned) being more brightly lit than the other side (where a viewer is). Some of the light incident on the glass from the criminal's side is reflected by the front and back surfaces of the glass. If the other side is relatively dark, then the criminal sees only the reflected light and thinks the glass is a mirror. The viewer, on the other hand, receives ample light transmitted through the glass and can clearly see the criminal. The mirror effect is enhanced if the viewer's side of the glass is coated with a very thin layer of metal that would increase the amount of reflected light to the criminal but still allow enough light for the viewer.

5.13 Even though the moon is in the earth's shadow, sunlight can still illuminate it if the sunlight is refracted into the shadow area by passing through the earth's atmosphere on the edges of the earth. However, such refraction removes the blue end of the visible spectrum for the same reason that the sky is blue (FC 5.59) and leaves only the red end of the spectrum. Hence, the sunlight that is refracted sufficiently to illuminate the moon is red. The same color subtraction is responsible for the red skies during sunrises and sunsets (FC 5.58).

5.14 Although mirages are normally due to refraction of light (see FC 5.7), this particular illusion appears to have been a mirage due to reflection. The girl probably saw a reflection of herself on the thin mist. Nothing more than this suggested cause has been published on reflection mirages and their physical possibility can only be guessed at now.

5.15 There is no one equation giving the number of images possible in the two mirrors as a function of the angle between the mirrors and the angular location of the object with respect to the mirrors. The most complete work done on the problem is that by Chai (989).

5.16 The green flash is due to the separation of the colors in the sunlight by the earth's atmosphere, similar to the dispersion of light by a prism. As a ray from the sun enters the earth's atmosphere, it is refracted such that it is slightly closer to being vertical than before. As a result, the sun appears to be somewhat higher in the sky than it really is. The shorter wavelengths of light (the blue end of the spectrum) are refracted more than the longer wavelengths (the red end), and, as a result, there should be a blue image of the sun slightly higher than a red image, with images of intermediate colors somewhere in between. However, the blue is lost by atmospheric scattering (see FC 5.59) and the highest image is the next color, green. As a result, green is the last color seen just as the image of the sun dips below the horizon.

5.17 The unmixed sugar creates a variation in the refractive index with depth with the maximum being at the bottom where there is more sugar. As the laser beam enters this solution, let us say initially at a slight downward tilt, the beam is continuously refracted to a greater tilt as it encounters progressively greater values of the refractive index. Eventually it reflects from the bottom surface. As it moves upward it again encounters a continuously changing refractive index and again is continuously refracted. This same type of refraction, but with sound instead of light, is discussed in the answers to FC 1.29 and 1.38.

5.18 The light rays from the sun are refracted by the atmosphere. The closer the sun is to the horizon, the more this refraction is. Consider the sun when its lower edge appears to be on the horizon. Were it not for the refraction, the sun would just then actually have its lower edge a little **more** than half a degree below the horizon. The upper edge, in the meantime, appears to be slightly **less** than half a degree from where it would be if there were no refraction. As a result, the vertical width of the sun appears to be somewhat less than it would be with the sun overhead (actually about 6 arc minutes short). The horizontal width suffers very little shortening due to refraction (about half an arc second). Thus, when the sun is on the horizon it appears to be an ellipse. (Please don't confuse this refraction effect with the optical illusion of FC 5.134.)

5.19 The blue ribbon is due to the reflection of the blue sky by the waves on the horizon. According to the answer to FC 5.20, the average contribution to the reflected light by the horizon waves comes from the sky about 30° up from the horizon. During much of the day that portion of the sky is a deeper blue than the rest of the sky, and therefore the ribbon should be a deeper blue. The reflection polarizes the light parallel to the water (see FC 5.49).

5.20 Certainly the waves do not all have a slope of 15°, but the average contribution of the light scattered by the waves on the horizon comes from the sky about 30° up from the horizon. So, the overall effect is the same as if all waves did have 15° slopes. Waves with small slopes are more probable than waves with larger slopes, but they reflect only a small part of the sky. Waves with somewhat larger slopes are less probable, but they reflect larger portions of the sky at greater angles to the horizon. Waves with relatively large slopes are so improbable that they contribute almost nothing. The overall effect is that the part of the sky at 30° from the horizon is most strongly reflected by the waves on the horizon and that portions of the sky at smaller angles have much less reflection and are not seen.

5.21 The tilt of the random waves on the water spreads the reflected image of the light source (sun, moon, or artificial light). The spread to the left and right is less than the spread between the observer and the horizon because of the geometry involved in reflecting light to the observer. The ratio of the width to the length of the bright area is sin θ where θ is the angle of elevation of the light source. The dark triangle above the horizon is a contrast effect. By blocking off the luminous area from your field of view, you eliminate that illusion.

5.22 The cloth can glisten if it has a regular pattern of threads running parallel to each other to give a furrowed surface. When such a cloth is viewed at certain angles, the reflection of incident light is relatively large. At other angles, the reflection is less. So, if the cloth is moved around in the light, sometimes it reflects well, sometimes not. In other words, it glistens. The orientation giving the most reflection is when a line perpendicular to the furrows bisects the angle between the incident light ray and the light ray reflected to the observer's eyes.

5.23 The eye produces a real image of the nail on the retina that is inverted from the nail's orientation. The nail looks right side up, however, because the brain effectively inverts the real image in its interpretation of the scene. The nail also puts a shadow on the retina whose orientation is the same as the nail. But since the brain inverts the scene on the retina, that shadow appears to be upside down.

5.24 The optimum hole radius is approxiately $\sqrt{0.6\,\lambda f}$ where λ is the wavelength of the light and f is the distance from the hole to the screen or film. Larger holes give less resolution in the photograph. Smaller holes produce diffraction patterns. (Diffraction is an interference effect due to the wave nature of light. Similar diffraction, but with sound instead of light, is in

FC 1.42 and 1.43.) The pinhole camera does suffer chromatic aberration because the optimum distance to the film for a given pinhole depends inversely on the wavelength of light, which varies from about 0.40 microns (blue) to 0.65 microns (red).

5.25 The images are pinhole images made by the tiny holes in the leaves. They are always present during the day but are usually lost in the overall glare of light. During an eclipse that glare is reduced somewhat.

5.26 Light falling on the dew drops is strongly reflected back along the path to the sun, that is, retroreflected. Part of the reflection is at the front surface of the drop; part is at the back surface at the point on the axis through your eyes and the sun. Light incident at other angles on the drops can also enter the drop to reflect on the backside.

5.27 Reflectors that return the light to its source, even if the light source is not on the reflector's central axis, are called retroreflectors. They can be spheres (FC 5.26), triangular prisms, or incorporate mirrors and lenses. A perfect retroreflector would be practically useless since the eye is rarely exactly at the light source. But most retroreflectors are sufficiently imperfect that the cone of light returned toward the source is wider than the cone of light intercepted by the reflector. A simple retroreflector is a corner of three mutually perpendicular mirrors. A light ray entering the corner from any direction will reflect off all three mirrors in succession and then be returned opposite its initial direction.

5.28 The drops focus the light, placing an image of the sun on the leaves, which burns the leaves.

5.29 Chance orientations of the water waves reflect light to your eyes. The illusion that there are streaks of light radiating from the head of your shadow is due to the required wave orientation for the reflection to reach you and the constantly changing wave pattern.

5.30 Cats and other animals have retroreflecting eyes that are noticeable in otherwise dark surroundings. The eye incorporates a lens and a curved mirror that returns the light in a cone which then passes the source of light. In the case of carnivores, there is a layer of zinc cysteine crystals behind the retina that provides the high reflectance.

5.31 The horizontal line marks the height at which falling snow melts. The snow reflects more light above the line than the water drops do below the line.

5.32 Light does emerge in a large range of angles from a water drop, but the most intense light emerges at the rainbow angles (in ray theory you can say that there is the densest clustering of the emerging rays at the angle). Since the different visible wavelengths suffer different amounts of refraction (blue is refracted more than red), the exact angle at which the emerging light is brightest is slightly different for each color. Hence, at the rainbow angle the colors are not only bright but also slightly separated so that they can be distinguished. (Also see the answer to FC 5.44.) However, the dispersion separating the colors is not prismatic. The first clue to its true nature is in the formation of the supernumeraries (FC 5.34).

The color sequence in the secondary bow (which is higher in the sky than the primary bow and somewhat rarer) is reversed because the participating light rays reflect twice inside the drop. As a result, they emerge at a different angle than the rays participating in the primary rainbow. Since the blue is refracted more than the red, the drops contributing the blue to the secondary rainbow must be at a slightly greater angular elevation than the drops contributing the red. The exact opposite is true for the primary rainbow because only one reflection is involved there. Thus, the color sequence is reversed.

More than two rainbows have been seen in the lab. [For example, see my paper in the *Amer. J. Physics,* **44**, 421 (1976).] A few people have reported seeing the third order rainbow (corresponding to three internal reflections) when the sun is low and below some dark clouds. In general the higher order rainbows are not seen because they are dimmer than the glare reflected from the surface of the drops, the glare transmitted through the drop with no internal reflections, or the background sky light.

5.33 The uneven distribution of the red is due to the vertical flattening of the falling drops by the air flow around them. Since the horizontal cross sections of the drops remain circular, the colors on the vertical legs are the expected rainbow colors because the light traverses the drops through such a circular cross section. On the top of the arc the light must go through a flattened cross section, which displaces the red inward and downward in one's view of the bow. As a result the contribution of red is diminished. Smaller drops are less affected by the air flow and therefore can contribute to all parts of the rainbow.

5.34 A correct calculation of the rainbow intensities and color distribution abandons the technique of tracing light rays through the drops in favor of treating the light as a wave. Even if the incident light wave is a plane wave, the emerging light wave is not. As a result the emerging light creates an interference pattern. The major peaks (the brightest areas) in the pattern are just the bright colors in the usual rainbow, so nothing substantial has been changed from the previous ray approach. Some subtle effects are noticed, however. The actual angular locations of the colors are now more accurate. The change in colors with change in drop size is finally accounted for (FC 5.44). But more important, the interference of the emerging waves explains the occasional faint bows just below the primary bow and just above the secondary bow. These extra bows are the other peaks in the interference pattern. They are seen less often only because they are less intense than the major peaks in the pattern and because their visibility depends on uniformity of drop size.

5.35 From all of the sky in the direction of the rainbows there is the general background sky brightness and some glare from the reflection of sunlight from the outside surface of the drops. Below the primary rainbow there is also light reflected once inside the drops. The brightest of this light is at the primary rainbow angle, but other such singly reflected light can exit any drops at a lower angle in the sky than the primary bow. However, such singly reflected light cannot exit from drops at a greater angle than the primary bow. A similar but reversed case holds for the light reflected twice inside the drop. The brightest of this twice reflected light exits at the angle of the secondary rainbow. Other twice reflected light can exit from drops at greater angles in the sky, but none can exit from drops at lesser angles. So, in addition to the background light and glare, there is an additional light below the primary bow and above the secondary bow but not between the bows. That band between the bows is therefore relatively dark.

5.36 The rainbow is polarized parallel to the bow at any given point because of the refraction and reflection of the light by the water drops.

5.37 The lunar rainbows are rare not only because moonlight is much weaker than sunlight, but also because of the weather and the time the moon is in the proper position and condition for making a rainbow. The time of day with the highest frequency of thundershowers is late afternoon and early evening (FC 3.41). Thus the moon has less opportunity of making rainbows. Also, the intensity of the moonlight changes as the moon changes phase, making the chances of a rainbow even slimmer.

5.38 The drops contributing to the rainbows are not at any particular distance from the observer. Only their angle from the line extending from the sun and through the observer matters. The drops can be anywhere from a few yards to several miles distant from the observer. If the only drops participating are just a few yards away, such as with a nearby garden sprinkler, then each eye sees its own rainbow displaced from the other.

5.39 The rainbow pillar is a leg of the primary reflection rainbow. The normal primary bow is formed from the direct sunlight. Near a body of water light reflected from the water can form another rainbow in the sky. Although the geometry required for such a reflection rainbow is necessarily the same as for the normal rainbow, its orientation in the sky is different from the normal rainbow because of the reflection. Were the whole reflection rainbow visible, it would have its center higher in the sky. Thus, near the horizon its leg is at a steeper angle to the ground than the leg of the normal rainbow. Some intensity of the sunlight is lost on reflection from the body of water, so the reflection rainbow is weaker and rarer than the normal rainbow.

5.40 In contrast to the preceding phenomenon, the reflected rainbows are merely the mirror images of the normal rainbows. Although Minnaert (954) describes the two arcs as being identical, Humphreys (164) correctly shows that the reflected bow appears flatter because less of the arc is seen than for the normal rainbow above it. The difference in appearance stems from the requirement of scattering angles if light emerging from water drops is to contribute a rainbow that then reflects from the water surface to your eyes. The water drops meeting such angle requirements are at a lower angular elevation in the sky than those giving the direct rainbow.

5.41 The normal dewbow is just a rainbow from the water drops on the grass. The primary rainbow is roughly 42° from an axis running through the sun and the observer's eyes (FC 5.32) and would be a complete circle if the ground did not interfere with the suspension of water drops in the field of view. If the ground is covered with water drops, however, then the part of the rainbow below the horizon can be seen. The angle of 42° still holds, but because the drops are limited to a horizontal plane rather than filling all of the space in front of the viewer, the shape appears to be hyperbolic. In the normal dewbow the incident light is essentially parallel rays from the sun. The street light gives diverging rays, and although the angle of about 42° is still required for the bow, the position of the drops contributing to the bow takes on the initially strange shape because of the diverging range of incident light rays available.

5.42 The sun dogs are due to the refraction of light by falling hexagonal ice crystals that have their central axis (which is parallel to the six faces) vertical. Although the crystals refract light into a large range of angles, the brightest of the emerging light is at the angle that least deviates the sunlight from its original direction. That angle of least deviation is about 22° if the sun, the ice crystal, and the observer are all in a horizontal plane. The observer then sees the bright light from the crystals 22° to each side of the sun. As the sun rises, however, the crystal's axis is no longer perpendicular to the light rays and the angle between the sun dogs and the sun increases somewhat. Eventually the sun is so high that the brightness is eliminated. The sun dogs are colorful because the ice separates the colors in the same way as does a prism.

5.43 The 22° halo is produced by the same type of refraction from falling ice crystals as in FC 5.42 except that the central axes of the contributing crystals are randomly oriented in a plane perpendicular to a ray of incident sunlight. Thus, at any point 22° from the sun there are some crystals that happen to be oriented properly to give bright light. The collection of these contributing crystals forms the halo. Colors are again due to a prismlike separation of the colors. Since blue is refracted the most of the visible colors, it is on the outside of the halo.

5.44 An explanation of rainbows involving just the ray theory and prismatic separation of colors (FC 5.32) cannot account for white rainbows. The interference theory of rainbows used in FC 5.34 is needed. The colors in the normal rainbow are the major peaks in the interference of the light emerging from the water drops at the rainbow angles. As drops smaller than a millimeter in diameter are considered, the widths of those peaks increase and eventually overlap sufficiently to eliminate any distinguishable colors. The light exiting at the rainbow angle is still relatively bright but now has no color and thus gives a white rainbow.

5.45 Reflection from the outside surfaces of falling hexagonal ice crystals give the pillars of light above and below the sun. The crystals can be short along their central axis as compared to their width, in which case they are called plates. Or they can be longer than their width, in which case they are called needles or pencils. Both can give sun pillars. For example consider the plates. The air flow around them forces them to be horizontal to maximize their air drag. If the plates are higher in an observer's field of view than is the sun, sunlight reflects from the bottom of the plates and gives the observer a relatively bright area in that portion of the sky above the sun. If the plates are lower in the field of view, reflection is off the top surface.

5.46 Recent work by Greenler (1034, 1065) provides computer simulations of the 22° halo and the sun pillar. To untangle all of the observed and conjectured arcs, a complete simulation of light scattered from falling ice crystals—both plates and pencils (FC 5.45), and both oriented and spinning—is needed for the whole sky and for all positions of the sun. Some of the present, although perhaps controversial and erroneous, explanations for a few arcs and halos are the following (the letters refer to Figure 5.46):

(a) and **(b)** 22° halo and its sun dogs: See FC 5.43.
(c) 46° halo: the refraction of light to make this halo is similar to that in the 22° halo with one exception. The ray passes through a 90° corner, rather than a 60° corner as

with the smaller halo, by passing through one of the end faces and one of the six side faces of the crystal. Again, the brightest light is with the geometry giving least deviation to the rays. The light so scattered by appropriately oriented ice crystals gives the 46° halo.
(d) Circumzenith halo: sunlight enters and exits through two adjacent faces that are perpendicular to each other. To create the halo, the sun must be lower than 32° from the horizon.
(e) Parhelic circle: light is reflected from the vertical sides of the falling crystals.
(f) Sun dogs to the 46° halo: these very rare bright patches are due to the same scattering as the 46° halo but are limited to those crystals having their central axis horizontal.
(j) Lowitz arcs: to produce these arcs, hexagonal ice crystal plates must spin about an axis that lies in the plane of the plate. Light entering one of the hexagonal sides then exits through another. The maximum brightness of this refracted light is for the geometry least deviating the incident ray, and the arc is the light from a collection of the crystals meeting this requirement.

5.47 Crown flash is due to the mirrorlike reflection of light from falling hexagonal ice crystal plates. (They are also responsible for the sun pillars in FC 5.45.) The electric field in the thundercloud makes electric dipoles of these crystal plates (i.e., makes one side positive and the other negative) and the dipoles align themselves in the field. Normally the plates fall with their broad side down so as to maximize the air resistance. But the electric field of a lightning stroke momentarily changes this orientation and hence the relative brightness from a particular part of the cloud. If the change in electric field propagates through the cloud, then the brightness could also.

5.48 One of the first suggestions was to cover the headlights with polarized filters oriented 90° to the polarizing filters placed over the windshield. With this orientation the driver would not see the light from an approaching car because the polarized light from its headlights would not pass the filter in front of him. Such a situation would also be dangerous. Orientations somewhat different from the 90° would be better so that some of the oncoming headlight could be seen. One of the drawbacks to the scheme (perhaps a fatal one) is that the filters also absorb some of the light from the surroundings, as for example, the light from a streetlight. So the overall view would be darker. Another drawback is that the tilt of the windshield would matter and thus would have to be standardized.

5.49 Direct sunlight is unpolarized, that is, the oscillations of its electric field are perpendicular to the direction of travel but along no particular axis. The reflection of sunlight from a surface will polarize the reflected light parallel to the surface, that is, the electric field oscillations are still perpendicular to the direction of travel but are preferentially oriented parallel to the surface. The extent of this polarization depends on the material and the incident angle of the light. If you are driving toward an afternoon sun, for example, the light reflected from the road is strongly polarized parallel to the road. Polarized sunglasses reduce this glare by blocking that sense of polarization and passing light with vertical polarization. On a microscopic scale, this blocking means that the long molecules in the filters are oriented horizontally and will absorb light whose polarization (and thus electrical oscillations) are also horizontal. Much of the glare is thus eliminated, but the general illumination of the surroundings is not as reduced.

A fisherman can reduce the glare of sunlight reflected from the top surface of the water and still see the light reflected from the fish. Of the unpolarized incident light, the parallel polarization has been preferentially reflected. The light entering the water then must be preferentially polarized in the opposite sense—perpendicular to both the direction of travel and to the parallel polarization. Once reflected from the fish, this light will be able to pass through the fisherman's sunglasses. Thus, the man sees the fish and not the surface glare. This explanation is not completely correct if the fish is more than about 5 ft. deep, because thereafter the scattering of the light by small particles suspended in the water polarizes the light horizontally (FC 5.55).

5.50 The polarization of the sunlight scattered by the atmospheric particles is derived from the same scattering physics as in the popular explanation for the blueness of the sky (FC 5.59). The unpolarized incident sunlight oscillates the electrons in the air molecules (nitrogen, oxygen, etc.), which reradiate the light. Consider, for example, a sun on the horizon and a scattering atom directly overhead. Since the direct sunlight is unpolarized, the electrons in the atom can oscillate along any axis in a plane perpendicular to the direction of travel of the sunlight. You can consider such oscillations as being either of two cases: where the electrons oscillate vertically or where they oscillate horizontally as you would view this overhead atom. The vertical oscillations do not radiate light vertically so you do not see that contribution. Instead you see only the light radiated by the horizontal oscillations. Such light is polarized along the same sense as the electron oscillations: north-south if the sun is due west. In other words, the light scattered from that part of the sky is polarized. Similar considerations can be made for the rest of the sky and for any elevation of the sun. Clouds are not polarized because of the multiple scattering of the light traversing them. Multiple scattering is also the cause of the neutral points in the sky's polarization pattern.

5.51 The ice is double refracting, that is, a beam entering the ice is split into two beams each experiencing different indices of refraction and having perpendicular senses of polarization. Instead of the arrangement in the problem, first consider ice between two polarizing filters. Light passing the first filter enters the ice, is split into two beams, which then propagate through the ice at different effective speeds (because of the difference in the refractive indices they experience). When they emerge, the two beams may be in or out of phase depending on the wavelength of the light, the length of the crystal, and the difference of the refractive indices. The sense of polarization of the emerging beam will depend on this phase difference. Suppose that when, say, the yellow of an initially white beam emerges that its sense of polarization happens to be perpendicular to the second filter's polarization sense. The second filter will therefore block the yellow light, and an observer will see the other colors in the visible range in the light from that filter. In the problem there are no polarizing filters, but the sky provides polarized light and the reflection of the light from the water pool provides the second polarizing selection.

5.52 The cellophane can be considered as having two special axes. Light polarized along one of these axes will experience a certain index of refraction. Light polarized along the other will experience a different index of refraction. If the incident light is polarized along a direction lying between these two special cellophane axes, then the light is effectively split into the two polarization senses of the cellophane and propagates through the cellophane at different effective speeds because of their difference in the refractive index. Upon emerging, these two polarization senses of the light are out of phase. The result is that the net polarization sense of the light is rotated by the cellophane. Normally two perpendicular polarizing filters will not transmit light. But with the cellophane inserted between them, the polarization sense of the light passing through the first filter is rotated by the cellophane. Then part of the light's polarization sense happens to lie along the polarization axis of the second filter, and some of the light gets through the second filter.

The food wrap does not have such special axes until it is stretched. The stretching untangles the spaghetti of the long molecules, orients the molecules along the direction of stretching, and thus makes them polarizing filters. Light whose electric field oscillates perpendicular to these oriented long molecules pass through the wrap, whereas the light having its electric field parallel to the molecules does not pass as readily.

5.53 The spots indicate the stress points in the bonding of the glass plates and thus behave like the stretched food wrap in the preceding problem.

5.54 The sense of polarization of the light entering the syrup is rotated by the helical structure of the syrup molecules. The rotation depends not only on the type of syrup but also on the wavelength of the light, the blue end of the visible range being rotated more per unit length of syrup than the red end. What colors are seen through the second filter depends on which of the colors that filter happens to block when the light emerges from the syrup. For example, suppose the second filter blocks the sense of polarization that yellow light happens to have when it emerges from a particular container of syrup. Then an observer will see not the white of the initial light, but the other colors in the visible range besides yellow.

5.55 The polarization detection

lies in the ultraviolet detectors of the insect eyes. There, the rhabdoms, which act as light guides in the photoreceptors, are twisted: some one way, the others the other way. The direction for maximum sensitivity of the polarization of the incident light differs for these two senses of twisting by about 40°. Thus, two rhabdoms acting together could determine the polarization sense of the incident light. That determination, along with an intensity determination by an ultraviolet detector insensitive to the polarization, orients the insect with respect to the sky.

5.56 A dichroic crystal can be considered as having two polarization axes. If the incident light is polarized parallel to one of these axes, then the crystal is clear. If the light happens to be polarized parallel to the other axis, then the crystal is dark blue. By orienting the crystal and observing its color, the Vikings could detect the polarization of the sky. With experience they could infer the position of the sun, even if the sun was below the horizon. Cloud cover, however, destroys the polarization of the sky light and would make the crystal useless.

5.57 A blue absorbing pigment in the macula lutea (the depression in the eye's rear) absorbs according to the polarization of the incident light. For example, blue vertically polarized light is absorbed horizontally to leave a horizontal yellow hourglass (yellow is blue's complementary color).

5.58 and 5.59 The basic color determination of the sky is in the wavelength dependence of the scattering of sunlight by the atmospheric molecules according to the Rayleigh scattering model. The electric field of the incident sunlight oscillates the electrons in these molecules, which in turn radiate light. The overall effect is to scatter the sunlight. Light with shorter wavelengths (the blue end of the visible range) is deviated more from its original direction than is light with longer wavelengths (the red end). When the sun is near the horizon the sky above an observer is therefore largely blue. The sky more than 90° from the sun is less blue because it is illuminated with sunlight which must traverse a long path through the atmosphere and is therefore somewhat depleted in the blue. The sky near the sun on the horizon is red or yellow because it too is illuminated with light whose long traversal of the atmosphere depletes the blue. Dust from a variety of sources (e.g., volcanos, forest fires) can not only scatter additional light, but can also display a different wavelength dependence than the Rayleigh scattering. Sunsets and sunrises after a major volcanic eruption can be brilliant (and also lead to the blue sun and moon of FC 5.84). The particular hues seen in any particular sunset are due to a combination of the normal Rayleigh scattering and the dust scattering.

5.60 The purple light is due to dust at an altitude of about 20 km in the atmosphere. Some of the sunlight passes through a layer of the dust, comes out underneath the layer, and then because the layer is curved around a spherical earth, reenters the layer. Its first passage through the layer scatters out most of the short wavelength (blue and green) light, so the light reentering the layer is red. Some of this reentering light is then scattered by the dust to an observer. That observer also receives blue light from sunlight scattered by the atmosphere above the dust layer. (In other words, the blue light of FC 5.59.) The combination of red light due to the dust and the normal blue light gives an overall purple for the observer. The second purple light is believed to be due to a second dust layer at an altitude of 70 to 90 km.

5.61 The enhanced zenith blueness comes as a surprise. According to the Rayleigh scattering of FC 5.59, the zenith sky should be blue-green and then yellow as the sun sets. The missing element in a simple Rayleigh scattering picture is the absorption by the atmospheric ozone in the red end of the visible spectrum. With the red end removed, the blue end is left and appears richer. The geometry favoring such a blue enhancement occurs when the sun is about 6° below the horizon and the light scatters from directly overhead the observer. The blue is further increased by dust in the atmosphere because the dust absorbs more of the red and yellow from the direct sunlight than the blue (see FC 5.84).

5.62 Next to the earth's shadow the sky is illuminated with sunlight from which all of the short wavelength (blues and greens) have been subtracted (FC 5.58).

5.63 The particulate matter (industrial pollution, smog) with radii smaller than about 0.2 micron scatters the blue light out of the light reaching the distant observer.

5.64 The sky is bright because of the scattering of sunlight by the air molecules. But there is a problem. For every molecule scattering light to an observer there is, on the average, another molecule on the line of sight and half a wavelength closer to the observer. The light scattered from these two molecules arrive exactly out of phase at the observer and therefore cancel each other. Since the argument can be repeated for any portion of the sky except directly toward the sun, the sky should be completely dark except in that special direction and except for the stars and planets. The argument has a slight flaw, however. Although the molecules can be paired this way on the average, they cannot be continuously paired because of the fluctuations in the distribution of the molecules. Were there no fluctuations, the sky would be dark.

5.65 Apparently nothing beyond a description of the effect has been published. The yellow glasses may be of advantage if the haze is composed of relatively small particles, smaller than about 0.2 micron in radius. Such small particles scatter the short wavelengths of light (the blue end) more than the long wavelengths (the red end). Thus the reds and yellows may illuminate the ground in a more direct beam whereas the blues and greens, having suffered more scattering in the haze, are more diffused. By eliminating the blues and greens, an observer might be able to see more of a shadow of an object in the field of view.

5.66 Instead of being easier to see, stars are harder to see through a shaft. Although most of the sky is blocked off, the sky surrounding the star's image is still just as bright as without the shaft. The shaft certainly cannot make the star itself brighter. Experimental work on viewing a small luminous test area in a large surrounding dark area indicate that the threshold in distinguishing the test area is decreased if the lumination of the surrounding area is increased until it is about the same as the test area. Instead of being more easily seen, stars are therefore harder to see through a shaft because their threshold of visibility is increased when the sky is blocked off.

5.67 If the water is pure and deep, it is blue due to the surface reflection of light from the blue sky. Shallower water has greater green because of reflection from the bottom. Contamination can add a variety of hues to the water because of selective absorption or, in the case of micron-size particle suspensions, because of the scattering of the light. This latter effect is similar to the scattering in FC 5.87 and 5.88.

5.68 The greenish tinge is added by the reflection from green vegetation. A similar upwelling of light adding features to an overcast sky is responsible for the sky maps of FC 5.72.

5.69 The illumination of the "dark" part of the moon (i.e., the part of the moon in a shadow from the direct sunlight) is due to earthshine, the light reflected from the earth's atmosphere and surface.

5.70 The scattering of light off objects much smaller than the wavelength of visible light follows the Rayleigh scattering model in FC 5.59. The water drops in the clouds are usually larger and merely reflect the sunlight from their outside surfaces. No color separation results from such reflection, and the scattered light remains white. (An exception is in FC 5.73.) Very dense clouds are black because little of the sunlight is able to penetrate them, being either absorbed by the water or reflected upward.

5.71 Sunlight scatters from water molecules as from any other molecule in the atmosphere, and so individual water molecules contribute to the sky brightness (FC 5.58). The scattering from water drops is greater than from an equal number of individual molecules because of their close spacing in the drop where adjacent molecules are about 1000 times closer than the wavelength of visible light. Consider two such adjacent molecules. When the sunlight oscillates their electrons, the oscillations are in phase because the electrons sample essentially the same part of the incident light wave. The electric fields radiated by those electrons are also in phase and constructively add to give twice the electric field as from a single atom. The intensity of the radiated light is the square of the amplitude of the radiated electric field and therefore must be four times that from a single atom. If these two molecules are well separated (much more than the wavelength of the light), there is no average constructive interference and the radiated intensity is only the sum of the intensity from each, that is, only twice the intensity from a single atom. The clouds are bright because of the constructive interference of the closely spaced water molecules in the water drops.

5.72 The maps are formed from the selective reflection of sunlight from the ice and water surface and then from the clouds. For example, the direct sunlight is reflected more from the solid ice than from the water. Since both reflect well, the difference in the reflections can be seen on the bottom of the clouds.

5.73 The mother-of-pearl clouds are composed of small drops whose radii (0.1 to 3.0 microns) are near or somewhat larger than the wavelength of visible light. Such drops scatter light according to the Mie model rather than the Rayleigh model of FC 5.59 for smaller drops or the simple reflection from larger drops. The diffraction of the light around the drops depends not only on the drop radius but also on the wavelength of the incident light and is therefore responsible for the beautiful colors. The clouds can be viewed up to two hours after sunset because of their high altitude: they are still illuminated for that long after a ground observer has entered twilight.

5.74 The interference fringes result from the scattering of the light by the dust particles on the front surface of the mirror. Consider two rays. One is scattered by a dust particle before entering the mirror and being reflected in the normal way. The other is first reflected in the normal way and then is scattered by the same dust particle on leaving the glass. Since the paths taken by each were slightly different, the rays can have a variety of possible phase differences when observed together, depending on the angle of the rays and the wavelength (color) of the light. Thus, the light shining on the dusty mirror produces an interference pattern for the observer, with the colors separated because of the wavelength dependence in the phase difference of interfering rays.

5.75 The light beam dims not only because of the spreading of the beam but also because of the attenuation of the beam by the scattering of the atmosphere. (Were there no scattering, the beam would be invisible.) The attenuation exponentially diminishes the beam's intensity, giving a rather abrupt end to the beam.

5.76 Both the zodiacal light and the gegenschein are due to the scattering of sunlight from interplanetary dust, probably derived from the astroidal belt. The dust giving the zodiacal light lies inside the earth's orbit and is visible only under the special circumstances described in the problem. The gegenschein is sunlight backscattered from dust outside the earth's orbit.

5.77 The windshield wiper rubs circular groves of gummy dirt on the windshield that then reflect light to the driver. The brightest reflection occurs when the incident light ray is perpendicular to a tangent to the grooves. The collection of these brighter reflections forms the streak of light along a radius of the curve through which the wiper moves.

5.78 The brown is primarily due to selective absorption by the nitrogen dioxide in the hazes.

5.79 The glory results from light backscattered to its source by small particles whose radii are near or somewhat larger than the wavelength of visible light. The scattering is described by the Mie theory rather than the Rayleigh theory for smaller particles (such as in FC 5.59) or by the normal reflection and refraction models for larger particles. The light that is returned in the direction of the light source enters a drop on an edge and exits from the edge on the opposite side of the drop after suffering both a reflection within the drop and a skimming along the drop's surface. That skimming, which is described as being a surface wave, is not part of the standard ray optics used in modeling a rainbow (FC 5.32). The return angles for different incident colors are slightly different and thereby produce the distinct colored rings around the shadow of the observer's head. Since the angles involved in a particular pattern depend on the size of the drops, the colors are lost if the drops in the cloud have a large range of sizes.

5.80 The solar and lunar corona are the diffraction patterns of the light from the small water drops in the line of sight. The diffraction is handled best with the Mie theory as in (FC 5.79) but can be approximated with the conventional diffraction of light around a small sphere (as in FC 5.96 and 5.98). In the latter interpretation, the light rays passing opposite sides of a water drop interfere with each other on the other side of the sphere to create bright and dark rings corresponding to constructive and destructive interference. The angular positions of the bright and dark rings depend on the drop size and the wavelength of the light. If the drops are uniform in size, then rings of different colors can be distinguished, the longer wavelengths (red) on the outside and the shorter wavelengths (blue) on the inside of a ring.

5.81 The frosty glass corona results from the scattering of light as in the preceding problem but with a few changes. Instead of being randomly spaced spheres, the drops are now fairly uniformly spaced, flat drops.

5.82 The Bishop's Ring is a diffraction pattern from small particles, usually dust from volcanic eruptions, in which settling has sorted the suspended particles to a uniform size. As in FC 5.79 the Mie theory of scattering is needed to calculate the intensity of the scattering pattern for these particles whose radii are about the wavelength of visible light.

5.83 Colored rings can be seen surrounding street lights even if the night is perfectly clear. Similar to the corona in the previous problems, these corona result from diffraction of light around small objects which are about the wavelength of visible light in size. The difference in this corona is that the objects are inside the eye. Some of the possible diffracting objects responsible for these entoptic halos are the radial fibers in the crystalline lens or the mucus particles on the corneal surface. Similar entoptic diffraction patterns are in FC 5.96.

5.84 In contrast to the Rayleigh scattering in FC 5.59, the blue moons are due to the scattering of light by atmospheric aerosols whose radii range from 0.4 to 0.9 micron (which contains the wavelength range of visible light). Particles in this size range scatter the long wavelength end (red) of the visible range more than the short wavelength end (blue). As a result, a normally white moon seen through such an aerosol will have the red end of the spectrum scattered out, leaving the observer with the blue end and thus a blue moon. These aerosols are occasionally produced by volcanic eruption or combustion as in large forest fires. The selection of this size range can result from the particles settling in the atmosphere or from condensation increasing the size of small nuclei until they are within this size range.

5.85 In spite of several studies, the value of yellow fog lights is not clear. If the particles are smaller than about 0.2 micron in radius, the blue will be scattered more than the red end of the visible spectrum. Yellow light would therefore better penetrate the fog because it has a longer wavelength than blue and green. However, for the particular size range described in FC 5.84, the result could be the exact opposite. For even larger particles as in a true fog, the yellow would be of no advantage. One problem in these general conclusions is that the absorption of the light by a particular type of suspended particle could be important.

5.86 The blue results from the scattering of light from very small particles, smaller than the wavelength of visible light. They can be macromolecules (large molecules) of terpenes released by the vegetation. Or they can be particles released from the tips of the vegetation (e.g., the tips of a leaf) where the electric field is relatively high (FC 6.33) and brush discharge (FC 6.46) might occur. The scattering is adequately modeled by Rayleigh scattering (FC 5.59) in which the blue is scattered more from the direct sunlight than are the other colors. If the particles are closer to being the size of the wavelength of light, then the Mie scattering theory is needed to predict the exact scattering intensities and color.

5.87 In order for you to see your own shadow in the shallow water, you must be able to distinguish the light reflected from the top surface of the water. In clear water that relatively feeble light is lost in the light reflected from the water's bottom. With mud in the way, the bottom light is partially or totally eliminated, allowing you to distinguish the light and dark areas from the surface reflection. To see the shadows of other people, there must be more elimination of the bottom light.

The colorful edges on shadows result from the scattering of light by small particles suspended in the water. They, like the atmospheric particles of FC 5.59 and the other particles in FC 5.88-5.90, scatter the shorter wavelengths (blue). Consider viewing a shadow near your own shadow. The near side is blue because this particle scattering is distinguishable against the dark background of the shadow. The far side of the shadow is red because light from that side has had the blue component removed by the scattering.

5.88-5.90 In each of these cases, light is scattered from very small particles. As in the arguments for the blue sky (FC 5.59), the Rayleigh theory of scattering is applicable for particles smaller than the wavelength of visible light. In such cases blue light is scattered more than red light. Thus, the suspended milk globules and the smoke particles appear to be blue when seen while looking from the light source or from the sides, but appear to be red or yellow when seen while looking toward the light source. (In the campfire smoke case, the smoke is first seen because of sky light coming from behind the viewer and then later seen because of sky light in front of the viewer.) If the particles are near the wavelength of visible light or somewhat larger, then the Mie theory of scattering is needed. (The Mie theory is used in FC 5.73 for direct backscatter.) When cigarette smoke is inhaled, condensation in the mouth increases the drop radii such that they are comparable to the wavelength of light, and yellow is preferentially scattered instead of blue.

5.91 The colors in a soap or oil film are due to the same type of thin film interference that is responsible for the blue Morpho wing in FC 5.94. Briefly, light reflected from the first surface of the film interferes with the light reflected from the second surface. Whether the interference is constructive or destructive depends on the wavelength of the light, the refractive index of the film, and the path length of the light inside the film. If the film is less than about one fourth the wavelength of light and has air on both sides, as with a soap film held on a vertical wire hoop, the interference is destructive and the film appears dark to an observer on the same side of the film as the light source. As one considers progressively thicker films, the interference gives progressively longer wavelength colors for those experiencing constructive interference in reflection. If a thin soap film is held vertically and illuminated with white light, the top portion may be thin enough to be dark. Further down the film are the colored bands corresponding to the constructive interference. But, mysteriously enough, just below the dark portion is a white strip. For the film thickness there, partially constructive interference comes from the entire range of visible light and therefore gives the observer a white light reflection (Bayman and Eaton, personal communication).

5.92 The source of these rings have apparently not been investigated but they seem likely to be the entoptic halos of FC 5.83 caused by mucus particles or small water drops on the surface of the eye.

5.93 Liquid crystal, which is a material somewhere between being a liquid and being a solid, comes in three basic types, depending on its molecular arrangement. The smectic type has its molecules aligned in the same direction and in parallel layers. The nematic type has similar alignment in a single direction but lacks the layering. The third type, the cholesteric one, is responsible for the colors in the liquid crystal toys now being sold in the United States. This type contains layers of molecules in which the molecules of any one layer are aligned in the same direction parallel to the plane. The alignment direction shifts from layer to layer such that a vector giving the alignment would rotate in a helical path as one considered deeper and deeper layers. The pitch of this helical path determines the wavelength of the light strongly reflected by the cholesteric liquid crystal. Other wavelengths merely pass through the crystal. By applying pressure to the crystal (the toy comes packaged in a flexible plastic container) or by changing the temperature of the crystal, one can alter the pitch angle and therefore vary the color selectively reflected by the crystal.

5.94 The rich blue from the top surface of a Morpho butterfly wing is due to thin film interference of light from thin terraces that lie almost parallel to the wing and on a support structure sprouting almost perpendicular to the wing. Of the white light striking such a thin terrace, part is reflected (call it ray A) and part is transmitted into the terrace. Of the latter, part (call it B) is reflected from the bottom surface of the terrace and emerges upward along with A. Rays A and B can interfere with each other constructively or destructively depending on their phase difference. That phase difference depends on the wavelength of the light, the thickness of the terrace, the refractive index of the terrace, and the angles of the light entering and leaving the terrace. For light incident perpendicularly to the terrace, the wavelength corresponding to blue light will produce emerging rays A and B in phase so that they constructively interfere. Rays A and B of all of the other colors of the incident white light more or less destructively interfere and do not contribute noticeable light to the observer. As a result the observer sees blue. As the observer changes either the angle of view or the angle of incident light, the path length of the light inside the terrace changes, thereby varying the phase difference between rays A and B. The wavelength giving maximum constructive interference changes as a result, and a slightly different blue hue is seen.

5.95 The dark lines between your fingers are the dark fringes of the diffraction of light through the space between the fingers. The light passing through part of the slit, say adjacent to a finger, interferes destructively with light passing through a different part of the slit, say further from that finger, to give a dark line at the observer.

5.96 The eye floaters are the diffraction patterns of light passing spherical blood cells floating just in front of the fovea on the retina. (The fovea is a pit of densely packed cones directly opposite the opening of the eye.) Blood cells loosened by old age or violent blows to the head swell to spheres by osmotic pressure when they float in the watery matter in front of the retina.

5.97 If the photograph is exposed just right, it might record spikes of the stars because of their twinkling (FC 5.102). Spikes can also be present because of the diffraction of the star light by the straight sections of the camera's aperture. These apertures are not perfectly round but are composed of many straight edges so that the aperture width is adjustable. The human pupil is not perfectly round either, and diffraction around the straight edges creates spiked stars there too. The spikes must always appear in pairs.

5.98 The interference rings shown in Figure 5.98 are the diffraction pattern of the light by the sphere. Light passing one side of the sphere interferes with light passing the other side to give bright and dark rings on a distant screen. The center of the pattern is equidistant from the two sides, and the light rays from the two sides arrive in phase to give constructive interference.

5.99 and 5.100 The cause of shadow bands is not well understood. Probably the best current explanation is that they are interference patterns of light that traverses air cells of varying densities. Such cells in the upper atmosphere could be naturally occurring turbulent cells.

5.101 The lake acts as a single slit, and the observer flies through the maxima and minima of its diffraction pattern.

5.102 Stars twinkle because the air is turbulent, which ultimately is due to uneven heat distributions in the atmosphere. Small turbulent cells, several centimeters or more across, are constantly present to refract the passing starlight first one way, then another. This small shimmy is noticeable with the small star images but less noticeable with the larger images of the moon and the planets.

5.103 The ultraviolet light is absorbed by the organic molecules in the pigments and alters their molecular bonding, eventually eliminating most of the color properties of the pigment. This ultraviolet fading of colors was discovered to be one of the most serious threats to paintings hung in modern museums that had begun to install common fluorescent lamps for uniform illumination. The lamps also emitted appreciable ultraviolet light. Now, either the lamps or the paintings are filtered to remove the ultraviolet light, or the museum has returned to incandescent bulbs.

5.104 Light has momentum and can therefore exert a force. The laser used in this type of experiment provides an intense beam of light that exerts sufficient force on the sphere to raise it. The stability is due to the refraction of light by the sphere. The laser light is most intense in the center of the beam. Consider a sphere somewhat off center but still in the beam. Light entering the sphere near the beam's edge refracts into the sphere, propagates across the sphere, and then refracts out of the sphere toward the center of the beam. That light beam has suffered a net deflection, and therefore must exert a force on the sphere. Light entering on the side near the beam's center suffers similar deflection but toward the edge of the beam. Both deflections provide lift to the sphere. Both also provide sideways forces. But the light deflected toward the center is less intense than the light deflected toward the edge. Thus, the net sideways force is toward the center. If the sphere wanders from the center, this net sideways force brings it back.

5.105 The dark and bright bands are the diffraction pattern from the screen. You can see similar but more colorful patterns by viewing a light through an umbrella's fabric.

5.106 The range over which a star radiates light depends on its surface temperature (to the fourth power when expressed in Kelvin degrees). The higher the temperature of the star, the lower the wavelength at which the peak of its radiation occurs. A cool star may have an insignificant amount of radiation in the visible range. As one considers progressively hotter stars, the radiation range enters the visible range from the red end. Thus, a star may have only red or red-yellow radiation in the visible if its temperature is just right. A hotter star could have its peak in the center of the visible and then emit all the colors approximately uniformly. Such a star would be white, as is the sun. A still hotter star could have its peak in the ultraviolet and emit more blue light than the other colors in the visible and therefore appear somewhat blue.

5.107 As yet the lights associated with tornadoes are not explained. In fact, tornadoes themselves have not been explained (FC 4.68). Most likely the light results from the electrical discharges present in the tornadoes.

5.108 The light is from molecules excited by charge differences on the crystal planes as the crystals are fractured in the pounding and scrapping of the stirring.

5.109 Solar ultraviolet light is responsible for both tanning and sunburning. If you are exposed to large or lengthy doses of ultraviolet light, both the dermis and epidermis of the exposed skin may be damaged. As a result the capillary vessels dilate and bring more blood to the skin surface, both reddening and warming the skin. Shorter exposures to the ultraviolet light tans light-colored skin by first oxidizing a pigment normally without color and then by activating (perhaps indirectly by deactivating an inhibitor) tyrosinase. This activation increases the amount of melanin, a brown or black pigment. The melanin protects the nuclei of skin cells by forming a layer over the cells which filters out the ultraviolet.

Suntan and antisunburn lotions come in three main types. Some (with zinc oxide or titanium oxide) screen all the ultraviolet and visible light and provide no tanning but do protect sensitive skin. Others (e.g., benzophenones) absorb all of the ultraviolet and also provide no tanning. The third group (containing substances such as aminobenzoic acid) provides both tanning and protection against sunburn by selective absorption. The ultraviolet wavelengths extend from about 0.28 microns to about 0.40 microns. Shorter wavelengths do not pass through the atmosphere, and longer wavelengths are in the visible. Of this range, the wavelengths between 0.29 and 0.32 microns are the most efficient in sunburning, whereas the wavelengths between about 0.31 and 0.40 microns are the most efficient in suntanning. The point of the third class of lotions is to filter out the wavelengths below about 0.31 microns.

Sunburn and tanning are less likely in the morning or late afternoon because the sunlight must traverse a greater path through the atmosphere and more of the ultraviolet light is absorbed. Glass also absorbs the ultraviolet wavelengths. Sunburn is more likely on high mountains because of the shorter pathlength of the light through the atmosphere. It is more likley at the beach because of the ultraviolet reflections from the sand.

5.110 and 5.111 In each of these luminescent cases two materials are responsible for the light production. These materials are given the general names of luciferin and luciferase, but what they are differs among the organisms. The luciferase is just an enzyme, a biological catalyst for the reactions in the light production. Marine creatures can be luminescent in three ways. They may have cells designed specifically for that purpose, perhaps even advanced enough to resemble lanterns. They might instead issue luminescent clouds. Or they actually may not be luminescent themselves but play host to luminescent bacteria. The dinoflagellates turn the sea red, yellow, or brown during the day by their natural color, but at night they glow blue when disturbed. Some internal clock regulates that light. Dinoflagellates kept under continuous dim light still will have their maximum light output about 1 A.M. when disturbed. Such rhythm may last for several weeks under the dim light. Fireflies set off a series of chemical reactions to produce light. Their conversion of energy to light is 100% efficient in that one photon is emitted for each luciferin molecule oxidized in the chemical reactions. The light is said to be "cold" light because, in contrast to incandescent bulbs, candle fire, red-hot pokers, and so on, the firefly light does not result from high temperatures and rapid thermal agitation of the molecules. Bacterial light, which is responsible for much of the luminescence reported for food, is a part of the natural process of that bacteria in obtaining energy from nutrients.

5.112 This type of glass contains small crystals that react to the illumination. For example, if the crystals are silver bromide, then the light transforms the silver ions to silver atoms that then darken the glass. The silver atoms are still trapped near the bromide, however, so as soon as the light is dimmed, the two recombine, and the darkening is reversed.

5.113 The black-light poster fluoresces by absorbing ultraviolet light and then radiating visible light. Under an ultraviolet light the poster seemingly glows without any stimulus. Whiter-than-white soaps do almost the same thing. They convert the natural ultraviolet to blue light and thus add to the visible light radiated by the materials containing the soap product. In the advertising jargon, the resulting "white" (meaning visible) light is more than the natural visible light.

5.114 Electrons emitted by an electrode collide with an atom of mercury vapor, exciting one of the outer electrons in the atom. That excited atom quickly deexcites to return the electron to its former energy level. As a result of the deexcitation, the atom emits ultraviolet light that is absorbed by phosphor crystals coating the inside of the tube. The deexcitation of the crystals results in the visible light we see. The crystals should emit some light for at least two cycles of the 120 cycles a second at which the maximum discharge occurs (the rates can differ with the country). If the crystals deexcited instantaneously after being excited, then the tube would have an intolerable flashing.

5.115 The speckle is an interference pattern of incident parallel rays (spatially coherent light) scattering from the very small structures on the diffuse scatterer. Spatially coherent means that the phase of the light emitted by one portion of the light source is correlated with the phase of the light emitted by another portion. Such a correlation is needed to maintain a stationary interference pattern. To some extent direct sunlight is spatially coherent and can produce these speckle patterns. The apparent motion of the pattern when the observer moves is due to the parallax present because the observer's eyes are not focused on the scattering surface. For example, if the scatterer is several meters away from the observer and if the observer is nearsighted, then the observer's eyes are focused in front of the scatterer. As the observer's head moves to the left, for example, parallax causes the speckle pattern to appear to move to the right.

5.116 In each of these two cases humming creates a stroboscopic image of the TV screen or rotating disc on the retina. Each source has periodic changes in appearance. The disc turns. The TV has a recurring image because of the line-by-line, horizontal sweep of the electron beam exciting the screen. A correct humming frequency oscillates the head, and thus the eye, appropriately to bring the recurring image back to the same place on the retina. The image then looks frozen. If the humming is not at such a frequency, then the head and eye oscillations are out of synch with the TV or disc, and the recurring image appears to migrate. For example, if the humming is at a frequency slightly too high for a frozen image, then the pattern on the disc will appear to move backwards against the sense of rotation of the disc.

5.117 The elliptical motion of the pendulum is seen because the eye covered with the darkened filter experiences a delay by several milliseconds in its perception of the pendulum's position. In the brain's interpretation of the pendulum positions perceived by the two eyes, the pendulum is placed either closer or further away than its true position. Thus, the pendulum swing is interpreted as being two dimensional rather than one dimensional. For example, suppose the pendulum is swinging to the right while viewed with the left eye having the darkened filter. The right eye perceives the true position of the pendulum, whereas the left eye perceives it as where it was several milliseconds previously. You mentally extrapolate the light rays from these two positions backward until they converge so that they make sense as having come from a single object. This extrapolation means the pendulum will appear to be further away than it really is.

When the pendulum swings back to the left, a similar lag in perception occurs for the covered eye, and the brain interprets the pendulum as being closer than it really is. Overall, the pendulum appears to swing in an ellipse as in the right-most sketch of Figure 5.117. The cause of the visual latency is not well understood. One analogue for the visual system is a series of delay line filters in which the time resolution of the system is improved by an increased feedback signal when the eye is subjected to greater illumination. A decrease in illumination reduces the feedback and thereby worsens the time resolution.

5.118 The entire screen of the TV is not continuously lit, but is constantly swept horizontally line by line with the electron beam in the TV tube. The sweeping light can act as a stroboscope in lighting the top, causing you to see a frozen image of the top surface, or an image that turns one way or another, all depending on how the sweep frequency compares with the rotation rate of the top.

5.119 Rods (which are primarily used in low levels of illumination) are packed the densest compared to the cones (which are for greater levels of illumination) toward the periphery of the retina. Staring directly at a star places its image on the fovea, in which there are no rods. Jumping your eyes away from the star sweeps the star's image across the greater packing of rods and enhances the perception of the star.

5.120 Blue arcs are still under research. Although their cause is poorly understood, they appear to result from neurones excited by those axons directly stimulated. Suppose a stimulating light excites a particular set of photoreceptors that are linked to a particular set of ganglion axons. That set of axons can then excite other nearby neurones, which then stimulate the photoreceptors to which they are linked. The result is an apparent arc of stimulation stretching away from and around the fovea (the central pitlike region on the retina) with one end on the point under direct stimulation. The cause of the blueness has not been discovered.

5.121 Apparently the production mechanism for phosphenes is not understood at all, because there has been almost no work published on modeling the phenomenon. The physical source has not even been identified conclusively, although some research has shown direct electrical stimulation of the occipital lobe at the rear of the brain can result in phosphene images.

5.122 All the lamps should turn on at the same time because the lag of electricity is imperceptible. However, the street lamps at an intersection appear to turn on sooner because they collectively provide more light to the observer and therefore suffer less visual latency (FC 5.117) than the street lamps between the intersections.

5.123 and 5.125 The elaborate network of blood vessels on the retina creates distinct shadows on the retina, but those shadows are rarely seen because the brain ignores any constant image from the eye. The vessels remain fixed with respect to the retina, their shadows remain constant, and thus you do not "see" the shadows. Exceptions occur in two cases. Upon first opening your eyes in the morning, the sudden casting of these shadows is a change and thus will be seen briefly until the brain fades out the information as being from a constant image. The other exception is that the blood cells carried by the retinal capillaries can cast shadows that are seen because the cells jerk their way through the capillaries. These cell shadows are the jerky specks in your field of view when you stare at a featureless illuminated area.

5.124 The cause of these geometric designs is not well understood and is currently being researched. They could develop at the retina or in the neural pathways. They could also depend on the interaction of the information of one eye with that of the other eye. One type of pattern, roughly geometric but lacking precision and complexity, appears to be dependent on monocular vision. The more complex patterns, on the other hand, appear to result in part from binocular vision.

5.125 See FC 5.124.

5.126 Vision in bright light is by the cones on the retina, whereas in low illumination vision is by the rods. These do not have the same spectral response. The peak response of the cones is in the yellow (corresponding to a wavelength of light of about 0.56 micron) with a much lower response in the blue. The rods respond best in the green (wavelength of about 0.50 micron for peak response) and have a much lower response in the red. If you observe red and blue while the room illumination dims from being initially bright, your vision shifts from using the cones to the rods, and the relative response to the blues and reds changes drastically.

5.127 The bright and dark bands, Mach bands, are illusionary. They can be photographed in the sense that an observer sees the illusion as readily in a photograph of a shadow edge as with a real shadow edge. The current theory for the band production involves an inhibitory effect in the neural network of photoreceptors being stimulated by the incident light near the shadow edge. In short, the signal of a firing photoreceptor inhibits the signal of a neighboring photoreceptor also being stimulated. Consider a shadow edge on the retina. On one side, say the left, the retina received a uniform bright illumination. On the other side, the right one, the illumination is uniformly dimmer. In a short intermediate region, the illumination drops from the bright level to the dim level. On the bright side of this intermediate region a bright Mach band appears whereas a dark one appears on the dim side. In the uniformly lit region all of the photoreceptors are inhibited by their neighbors. So, the signal level from this area is less than would be expected if there were no inhibition. A photoreceptor on the edge of the intermediate region is less inhibited because on one side, in the intermediate region, there is less illumination. Hence a bright band is perceived there. A photoreceptor on the other edge of the intermediate region, that is, bordering the uniformly dim area, is more inhibited than the receptors in the dim area because of the illumination in the intermediate region. Hence, a dark band is perceived there.

5.128 The Land effect is not well understood in its complete process but is currently modeled by supposing that there are three types of cones on the retina that are distinguished by where in the visible range their peak response lies: one type each for the short-wavelength, middle-wavelength, and long-wavelength ranges. When you observe a color scene, each set of cones somehow measures the reflectance of light from the scene in those three wavelength ranges, compares those reflectances, and then creates your color sensations. Color results from the black-and-white slides described in FC 5.128 because the reflectance information at two different wavelengths is apparently sufficient to trigger color responses. Hence, the colors you perceive may be almost independent of the wavelength of the light you receive.

5.129 The colored edges of the light source are due to the chromatic abberation of the eye. In the arrangement of Figure 5.129, the finger blocks the right side of the viewing eye such that the observed light rays from the window enter the eye only on the left side of the eye. Upon entering the eye, the red rays are refracted slightly less than the blue rays. Although the aberration is normally not noticeable, the partial blocking of the eye with the finger allows the viewer to distinguish the different images of the window edge for different colors. Consider light radiated from the right side of the window and entering the left side of the eye. The small difference in refraction for red and blue light results in the red image of the window edge being slightly to the left of the blue image on the retina. The blue image is lost in the white light coming from the window somewhat away from the edge. Thus, the observer sees a reddish window edge. A similar separation of colors occurs for the left side of the window, but this time the red image happens to be lost in the white light from the rest of the window, and the viewer is left with a bluish edge.

5.130 Previous to current research, the colors were thought to result from the difference in times needed to turn on the color responses in the visual pathway after the observer viewed darkness. In particular, the theories concluded that red turned on slightly sooner than blue, and thus the leading edge of the white area was red. However, recent research indicates no such difference in turn-on times for the different colors. One of the current theories about the color response to the disc is that the pattern of its intensity variations either mimics or creates a photoreceptor firing pattern that mimics the brain's color coding. In other words, the proper pattern of intensity changes sends a Morse-code-like signal to the brain telling the brain it sees a particular color for that particular code. As in FC 5.128, seeing red does not necessarily mean that you are viewing light with a wavelength of about 0.6 microns. Color perception appears to be far more intricate.

5.131 Whereas the blues and greens from the fluorescent lamps turn off during part of the cycle in the 120 cycle-per-second stimulation of the lamp, the yellows and reds (if the lamp has much red) do not. As a result, a spinning black-and-white disc or a spinning coin will reflect colors to the observer that change with time. The difference in the color duration comes from the three types of emissions present in the lamp's output. The short-lived blues and greens come primarily from the mercury emission lines (which have very short lifetimes) and a phosphorescence also with a short lifetime. The yellows come mostly from a long-lived fluorescence from the same phosphor.

5.132 The picture on a TV screen is not created whole but is quickly produced by the electron beam in the TV tube moving horizontally line by line downward until the bottom of the screen is reached. The sweep is so fast that you don't perceive it. If you swing your eyes to the right, each horizontal sweep leaves an image on the retina for about 75 msec or so. Because your eyes move during the beam's line sweep, the lasting image of a particular line is slightly to the right of the line just below it because the top one was made slightly sooner. The overall lasting image of the screen is therefore tilted as shown in Figure 5.132. Multiple images are observed because during the full swing of your head the electron gun has filled the screen several times, each full image giving you an image to carry briefly off to the right.

5.133 Most of the stereoscopic illusions depend on imitating normal binocular vision that gives depth in our normal viewing. In the stereoscope (the device now bought only for children but that once provided much evening entertainment for all) contains two photographs shot with camera positions separated by a few centimeters and with an appropriate angle adjustment to mimic our normal viewing. When you examine the photographs with the stereoscope, the different images are fused to make a single image with apparent depth. Three-dimensional movies depended on providing a similar shift in perspective for each eye. For example, the two images might be projected, one in blue and the other in red. The audience would then wear glasses with blue cellophane over one eye and red cellophane over the other. Again the two images are fused to provide a single one with depth. Different polarizations of the projected light could be used instead of the different colors. The glasses would then have the left side passing a different polarization than the right side. Three-dimensional postcards employ a single picture but provide different images for the left and right eye by a grid of prisms or furrows in a plastic sheet overlaying the photograph. Because of the tilted surfaces of the plastic overlay, the left eye receives a different perspective of the photograph than does the right eye, resulting in depth when the two are fused.

The illusion that the red letters on a blue background are in front of the background appears to be due to chromatic aberration of the eye. With the object viewed so that its light rays enter at some angle to the central axis of the eye, the blue rays are refracted more than the red rays. This difference means that only one of the colors can be brought to focus on the retina. The image in the other color will be a blur to one side of the sharp image. For example, consider viewing equidistant red and blue points on a card somewhere in front of your eyes. Suppose the red point is in focus. Its image lies further from the center of the head than does the blue blur from the blue point. Such a displacement in normal binocular viewing is interpreted as indicating the red point is closer than the blue point.

5.134 Exactly why the illusion is seen is not clear, even after much debate in the literature on the subject. The enlargement has nothing to do with atmospheric conditions (indeed, refraction decreases the size of the moon, as discussed in FC 5.18). Instead, the illusion appears to depend on the space between the moon and the horizon. For a very large space, corresponding to a large angular elevation of the moon, the moon appears to be its proper angular size of 0.5 arc degree. As the moon descends and the space between it and the horizon (or the objects on the horizon) decreases, the moon appears to grow.

5.135 The rays are actually parallel. Their apparent meeting at some point in the distance is an illusion. A similar illusion can be seen by standing in the middle of a very long, straight train track. If you pretend not to know about the distant track, it will appear to converge at some point on the horizon.

5.136 If you bisect the moon with a long stick, it will cut through the sun just as it should. But without such a reference, the mentally extrapolated line bisecting the moon will miss the sun. The illusion results from your perception of the sky as an overhead spherical dome.

5.137 The apparent bending of the searchlight beam is an illusion that depends on the illusion of the sky being an overhead dome just as in the previous problem. Holding a straight edge along your view of the beam will convince you the beam is really straight.

5.138 Apparently an explanation of why the illusion occurs is not available. It is a surprising illusion in that previous to its published description, an elevated object equidistant with a similar level object was thought to always appear to be more distant.

5.139 Whiteout can come in two different cases. A ground blizzard may whip up all the loose snow to limit visibility to just a few feet. With such restricted visibility, a person can become lost in the swirling snow after walking just a few feet. Another type of whiteout is the elimination of visual clues when the ground is covered with snow and the sky is filled with white clouds. The lighting becomes so diffused that no shadows are cast, and both the snow and the clouds appear to vanish. One might then have the impression of walking or skiing over a vast white emptiness. Permanent blindness might also result from constant viewing of snowfields under intense light because the visible and ultraviolet light can destroy part of the visual process. (W. C. Burkitt, personal communication.)

5.140 For an example, consider an astronaut viewing the earth from an orbit about 800 km (500 mi) above the surface. If he uses only his eyes rather than a telescope, signs of intelligent life are very difficult to detect. His eyes are diffraction-limited in their resolution, that is, the limit on how small an object he can resolve is imposed by the diffraction of light through the circular opening to the eye. (Some animals are limited in their resolution because of the photoreceptor spacing on their retinas. An image smaller than that spacing cannot be resolved.) For a typical human eye the smallest object that can be resolved occupies about 0.0005 radian in the field of view. This limit means that the astronaut can just barely resolve objects of about one kilometer on the earth's surface. Very few man-made features can be recognized with such resolution. Primarily the telltale signs of intelligent life are geometric structures, such as long straight superhighways. In an examination of thousands of photos from the Tiros and Nimbus meteorological satellites having resolution limits of about 0.2 to 2.0 km, only a highway and an orthogonal grid of some Canadian loggers were recognized.

5.141 A point source of light in a dark room appears as a light streak in the Christmas tree ball because the eye can accept reflected light rays in a significant range of angles. Mentally extrapolating those rays back into the ball gives the impression that the light originates from a line source. When the room lights are turned on, the pupil size gets cut in half and that acceptance range is decreased enough that the apparent line source is shrunk to a point.

5.142 The Moiré patterns appear according to how elements in the two overlaying grids, cloth, and so on, fall in step. For example, the comb and its reflection will periodically have teeth exactly overlapping, partially overlapping, and then not overlapping at all. The resulting Moiré pattern appears to be an enlargement of the teeth in the comb, or at least an enlargement of the same type of periodic structure.

6.1 Roughly speaking, the effects of current through the human body are the following:

less than 0.01 amp—tingling or imperceptible
0.02amp—painful and cannot let go (see FC 6.3)
0.03 amp—breathing disturbed
0.07 amp—breathing very difficult
0.1 amp—death due to fibrillation
more than 0.2 amp—no fibrillation, but severe burning and no breathing

The intermediate range of 0.1 to 0.2 amp is strangely enough the most lethal range for common situations, because this level of current initiates fibrillation of the heart, which is an uncontrolled, spastic twitching of the heart. The resulting disruption of blood flow quickly results in death. Above 0.2 amp the heart is merely stopped, and normal first aid procedure can restart it. But another, controlled electrical shock is the only way to stop fibrillation. Hence, the 0.1 to 0.2 amp range is more deadly than larger currents.

The current passing through a victim is usually determined by the skin resistance, which ranges from about 1000 ohms for wet skin to about 500,000 ohms for dry skin. The internal resistance is smaller, being between 100 and 500 ohms. Touching voltages higher than about 240 V usually results in current puncturing the skin. Often a person grabs a wire that has sufficient current to contract his hand muscles onto the wire. That level is initially not lethal, but the skin resistance drops with time until the lethal level of 0.1 amp is finally achieved. If you find someone "frozen" to a live wire but still alive, you should remove that person as quickly as possible without endangering yourself, or eventually the person **will** die.

6.2 An electric potential lies between the two metals where they contact, owing to the difference in the energy levels of the conduction electrons on the metals. When the frog leg touched the railing, a complete circuit through the railing, support, and leg was made, and electricity (those conduction electrons) flowed through the circuit. The current excited the muscles in the leg, forcing the leg to contract.

6.3 The wire does not capture your hand with an electrical force. Instead, the current through the hand muscles causes the hand muscles to contract, clamping your hand around the wire. Electricians working with live or potentially live wires often use the back of their hands or fingers to move the wires. If the touch does draw current, the muscle contraction then throws the hand away from the wire.

6.4 When activated by a nerve signal, the biological cells called electroplaques suddenly allow an ion flow—a current—across their membranes. The electric fish have a series of the cells from head to tail, and the combined electric potential from each during the ion flow (about 0.15 V between the interior and exterior of the cells) creates a voltage difference between head and tail. Many such series of cells are "wired" in parallel in the fish to provide sufficient amperage to flow externally from its head to tail to stun or kill its food or enemy. For example, the **Torpedo nobiliana** (a saltwater giant ray) has about 1000 electroplaques in series and about 2000 series in parallel. Were all the electroplaques placed in series, not only would the fish be rather long, but also the current through the series would be large enough to destroy the cells. By using a parallel wiring of the series, the current through each cell is kept low without sacrificing the external current. Fresh-water electric fish have greater numbers of electroplaques in series because they need more head-to-tail voltage to force the same current through the higher resistance of fresh water.

6.5 The microwaves are absorbed by the meat, mostly by the water in the meat, within a depth of one to several centimeters from the surface. The rate of absorption with depth depends on the frequency of the microwaves: the lower the frequency, the greater the depth. Most microwave ovens operate at a frequency of 2450 MHz, which penetrates and heats primarily the first two centimeters of meat. A microwave oven is designed to bathe the meat with microwaves from essentially all directions. Provided the meat is not too large, the amount of radiation reaching the center of the meat from all directions could be greater than the amount absorbed in the first centimeter on any one side. As a result, the center would absorb more than the outside layer of meat and would cook sooner. However, the result could be just the opposite if the piece of meat is large, or if the oven does not bathe the meat, or if the operating frequency is high and thus the penetration depth is small compared to the size of the meat.

6.6 The electrons move through the circuit at a relatively slow speed of 10^{-4} m/s, but the signal—the change in the electric field along the wire—moves at nearly the speed of light. It is the signal rather than the actual electrons from the wire in the switch that must reach the light. The signal may reach the filament in the bulb in as little as a nanosecond (10^{-9} s), hardly enough time to bother with. However, the filament must first be heated by the current through it before it can emit light. To emit visible light, the filament should be several thousand degrees Kelvin, and that temperature is typically reached 0.01 to 0.1 s after the switch is thrown.

6.7 When the two materials (e.g., shoes and rug or cat's fur and glass) touch, electrons from one of the materials tunnel through the electrical potential barriers on the surface to the other material. For example, when glass touches cat's fur, electrons tunnel from the glass surface to the fur's surface. Since neither are good conductors, this tunneling occurs only where the materials actually touch. To obtain more transferred electrons, more parts of the materials should be put in contact. Rubbing the surfaces together is the most convenient way of putting more parts of the materials in contact to obtain a greater transfer of electrons. The material losing the electrons is left positively charged; the other material becomes negatively charged. If the air is damp, the excess charge drains quickly to the airborne water drops. Smoke particles could also remove the charge. Without such discharge, normal contact of materials can produce fairly high potentials. For example, sliding across a car seat and stepping outside can leave you 15 kV higher than ground potential.

6.8 The apparatus must be arranged such that the water streams falling from the nozzles break into drops at about the level of the top cans. Initially, when the water first begins to fall, one can is charged negatively slightly more than the other can. Which can is more negative is sheer chance, because the initial charge difference is due to the charging by either cosmic rays or the earth's natural radioactivity. Suppose, for illustration, that the bottom-left can in Figure 6.8 is more negative. Then the top-right can will also be more negative than the top-left can because of the wiring. The right-hand stream is polarized as it falls through the top can. If the drops develop just then, the drops will be positive, the negative charge in the stream being repelled by the can around the stream. Since those drops fall into the bottom-right can, that can becomes more positive than before. Although the initial voltage difference between the bottom cans is trivial, some homemade Kelvin droppers develop a potential difference of as much as 15 kV.

6.9 The liquid stream breakup was first explained by Rayleigh, who showed that disturbances to the emerging stream would result in waves around the stream axis, these waves growing exponentially in amplitude until the stream disintegrated. Once broken up, the water pulls together by surface tension to form drops. The breakup can be avoided or delayed by the presence of a charged rod because of the resulting induced charge separation in the stream. For example, if the rod is positive, then the stream side nearest the rod is negative and that side furthest from the rod is positive because the rod's electric field has forced electrons to flow across the stream to be nearer the rod. With such a charge separation, the stream is less likely to break into drops. Suppose there is a separation into drops. Consider two freshly produced drops lying along a line radially extended from the rod. Because of the induced separation of charge in the drops, their adjacent side would be oppositely charged, and the two drops would attract each other. Thus, the drops do not form in the first place. If the rod is highly charged, then the stream's near side is so attracted to the rod that the stream is bent from its normal trajectory.

6.10 The charging of wire fences, airplanes, and similar metallic objects by blowing snow is the result of the same type of electron transfer as in FC 6.7. The metallic objects receive electrons from the snow particles and become negatively charged.

6.11 The tape separation appears to cause a separation of charge, the adhesive layer carrying away a different charge than the top surface of the tape to which it had just been attached. The glow is an electrical discharge between the two surfaces of tape that have just been separated.

6.12 The sugar is charged as in FC 6.10 and 6.7 when it is sifted. Since the falling sugar grains then have like charge, they repel each other, and some of the sugar is pushed to the side.

6.13 Contact between the tires and the road leaves the tires negatively charged. As the tires rotate and become uniformly charged, the negative electrons in the metal body and frame are repelled by the tires, leaving the body area near the tires positively charged. Sparking between a particular part of the car and some nearby grounded or oppositely charged body is then possible. The sparking would be merely a nuisance except in the case of gasoline trucks where gasoline fumes may be ignited. Years ago chains were dragged from the truck's body and on the ground in the belief that the chains would continuously discharge the truck. The chain **would** drain some of the electrons from the truck's body, but that would not leave the truck neutral and thus safe because it would then be positively charged and hence still susceptible to sparking.

6.14 Details of the charge separation in the splashing of water are not presently well understood. In the nineteenth century, however, Lenard showed that the larger drops floating in the air near the splashing were positively charged, whereas the smaller drops were negatively charged. Since the larger drops settle more quickly than the smaller ones, the air is left with negatively charged drops and a rather large electric field.

6.15 Apparently the effect of charge on a human being is not fully verified, much less explained.

6.16 The basic force preventing you from falling through your shoes, the floor, and the ground is the electrical repulsion between the atoms in each set of adjacent surfaces. (Also see FC 7.24.)

6.17 Four types of forces may hold a powder together. If the particles are less than about 50 microns, van der Waal's force (an attractive force between atoms) may be important. Electrostatic attraction of unlike charges on the powder grains can also hold the grains together. If the powder is damp, then the water can bond the grains by essentially a surface tension type of force. (But if there is too much water, the powder becomes just a slush.) And finally, if the particles are irregular in shape, their interlocking holds them together. Current research on powders and crumb formation attempts to explain the bulk cohesion of certain powders in terms of these forces on the microscopic scale.

6.18 Static electricity on the plastic wrap causes the wrap to cling to itself and to most food containers. For example, if the layer of plastic next to a metal wall has an excess of electrons, then it repels the electrons in the metal. The area next to the wrap is thus left positive and attracts the wrap. Since the wrap is not a good conductor of electricity, its static charge does not readily drain to the metal. As a result, the plastic clings. Some of the static charge originates when the plastic roll is being manufactured. Indeed, avoiding static charge during the production would be difficult. More charge separation can occur when you pull the plastic off the roll: a faster pull produces more charge. Humid air or a wet container drains the static charge and thus reduces the cling.

6.19 The ink in the dollar bill contains magnetic salts, probably iron salts, that are attracted to one of the magnet's poles.

6.20 The fluid in the leveler is diamagnetic, that is, when it is placed in a magnetic field, it produces a magnetic field in the opposite sense. The fluid is repulsed from the magnet, thereby forcing the bubble toward the magnet.

6.21 A changing current in the coil creates an accompanying changing magnetic field in which the ring lies. That field in turn creates a current in the ring such that the magnetic field of the induced current is opposed to the magnetic field of the coil. The second magnetic field supports the ring in the first magnetic field. If current is switched on, the sudden change of current in the coil induces a larger current in the ring and thus a larger magnetic field that may be sufficient to send the ring upward.

6.22 The alternating magnetic field induces currents in both the fixed sheet and the disc. Without the fixed sheet, the part of the disc over the magnet would fully have such induced currents, whose flow would be in such a direction that the magnetic field created by the flow would cause repulsion of the disc by the magnetic field from the magnet. Without the fixed sheet, the disc would therefore be repulsed by the magnet. With the sheet in place, however, the induced currents in it and in the unshaded portion of the disc attract each other (the magnetic field of one of the currents forces the other current closer), and the disc is continuously rotated.

6.23 The changing magnetic field from the rotating magnet induces currents in the aluminum disc. These currents in turn set up their own magnetic field. The interaction of the two fields creates a torque on the disc, causing it to turn in the same sense as the rotating magnet.

6.24 Why not build this simple device to see if it is truly workable? If the magnetic field is strong enough to start the ball up the inclined plane, won't it be strong enough to prevent the ball from sliding down the ramp underneath the plane?

6.25 The penetration depth of radio waves through the ionosphere depends on the frequency of the waves. The ionosphere is transparent to the relatively high frequency waves used in TV and FM radio transmissions. However, the lower frequency waves used in AM radio transmissions are reflected from the ionosphere. Thus, to receive a particular TV or FM program, the receiver must be near the transmitter to receive either a direct signal or at least a signal reflected from the environment (buildings, etc.). In the AM case, the receiver can be distant since it can use the signal reflected from the ionosphere. Occasionally the higher frequency signals are reflected from the ionosphere and can be received at surprising distances. Such reflections are likely due to the increased ionization in the ionosphere during meteor showers or during what is called sporadic-E conditions. The latter increase in ionization is not currently understood very well, but may be linked to increased radiation from the sun. The reflecting level of AM signals rises at night because the lack of sunlight decreases the ionization of the molecules on the lower side of the ionosphere. With a higher reflecting level, the AM signals travel further around the curve of the earth. To avoid an unmanageable mess of signals, the FCC requires most stations to cut their power or to leave the air during the night.

6.26 The receiving circuit resonates at a particular frequency that depends on the magnitudes of the capacitance (of the capacitor) and the inductance (of the coil). By varying the contact on the coil, the inductance can be altered, thus changing the frequency at which the circuit responds. The incoming signal is sinusoidal. Since the average power absorbed from a sinusoidal signal is zero, the listener would hear nothing were the signal not changed. The contact between the metal "whisker" and the crystal allows the current to flow in one direction only. Thus, the signal is rectified, because half the sinusoidal (say the negative part) is removed. With only half the sinusoidal wave in the circuit, the average power absorbed is no longer zero, and the listener can hear the signal.

6.27 The airplane reflects the TV signal to your antenna slightly later than the direct signal is received. The direct signal places an image on the screen; the later reflected signal places another, fainter image to the right on the screen—a ghost image—that changes as the airplane continues to move. The ghost is to the right because the electron gun producing the screen image scans from your left to your right.

6.28 The AM transmitters have vertical antennas. If the radio wave is directly received by the car's antenna, the electric field in that wave is polarized along the sense of the transmitting antenna, in other words, vertically. To gain the maximum signal strength, the receiving antenna should also be vertical.

6.29 In the FM region the appearance of the same signal at multiple places across the dial is due to the nonlinear response of the receiver in the presence of a very strong signal; the effect is called cross modulation. In the AM region, a signal can be overwhelmed by a stronger one normally at a different frequency if the receiver is near the transmitter of the second signal. Both the transmitter and the receiver work in a rather narrow frequency range. However, neither is exactly at a single frequency. For example, a transmitter may be primarily at 11,000 kHz, but it may also be transmitting a fraction of its power at 11,500 kHz. A distant receiver would not sufficiently amplify this very weak signal at 11,500 kHz for it to be heard. If a nearby receiver is tuned to another station whose primary frequency is at 11,500, it may also receive and sufficiently amplify the undesirable signal from the first station.

6.30 Low-energy solar electrons (with energies in the hundreds of electron volts) are swept into the plasma tail on the antisolar side of the earth, are somehow increased in energy (to several thousand electron volts), and then are directed into the atmosphere near the poles by the earth's field lines. Those lines enter and leave the earth at the magnetic poles, which are offset somewhat from the geophysical poles about which the earth spins. The energized solar electrons enter the atmosphere in an oval around the magnetic poles and excite nitrogen molecules and oxygen atoms by collision. Green light is emitted by deexciting atomic oxygen at altitudes from 100 to 150 km. Higher atomic oxygen produces a strong red light. Red light also comes from deexciting molecular nitrogen. These colors are observed in the ovals around the magnetic poles along the geomagnetic latitudes around 70°. With the north geomagnetic pole in Canada, this arrangement places auroras over southern Canada and northern United States. The same geophysical latitude over Siberia is at a lower geomagnetic latitude and has fewer auroras.

6.31 Lightning sends out electromagnetic pulses that are first heard directly as clicks on the detectors. Part of the pulse waves travel upward. Those waves traveling upward are concentrated into a beam in the ionosphere and are then bent over such that the beam travels along one of the magnetic field lines of the earth. When the beam reaches the opposite pole region, it is reflected by the stronger magnetic fields and returns along a field line to near the point of origin. But not all of the beam travels at the same speed: the higher frequency components travel faster. When the returning beam is detected, the higher frequencies are heard first, and then progressively lower frequencies arrive. Instead of the original click being repeated, this electromagnetic echo lasts longer and descends in pitch.

6.32 In a normal lightning stroke the charge distribution in the cloud is the following: small amount of positive charge at the base, large amount of negative charge in the lower middle, and large amount of positive charge in the top. The stroke begins with a discharge between the base and the lower middle, bringing electrons down to the base. This discharge proceeds from the base downward by a "stepped leader," which jumps in lengths of 50 m, pauses about 50 μs, and then jumps again. Each time, negative charge drains from the cloud to the bottom of the channel. Only the lower tip of the leader is visible, but the motion is so rapid at this stage and the following stages of the stroke, that the whole process appears to be luminous. The leader is crooked because the path of the descending channel is deviated by pockets of positive charge in the air. If a pocket is sufficiently strong, the leader may even be turned horizontal.

When the leader is near the ground, the electric field near sharp points is sufficiently high that electrical breakdown occurs and a positive return stroke starts upward to meet the leader. The point of meeting is highly luminous as the negative leader is neutralized and its electrons are brought to ground. This region of high luminosity and current propagates up the leader channel until it reaches the cloud, but the observer cannot resolve the rapid motion and sees a continuously luminous channel. The leader's trip down takes about 20 ms, whereas the return stroke takes only 100 μs. The light comes from the center of the leader channel, probably from a core no wider than a few centimeters.

6.33 The earth's electric field, which lies between the negatively charged surface and the positively charged upper atmosphere, should be discharged in 5 min or less because of the constant ionization of air molecules by cosmic rays and the earth's natural radioactivity. Some of the electrons from the ionization move to the upper atmosphere where at an altitude of about 50 km the conductivity is so good that the atmosphere is essentially a spherical conductor. The rising electrons will neutralize this positive conductor. Similarly, some of the positive ions from the ionization descend to the negative ground to neutralize it. Because the worldwide current resulting from the ionization is about 1800 amps, both the ground and the upper atmosphere should be discharged in a few minutes. They are not, because the worldwide lightning activity is constantly recharging the earth with electrons.

There may be a 200-V difference between the **heights** of your feet and nose, but your body is such a relatively good conductor that all of it is at essentially the same potential. Hence, no significant voltage difference exists across your body.

6.34 The cloud-to-air stroke terminates in a pocket of positive charge in the air. The ribbon lightning occurs when the wind is strong enough to move the ionized channel noticeably between strokes when more than just the leader and return strokes run between the cloud and ground. Bead lightning is not well understood. It may sometimes occur when the stroke is partially obscured by rain so that the observer is not blinded by the flash. Then, as the luminosity dies away, those portions of the stroke running along the observer's line of sight will last slightly longer because of the greater amount of light when such a portion is viewed on end. However, the bright beads may also occur for different, and as yet unknown, reasons.

6.35 The nature of ball lightning is still current research, and any explanation given here may be soon proven wrong. Probably the best explanation to date is that the ball is a plasma ball that is fed energy from external electromagnetic waves. Some electrical activity of a thunderstorm, lightning, or point discharge initiates ionization of the air or of a vapor gas (if something such as a metal conductor has been struck). The ionized gas remains integral because of its overall electrical neutrality, but grows to some equilibrium size because it absorbs energy from natural radio waves. Such waves are known to be generated either at the clouds or at the ground during intense electrical storms. The environment of the ball imposes constraints on the radio waves, creating standing waves, and the ball absorbs energy from an antinode in such a standing wave. An external source of energy like this is appealing, because the relatively long-living luminosity of the balls is otherwise very difficult to explain. If the light is only from an internal source of energy, and if a nuclear fireball is scaled down to the size of ball lightning in order to put an upper limit on such internal energy, the ball would glow for no more than about 0.01 s instead of the reported several seconds.

6.36 The H-bomb lightning strokes may have resulted from the charge produced when gamma rays from the burst scattered electrons from the air molecules. The leaders for the strokes apparently propagated upward from the instrumentation structures near the surface. Similar upward propagating leaders, which contrast with the downward leaders of normal lightning (FC 6.32), have been seen starting from tall structures such as the Empire State Building. Since the leaders run upward, so does the branching.

6.37 The hot lava hitting the seawater sends positively charged steam upward. After sufficient charge separation has occurred, the clouds of steam discharge back to the ocean, allowing electrons to flow upward through the ionization column. The upward flow of electrons is opposite the situation with normal lightning (FC 6.32).

6.38 Earthquake lightning is not well understood. Recently it has been attributed to piezoelectric fields created when seismic waves propagate through either the surface or low-lying rocks. (In the piezoelectric effect, electric fields are created in a material when the material is placed under stress. The diamond used in modern record players is an example of a piezoelectric crystal whose electric fields are created by the stress from the bumps in the record grooves.) Supposedly such electric fields would be large enough to cause atmospheric discharges on the surfaces. However, details and proof of this model are currently lacking.

6.39 The pointed wire was to provide a sufficiently high electric field that enough current would be attracted to allow Franklin to do his experiments. (The sharper the object, the greater the electric field is around it.) The silk ribbon was insulation between himself and the wet conducting twine. The key provided several sharp points for visible discharge of the electron current descending the twine. Often Franklin is pictured doing this experiment during a lightning storm. He was never that stupid. A lightning strike to the kite would have destroyed the kite, the twine, and possibly Franklin, regardless of a trivial piece of silk. In actuality, Franklin flew his kite before the full storm arrived.

6.40 The purpose of the lightning rod is to provide a safe route to ground for the descending current. The sharp point has a high electric field around it and can initiate the upward traveling channel that meets the downward traveling leader stroke (FC 6.32). Once contact is made, the electron current flows from the ionization channel and through the rod to the ground in which the rod is buried. The possibility of the stroke hitting the structure to which the rod is attached is thereby reduced. The lightning rod cannot appreciably discharge a passing cloud to avoid lightning because such discharge is too slow. The radioactive source proposed for the rods would have no effect and probably would prove to be dangerous if a lightning strike ruptured the source.

6.41 If the tree is thoroughly wet, the current descends through the water sheath and leaves the tree unharmed. If not, the current may enter the tree to descend through the sap. The rapid heating and expansion of the sap then blows the tree apart. Oak is more susceptible to explosions than many other trees because it has rough bark. If the lightning strike occurs early in the rainstorm, it may find only the top part of an oak wet, whereas a smooth bark tree would be wet to the ground. The oak would be blown apart, and the smooth bark tree left untouched. Forest fires are initiated by lightning strokes in which there is a continuous current running through the lightning channel between the main strokes, that is, between the first return stroke and the following ones. Since the continuous current is not always present, not all lightning strokes to trees would set the trees on fire.

6.42 The high frequency current of a lightning strike does not penetrate the metal walls of a car, airplane, and the like, but stays on an outside layer of the metal. Barring a puncture to the fuel and the consequent explosion, the occupants in such a metal enclosure will probably not even know that they have been hit.

6.43 At times the water droplets in the clouds are partially suspended by the local electric fields. A lightning stroke occurring then may decrease those fields, causing an enhanced fall of the droplets—a rain gush. As the electric fields regain their strength, the precipitation decreases.

6.44 The rapid evaporation and expansion of the moisture on your skin blows your clothes and shoes off. You may be otherwise unharmed if little of the current entered your body.

6.45 After reaching the ground, the lightning current spreads out and runs partially horizontal. If a cow stands as in Figure 6.45, an appreciable amount of the ground current enters the front legs and exits from the rear legs, electrocuting the cow. If you are caught outside during a thunderstorm, you should not lie down. If a strike hits nearby, the resulting electrical potential between your head and your feet may draw enough of the ground currents to kill you. Since you also should not stand up, the best position is to squat. That way you keep your head low while minimizing the contact area with the ground. With minimal contact area, the possible electrical potential from one side of the contact area to the other is least, and you will draw the least ground current.

6.46 and 6.48 Ground-level St. Elmo's fire and the Andes glow are both examples of corona discharge in which the electric field surrounding objects, usually pointed objects, is sufficiently high that electrical discharge can occur. The Andes glow is an especially strong type of corona, not fully understood yet. St. Elmo's fire can also occur on aircraft flying through rain or snow because of the charging discussed in FC 6.7, 6.10, and 6.14.

6.47 If a massive current enters a victim's body, the person will likely die because of internal burns. But the lightning may not penetrate the body if the person is wet. Then, most of the current descends through the water sheath on the body. (Wet trees struck by lightning may be completely unharmed. See FC 6.41.) In this case, the victim's breathing and heart may be stopped by the electrical shock, but quick application of artificial respiration can revive the person. Many times a lightning victim does not suffer a direct hit, but is hit by a side-splash from the object suffering the direct hit or is felled by the ground currents of the hit (FC 6.45). One reference suggests that most people dying from lightning do so only because rescuers prematurely give them up for dead. Hence, common first aid for electric shock should always be given a lightning victim.

6.48 See FC 6.46.

6.49 The pinwheel does not move because of something thrown off or pulled on, but because of the ionization of the air next to the point. Once the air is ionized in the high electric field of those points, the ions and the point have the same sign of charge and thus repel each other. The ionization and consequent repulsion occurs regardless of whether the pinwheel is charged negatively or positively.

6.50 The alternating high voltage induces an alternating current in nearby metal objects, which then may be discharged by an unwary person grounding part of the object.

7.1 As fascinating as the possibility of communicating with an extraterrestrial civilization is, one should remain somewhat skeptical about the occasional flood of "flying saucer" sightings—at least skeptical enough to retain the fundamental physics man has developed. Many of the sightings would have to violate those basic laws if they were truly sightings of machines. The gravity shielding scheme is untenable. If, as in H. G. Wells' story, a ship could be shielded from the earth's gravity but still exposed to the moon's gravity, the resulting acceleration would be ridiculously small, about a millionth of the acceleration of a freely falling body near the earth's surface. Besides, there is absolutely no evidence for antigravity or gravity shielding, and such effects might even be theoretically inconsistent with modern physics.

7.3 Numerous arguments have been published to resolve the paradox, ranging from limiting the extent of the universe to a finite radius, to red-shifting the light from very distant stars so severely that it is essentially nonexistent. [The red-shift is the Doppler shift of light from a source moving away from the observer and is similar to the Doppler shift in sound (FC 1.65).] Probably the best argument is a recent one (1587). The sky is not ablaze with light because a theory that assumes that all the stars in the universe are lit simultaneously must be wrong. The typical lifetime of a star can be taken as about 10^{10} years. Although this time seems very long, it is not when compared to the time about 10^{24} years necessary for the universe to reach thermodynamic equilibrium. "In other words, this means that the luminous emissions from stars are much too feeble to fill in their lifetime the vast empty spaces between stars with radiation of any significant amount."

7.4 The nature of noctilucent clouds is still controversial, but they are probably due to the condensation and freezing of water on dust particles at the mesopause, which is a relatively low temperature region near the altitude of 90 km. The dust could be cosmic dust (star dust), comet dust thrown off during a comet's passage by the sun, or dust from the astroidal belt. The clouds can be seen only during sunset because they are so faint. They can be seen then only because they are so high that they are still illuminated by the setting sun when the ground observer has entered twilight. The wavy structure is due to the passage of atmospheric gravity waves, periodic variations in the density and temperature of the air.

7.5 I have no answer to this question, and the literature is of no help in resolving the controversy. Experiments to show the statistical success of water witches is unconvincing and marginal and leaves the reader a believer only if he or she were already a believer. Unless someone publishes a very careful experiment in which the signal is clearly shown, water witching will remain controversial. For example, if the water witch subconsciously detects a very weak electromagnetic noise signal from running water, then that signal will have to be detectable on sensitive instruments and then correlated with the water witch's success.

7.6 The snow waves are probably the progressive lowering of snow that lays over a structurally weak layer of hoarfrost, a situation that is also responsible for some avalanches (FC 3.47).

7.8 The original, purely tongue-in-cheek calculations by David Stone (1424) indicated that such a jump by the Chinese would produce an earthquake with a Richter scale magnitude of 4.5. The jump would surely devastate part of China. But if the ground wave is resonantly amplified, destruction elsewhere would also be possible. To make resonance, the Chinese would have to jump every 53 or 54 min, the time a ground wave requires to circle the earth. To protect itself, the target nation would have to organize jumps whose waves would cancel the Chinese-generated waves. Since the target nation would have a smaller population, the jumps would have to be from proportionally greater heights. One writer to *Time* magazine argued that any jumps must be with stiff knees in order to impart the greatest energy to the waves. The difference between still- and loose-kneed jumping is not clear to me, since the energy imparted must be the gravitational potential energy in both cases. But were his argument correct, he points out that a resonant wave "would not be generated . . . because the only weapon derived from the actions would be the ear-shattering scream from 750 million badly maimed Chinese" (1425).

7.9 The protein molecules in the egg white are initially in a spaghettilike mess. Beating or heating the egg whites untangles those long molecules, and they then can attract each other sufficiently to give a firmer structure.

7.10 Apparently no complete explanation for the compression point has been published, although one could guess that it is caused by the flow of the adhesive layer beneath the tape toward the separation point.

7.11 through 7.13 These three phenomena are really the same. I'll explain the first one and leave the other two for you. Osborne Reynolds explained the whitening of beach sand in 1885 when he pointed out that the sand expanded when stepped on. Prior to the pressure, the sand grains were as closely packed as possible. Under the shearing from the footstep, the disturbance to the grains could only result in less efficient packing. In other words, the sand was forced to occupy more volume because of the shearing. Whereas the sand level suddenly rose, the water level could change only through capillary action, and that took some time. Thus, just after the footstep, the sand beneath the foot had risen above the water and was, for a little while, dry and white.

7.14 Recent publications indicate that the radiation level in high altitude jets is not of much concern. Radiation from the sun occurs principally during solar flares, and those are being monitored. The more serious danger was thought to lie in the heavy nuclei from our galaxy that could produce 1000 rads near the end of their trajectory if they stop in human flesh. This rate should be compared with the limit of 100 rads per week for radiation workers. However, the heavy nuclei flux at the flight altitudes is only a few percent of that in space and is thought not to be a serious problem after all. (Improperly shielded astronauts will have more to worry about.) Although the primary cosmic particles are apparently safe for airplane passengers, still uncertain is the effect of the secondary neutrons, low-energy protons, and alpha particles produced by the primary particles.

7.15 The flashes observed by astronauts and by researchers sitting in the beams of accelerated particles are likely produced by a variety of mechanisms. Either the cosmic ray particles or the man-made beams might be able to produce light by Cerenkov radiation in the eye, direct excitation of the retina, and fluorescence of the lens. (Cerenkov radiation is that radiation accompanying a particle whose velocity in a material exceeds the velocity of light in that material. The radiation forms a bow wave with the particle at the vertex, similar to the bow wave formed in the shock wave of a supersonic airplane.) Some of the light flashes are probably connected with the phosphenes discussed in FC 5.121.

7.16 The ability of using X-ray, infrared, and ultraviolet light to expose multiple layers of paintings lies in the different responses of the different paints and other materials used in the paintings. For example, infrared analysis of "The Marriage of Arnolfini" by van Eyck exposed an original sketch of Arnolfini's right hand that was done in charcoal on the white chalk background. That sketch was buried beneath the final painting of the hand. In the infrared photograph the charcoaled hand appeared because it greatly absorbed the infrared whereas the white chalk did not; the hand thus appeared dark in the photograph. Under ultraviolet light, different paints fluoresce differently, and alterations to original paintings can be detected in a similar manner.

7.17 Part of the energy from the particle and electromagnetic radiation emitted by the initial burst is absorbed by the air in the immediate vicinity of the burst. Those air molecules are highly excited or ionized, and the resulting deexcitation and recombination yields visible light. About half of the initial burst energy is released as mechanical energy (and develops a shock wave), about a third emerges as electromagnetic energy (infrared, visible, UV, X rays, and gamma rays) and the rest is given to particles. The shock wave rapidly compresses the air, heating it to incandescence. The temperature of the fireball's surface about 10^{-4} s after burst can be 3×10^5 K or greater. The fireball expands and cools, and eventually the shock wave breaks away from the ball and thus no longer causes incandescence.

7.18 Although Herbert did not have this in mind, there is an analogy in physics. If a metal plate is swung into a magnetic field, such as a metal pendulum swung between the pole faces of a horseshoe magnet, the kinetic energy of the plate is lost to Joule heating in the metal. This loss is due to the currents created in the plate by the change in the magnetic field experienced by the plate swinging through the magnet. For example, the field first increases as the plate nears the pole faces, and then decreases as the plate swings away. The currents created in the plate heat the plate in the same way that electric currents heat the coils in an electric oven. The original kinetic energy of the moving plate is eventually dissipated as this heating, and the plate stops.

7.19 Modern work on friction discredits the old theory about it originating entirely from surface irregularities and instead points to the adhesion between the surfaces (due to molecular attraction) as being the main cause of friction. In spite of this work, many physics textbooks still regard friction as being due only to hills and valleys in the surfaces jamming together.

7.20 Pure lead is soft; it can be cut with a fingernail. When the Cathedral's roof heated during a Washington summer day, it might have reached temperatures near 80°C, high enough that the lead became malleable enough to flow under its own weight. Less pure lead would be less malleable, and the roof was remade with an alloy of 94% lead and 6% antimony. The European structures did not have such a problem because their lead was unpure to begin with and because their summer days were not as hot.

7.21 Cracks begin at small, perhaps invisible defects that form when the material is first fabricated or that develop later under the wear and tear of use. These cracks greatly weaken the material structure, because they concentrate an applied force on the vertex of the crack. Such concentration means that a force normally too small to rip the material apart if there were no cracks, can now propagate the crack through the object. Some cracks expand because of corrosion. Foreign molecules may enter the crack, break the bonding of the material's molecules at the crack's vertex, and then react with these molecules. If the resulting new molecular structure occupies more volume than the original structure, the new structure pries the crack open.

7.22 The corrosion begins when electrons released by the wet nickel tunnel through the oxide layer on the chromium surface to reach an oxygen species. As a result, the nickel is slowly dissolved at the defect site. But the rate of the reactions is controlled by the electron flow. If only a few defects are in the bumper, the electrons come from only those few places, the nickel there is relatively quickly dissolved, and the iron layer is soon exposed and becomes rusty. If there are many (small) defects, then fewer electrons come from each, and the dissolving of the nickel and exposing and rusting of the iron at each defect are much slower, giving a longer life to the bumper.

7.23 The process of polishing a surface does not depend on transferring material from the hills to the valleys on the surface or on heating the surface. The first does not occur, and the latter may be undesirable if the surface is sufficiently heated as to develop waves. Polishing removes material from the hills on the surface. Under a heavy load, larger pieces of material are removed; under a light load, individual molecules are removed. Eventually the hills are worn down to the level of the valleys, and the surface is then smooth on a microscopic level.

7.24 Adhesives are adhesive because of molecular attraction between the surfaces of the adhesive and of the material to which it is applied. Any material could be an adhesive, although most are not useful because of their other properties. For example, liquid water could glue things together were it not for its low shear strength. Most adhesives are liquid, at least initially, because of the need for close contact between the glue and the materials being glued. In order for two surfaces to adhere, they must be within a few angstroms of each other. (An angstrom is 10^{-10} m, about the size of small atoms and molecules.) Most solid surfaces are too rough to allow more than a tiny amount of their surface area to be this close. A liquid glue can flow into those surface irregularities and thus provide the close contact. Another reason most surfaces, such as the broken edges of my coffee cup, do not adhere on contact is that they are dirty. Were they both clean and smooth, the surfaces might spontaneously adhere, as was feared in the early space missions. Such spontaneous adhesion can be observed in freshly separated layers of mica. If you rejoin the surfaces a few seconds after separation, the layers adhere. However, if you wait for several minutes, the air and its dust will contaminate the exposed surfaces to prevent any such adhesion.